An Aesthetic Critique of Digital Enhancement

Jackson Pollock, Echo: Number 25, 1951

An Aesthetic Critique of Digital Enhancement

Government of the Self and Desire

Sarah Bianchi

LEXINGTON BOOKS
Lanham • Boulder • New York • London

Published by Lexington Books
An imprint of The Rowman & Littlefield Publishing Group, Inc.
4501 Forbes Boulevard, Suite 200, Lanham, Maryland 20706
www.rowman.com

86-90 Paul Street, London EC2A 4NE

British Library Cataloguing in Publication Information Available

Library of Congress Cataloging-in-Publication Data Available

ISBN 978-1-66692-831-0 (cloth)
ISBN 978-1-66692-832-7 (ebook)

For Gerhard Schiffler

Contents

Preface

Everywhere we look, we are confronted with the buzzwords of self-optimization, be it at work or in sports.[1] Today, the call for self-optimization includes our personal lives, too, for example fitness and beauty. It even extends to cooking: pasta should be hand-made with the best possible dough. At the same time, health insurance companies inform us that cases of depression and burnout are on the rise in our societies.[2] Could there be a relationship between the two observations? These and similar observations motivated me to think fundamentally about the understanding of enhancement in the central field of digital change and its ambivalences from a philosophical perspective.

Much has already been said about the theme of enhancement in a narrow sense,[3] but little has been said about a broad understanding of enhancement as a specific way of life. Yet, it is this broader view that shows us we are *not* merely dealing with a current trend concerned with perfecting the human organism by the latest advances in biotechnology, but with an important theme that has *supertemporal* relevance. This supertemporal relevance goes *beyond* the traditional and narrow understanding of enhancement and leads to questions of how to give one's life form in relation to the dominant logic of perfection of our contemporary normative orders that give subjects their social status. These questions are much older than current technological trends; they can even be traced back to antiquity. The book at hand considers such a shift in focus as necessary because *only* this perspective enables us to diagnose one of the core problems in today's field of digital enhancement. This problem is the affect-sensitive notion of power, which has so far received little attention. Throughout the present book, affect-sensitive power is understood as the ways in and through which the logic of perfection, the common denominator of the field of digital enhancement, has subtle effects on subjects. From this angle, the present book focuses on the seemingly forgotten questions of possible spaces of freedom for the subject: to what extent is the subject *quasi-automatically pre-determined* by the subtle effects of given normative orders? To what extent does the subject *also* have the potential to distance herself from the subtle effects of the logic of perfection, and thus *transcends the contexts* marked by this logic and create spaces for feeling, thinking and living *differently*? This book responds to these questions by developing an understanding of aesthetic critique that entails the method of genealogical critique and the notion of enlightenment.

Such questions do not only refer to philosophical debates; they also belong to the field of art. While I was in New York, I used to go to museums whenever I had time; on one of these visits, I realized how closely the questions I deal with in my book are explored in works of art. One piece that particularly struck me was the drip painting *Echo* (1951) by the US-American abstract expressionist painter Jackson Pollock (cf. frontispiece), exhibited in the famous Museum of Modern Art (MoMA). Obviously, Pollock does not explicitly refer to the theme of enhancement. But if we understand the logic of perfection in a broader way as structure, then the relation becomes clear. The common interpretation of this painting is that Pollock is expressing the tension between structure and *anti*-structure in abstract terms.[4] For example, we see an ear in the top left-hand corner. But the ear is not depicted in its entirety: all we see is a few lines in black and white, which merely allude to the figural form of an ear. At the same time, the allusion is also deceptive: within the lines that could be the auricle of the ear, we see something that appears to be teeth, but could be also the lashes of an eye. Like a real "echo," to refer to the painting's title, Pollock shows that the effect of the structure on the formation of things has gaps, just as an echo that one creates in the forest does not repeat the entire sentence, but perhaps only the last word. To put it in more general terms, Pollock plays with the modes of representation and anti-representation. From this perspective, the painting attracted my interest. Here we see that, according to Pollock, structure does not have the power to dominate the representation of things. This theme is also of major interest in poststructural thinking, where it centers on the difference between form or structure *on the one hand* and content (or in this case things) *on the other*.

If we translate this scenario to the present book's context, then the "echo" takes place between the logic of perfection, i.e., the structure or form, and the subject's life formation, i.e., the content. From this angle, the question is shifted to the extent to which the "echo" of the logic of perfection leaves gaps that give the subject space to understand and experience the original call of the logic of perfection differently. The theme of being called by an order is a key part of poststructuralist thinking, which mainly goes back to Louis Althusser. If Althusser demonstrates how much given structures appear as an authority to the subject, like a policeman on the street calling a pedestrian, then Pollock expands this understanding. He shows how small and ordinary things can be called and thus be formed by echoes. Obviously, things cannot hear and thus cannot really be called, but we as the listeners are the people that are called. According to Pollock, however, these calls leave space for the listener—or the person seeing the canvas—to interpret things *differently*. So, Pollock's approach makes it obvious that the call does not have complete power over things, but that there could also be spaces of freedom for seeing things otherwise. The eyelash, for example, could also be understood as something else, to return to one of Pollock's figural, echo-like allusions.

The German abstract expressionist painter, photographer, and sculptor Gerhard Richter shows, from another angle, what Pollock expressed through the painting *Echo*. He writes: "Life is not what is said, by the saying, not the painting, but the process of its formation."[5] Here, Richter does not use the term "echo," but he stresses the contingent conditions in and through which pictures, even life *itself* emerges.[6] Just as the "echo" makes explicit the possible frictions in the relation between the subject and the normative order, understanding the contingent conditions of feeling, thinking and living reformulates these possible frictions. By understanding that things do not necessarily have to be ordered the way they have been ordered so far, the subject gains space to see the potential for different constellations, just as the subject that only listens to a fading "echo" gains space for different impressions. From this perspective, it becomes possible to understand that life is not the fixed form of something that was already said, but rather *the process* in and through which its ongoing formation takes place now and could take place differently in the future. And the potential to see these different formations, as we can say with Richter, depends on the perception of contingent conditions. To again transfer this view to the context of the present book, we can say that the perception of the contingent conditions formulated by Richter can theoretically conceptualize how to create the possible space between the subject and the normative orders. The aesthetic critique developed here has the task of theoretically conceptualizing the perception of these contingent conditions of the subject's life formation; in doing so, it can make plausible how the subject can create space for enlightened views on how to give life a form.

ACKNOWLEDGMENTS

Now it is time to say a few words on the present book's general context. This is my second book, or (as people pointed out during my research stay in California) my third book in the world if one counts my master's thesis in history on the German reception of the Dreyfus Affair (Peter Lang, 2012). In summer 2015, as project leader of the sub-project "Government of the Self and Desire: Critical Aesthetics of Digital Enhancement," I started to write the present book in the American program *The Enhancing Life Project* of The University of Chicago and Ruhr University Bochum, funded by the John

Templeton Foundation in association with Humboldt University of Berlin and Princeton University. As part of the project, my research focus concentrates on the following big question formulated by *The Enhancing Life Project*: How can we conceptualize the tipping point between the endangerment and the enhancement of life? I am very pleased that I was part of *The Enhancing Life Project* and that The John Templeton Foundation funded the project generously. The argument developed in this book is my personal opinion and does *not* represent the point of view of The John Templeton Foundation.

Within the broader context of *The Enhancing Life Project*, I had the honor to be a short-term visiting scholar at Stanford University, funded by the PSCS visiting scholar award from Stanford University, and also to be a visiting postdoctoral research associate at the Department of Philosophy at Princeton University supervised by Prof. Alexander Nehamas. From summer 2018 onwards, I continued my research for the book at hand as an assistant professor ("wissenschaftliche Mitarbeiterin," postdoctoral) in the project "Perspectives of Critical Theory," supervised by Prof. Dr. Martin Saar and Prof. Dr. Raymond Geuss at the Department of Philosophy at Goethe-University Frankfurt. I then continued the research as a visiting professor at the Department of Philosophy of the Technical University Dresden, and am now finishing the publication of the second book as a postdoc back in Frankfurt.

During the research period, I had the chance to present the book at various events on a national and an international level. I am very thankful for the helpful feedback that I received from audiences. I cannot name every occasion, but I would particularly like to express my appreciation for the feedback that I received at the following events: Initially, I presented chapters of this book in Chicago, Banff, Stanford, London and Berlin during international conferences and transatlantic research collaborations of *The Enhancing Life Project*. Then, I felt very honored to have the chance to present papers in Frankfurt at the departmental colloquium of the Institute of Philosophy, at the international critical theory workshop on Martin Saar's work in Frankfurt, at the Critical Theory Colloquium in Dublin (normally in Prague), at different international and national conferences of the Friedrich Nietzsche society in Naumburg and in Schulpforta, at the Nietzsche research colloquium by Prof. Dr. Sommer in Freiburg, at the conference on the theme of machine culture in Nietzsche in Weimar (which I co-convened with the former head of the Friedrich Nietzsche Kolleg, Dr. Rüdiger Schmidt-Grépály), at the international conference "Adorno and Media" in Karlsruhe, at the congress of the German Society of Aesthetics (*Deutsche Gesellschaft für Ästhetik*), and at the congress of the German Society of Philosophy (*Deutsche Gesellschaft für Philosophie*).

At this point, it needs to be said that one cannot write a book alone. And perhaps this is especially true for me. So, I would like to take this opportunity to thank all the people who have accompanied me in this research project. First of all, I must thank Prof. Dr. William Schweiker and Prof. Dr. Günter Thomas, the two principal investigators of *The Enhancing Life Project*, for initiating my book's project. Secondly, I wish to thank Prof. Dr. Martin Saar for everything I have learnt from him. Without his continuous support, I would not have been able to write the book at hand. When I attended one of his talks for the first time, at the Berliner Nietzsche Colloquium around 2011, I did not have the slightest clue that I would later have the honor to work at his chair. Special thanks to Prof. Dr. Rainer Forst, Prof. Dr. Raymond Geuss and Prof. Dr. Bernard E. Harcourt, who were present at different stages of this book. Moreover, I am very thankful for Prof. Dr. James Tully's so seminal comments on my manuscript. It is such an honor that Jim had time to read the entire manuscript and I hope that I could include his comments the way they deserve. Furthermore, I would also like to thank the other fellows of *The Enhancing Life Project*, in particular the vulnerability group ("the vulnerability ladies," as we used to say) in which I participated: Prof. Dr. Heike Springhart, Prof. Dr. Andrea Bieler, Prof. Dr. Kristine Culp and Prof. Dr. Pamela S. Anderson, who unfortunately died recently, as well as Dr. Ellen Ueberschär, who served as academic consultant of *The Enhancing Life Project*. Many thanks to Prof. Dr. Alexander Nehamas for being my supervisor during my research stay at Princeton University. I also wish to thank my former colleagues at Martin Saar's chair, Andrea Blättler, Corina Färber, Hannes Kuch, and Kristina Lepold, the members of Martin's colloquium and the members of the Political Theory colloquium, that was founded by Rainer, for fruitful discussions. And of course, last but not least, giving undergraduate and graduate seminars at Humboldt-University Berlin, Goethe-University Frankfurt and Technical University Dresden taught me a great deal, so I would like to thank the students in Berlin, Hessen and Saxony, too.

Moreover, as I am not a native speaker, the present book could have not been written without the great support of my editors (*Lektoren*). Dr. Thomas Cannaday and Audrey Fausser edited chapters at a very early stage. A late version was edited by Dr. Ciaran Cronin, Prof. Dr. Thomas Heinrich and *Textworks Translations*. I owe them a lot. The final version of the present book has been proofread by *Textworks Translations*. I am very thankful for their incredible help. Without them, it would have not been possible to finish the book in this form. I would also like to thank Jana Hodges-Kluck and Deanna Biondi for giving my manuscript such a warm welcome at Lexington Books and for transforming it into a real book. I am very pleased that my book is being published by Lexington Books, especially since Lexington Books is known for its Critical Theory Series, *Essex Studies in Contemporary*

Critical Theory. The original plan was to ask the editors to include my manuscript in the series, but unfortunately the series no longer exists. Nonetheless, this book is written in its spirit. And I also thank the unknown reviewer for the incredibly helpful comments. I tried my best to integrate them into the present book.

Furthermore, I could not have written this book without the support of my family and friends, in particular Inge, Reinhold, Vera, Anna, Linda, Franziska, Nada, Melina, Bitía, Sima, Chris, Michael, Jörn, Mariel, Klaus, Anne, Adrien, Susumu, and Katharina. This book is dedicated to my grandfather, Gerhard Schiffler, who always said "il faut corriger la fortune."

Frankfurt am Main
Summer 2023
Sarah Bianchi

NOTES

1. Maren Keller, "Selbstoptimierung: Die Arbeit am Ich. Fitter, gebildeter, schöner–die Soziologien Anja Röcke erklärt, warum wir uns selbst als eine Dauerbaustelle betrachten," *Der Spiegel*, April 25, 2022, available at: https://www.spiegel.de/wissenschaft/mensch/selbstoptimierung-warum-wir-uns-selbst-als-eine-dauerbaustelle-betrachten-a-0c64a4f6-0002-0001-0000-000200446780.

2. See Ashley Abramson, "Burnout and stress are everywhere," *Monitor on Psychology* 53, no. 1 (2022): 72, available at: https://www.apa.org/monitor/2022/01/special-burnout-stress.

3. Dagmar Fenner, *Selbstoptimierung und Enhancement: Ein ethischer Grundriss* (Wien et al.: utb, 2019).

4. Without author, "Audio Guide Museum of Modern Art," available at: https://www.moma.org/collection/ works/79251.

5. Gerhard Richter, *Text: 1961 bis 2007 Schriften, Interviews, Briefe* (Köln: Walther König, 2008), 222 (trans. by the author).

6. Whereas Richter understands the notion of contingency in the ordinary sense of "randomly," it is more productive from a critical perspective to understand the notion of contingency in the broader sense of "not necessary."

List of Abbreviations

WORKS BY MICHEL FOUCAULT

ABM Foucault, Michel. *Abnormal: Lectures at the Collège de France 1974–1975*. Ed. V. Marchetti and A. Salomoni, trans. G. Burchell. London/New York: Verso, 2016 [1974/1975].

AE Foucault, Michel. "An Aesthetics of Existence." In *Foucault Live: Collected Interviews, 1961–1984*, ed. S. Lotringer, trans. L. Hochroth and J. Johnston, 450–454. 2nd ed. South Pasadena: Semiotext(e), 1996 [1984].

AK Foucault, Michel. *The Archeology of Knowledge And the Discourse on Language*. Trans. A. M. S. Smith. New York: Vintage Books, 2010 [1969].

AOE German edition: Foucault, Michel. "Der 'Anti-Ödipus': Eine Einführung in eine neue Lebenskunst." In *Dispositive der Macht. Michel Foucault über Sexualität, Wissen und Wahrheit*. 118–175. Berlin: Merve, 1978 [1972].

CF Foucault, Michel. "The Confession of the Flesh." In *Power/Knowledge. Selected Interviews & Other Writings 1972–1977*. Ed. C. Gordon, trans. C. Gordon et al. 194–228. New York: Vintage Books, 1980 [1977]

German edition: "Ein Spiel um die Psychoanalyse. Gespräch mit Angehörigen des Département de Psychoanalyse der Universität Paris VIII in Vincennes." In *Dispositive der Macht. Michel Foucault über Sexualität, Wissen und Wahrheit*. 118–175. Berlin: Merve, 1978 [1977].

DE German edition: Foucault, Michel. *Schriften in vier Bänden: Dits et Écrits*. Ed. D. Defer and F. Ewald, trans. M. Bischoff, H.-D. Gondek and H. Kocyba. Frankfurt a.M.: Suhrkamp, 2001–2005.

DE 1/33 German edition: Foucault, Michel. "Die Prosa der Welt." In *Dits et Écrits*. Vol. 1: *1954–1969*. Ed. D. Defert and F. Ewald, trans. M. Bischoff et al., 622–643. Frankfurt a.M.: Suhrkamp, 2001 [1974].

DE 3/255 German edition: Foucault, Michel. "Sicherheit, Territorium und Bevölkerung." In *Dits et Écrits*. Vol. 3: *1976–1979*. Ed. D. Defert and F. Ewald, trans. M. Bischoff, 900–905. 2nd ed. Berlin: Suhrkamp, 2016 [1978].

DE 4/291 German edition: Foucault, Michel. "'Omnes et Singulatim': Zu einer Kritik der politischen Vernunft." In *Dits et Écrits*. Vol. 4: *1980–1988*. Ed. D. Defert and F. Ewald, trans. M. Bischoff, 165–198. Frankfurt a.M.: Suhrkamp, 2005 [1979].

DE 4/354 German edition: Foucault, Michel. "Die Rückkehr der Moral." In *Dits et Écrits*. Vol. 4: *1980–1988*. Ed. D. Defert and F. Ewald, trans. M. Bischoff et al., 859–872. Frankfurt a.M.: Suhrkamp, 2005 [1984].

DP Foucault, Michel. *Discipline and Punish: The Birth of the Prison*. Trans.A. Sheridan. 2nd ed. New York: Vintage Books, 1995 [1975].

German edition: Foucault, Michel. *Überwachen und Strafen: Die Geburt des Gefängnisses*. Trans. W. Seitter. 16th ed. Frankfurt a.M.: Suhrkamp, 2016 [1975].

French original: Foucault, Michel. *Surveiller et punir: Naissance de la prison.* Paris: Gallimard, 2013 [1975].

DTP Foucault, Michel. *Discourse and Truth & Parrēsia*. Ed. H.-P. Fruchaud and D. Lorenzini, trans. N. Luxon. Chicago/London: The University of Chicago Press, 2019 [1982/83].

ECSPF Foucault, Michel. "The Ethics of the Concern for Self as a Practice of Freedom." In *The Essential Work of Foucault (1954–1984)*. Vol. 1: *Ethics: Subjectivity and Truth*. Ed. P. Rabinow, trans. R. Hurley et al., 281–302. London et al.: Penguin Books, 2000 [1984].

German edition: "Die Ethik der Sorge um sich als Praxis der Freiheit." In *Ästhetik der Existenz: Schriften zur Lebenskunst*. Ed. D. Defert and F. Ewald, trans. M. Bischoff et al., 253–279. 5th ed. Berlin: Suhrkamp, 2015 [1984].

FWL Foucault, Michel. "Friendship as a Way of Life." In *Essential Works of Foucault (1954–1984)*. Vol. 1: *Ethics: Subjectivity and Truth*. Ed. P. Rabinow, trans. R. Hurley, 303–320. London: Penguin Books, 2000 [1984].

GE Foucault, Michel. "On the Genealogy of Ethics: An Overview of the Work in Progress." In *Essential Works of Foucault (1954–1984)*. Vol. 1, *Ethics: Subjectivity and Truth*. Ed. P. Rabinow, trans. R. Hurley et al., 253–280. London et al.: Penguin Books, 2000 [1984].

German edition: Foucault, Michel. "Zur Genealogie der Ethik: Ein Überblick über die laufende Arbeit." In *Ästhetik der Existenz: Schriften zur Lebenskunst*. Ed. D. Defert and F. Ewald, trans. M. Bischoff et al., 191–219. 5th ed. Berlin: Suhrkamp, 2015 [1984].

French edition: Foucault, Michel. "À propos de la généalogie de l'éthique: Un aperçu du travail en cours." In *Un parcours philosophique: Au-delà de l'objectivité et de la subjectivité*. Ed. H. L. Dreyfus and P. Rabinow, trans. F. Durand-Bogaert, 323–346. Paris: Gallimard, 1984.

GL Foucault, Michel. *On the Government of the Living: Lectures at the Collège de France 1979–1980*. Ed. M. Senellart, trans. G. Burchell. New York: Picador, 2012.

German edition: Foucault, Michel. *Die Regierung der Lebenden*. Ed. F. Ewald and A. Fontana, trans. A. Hemminger. Berlin: Suhrkamp, 2014.

GSO Foucault, Michel. *The Government of Self and Others: Lectures at the Collège de France 1982–1983*. Ed. F. Gros, trans. G. Burchell. New York: Palgrave Macmillan, 2011 [1982/1983].

German edition: *Die Regierung des Selbst und der anderen: Vorlesung am Collège de France 1982/83*. Ed. F. Ewald and A. Fontana, trans. J. Schröder. Frankfurt a.M.: Suhrkamp, 2012.

GSO2 Foucault, Michel. *The Government of Self and Others II: The Courage of Truth. Lectures at the Collège de France 1983–1984*. Ed. F. Gros, trans. G. Burchell. New York: Palgrave Macmillan, 2011.

HS Foucault, Michel. *The Hermeneutics of the Subject: Lectures at the Collège de France 1981–1982*. Ed. F. Gros, trans. G. Burchell. 2nd ed. London: Picador, 2005 [1981/1982].

HS1 Foucault, Michel. *The History of Sexuality*. Vol. 1: *An Introduction*. Trans. R. Hurley. New York: Vintage Books, 1990 [1976].

German edition: Foucault, Michel. *Sexualität und Wahrheit*. Vol. 1: *Der Wille zum Wissen*. Trans. U. Raulff and W. Seitter. 21st ed. Frankfurt a.M.: Suhrkamp, 2017 [1976].

HS 2 Foucault, Michel. *The History of Sexuality*. Vol. 2: *The Use of Pleasure*. Trans. R. Hurley. New York: Vintage Books, 1990 [1984].

German edition: Foucault, Michel. *Sexualität und Wahrheit*. Vol. 2: *Der Gebrauch der Lüste*. Trans. U. Raulff and W. Seitter. 13th ed. Berlin: Suhrkamp, 2017 [1984].

French edition: Foucault, Michel. *Histoire de la sexualité*. Vol. 2: *L'usage des plaisirs*. Paris: Gallimard, 2014 [1984].

HS3 Foucault, Michel. *The History of Sexuality*. Vol. 3: *The Care of the Self*. Trans. R. Hurley. New York: Vintage Books, 1988 [1984].

HS4 Foucault, Michel. *Histoire de la sexualité*. Vol. 4: *Les aveux de la chair*. Ed. F. Gros. Paris: Gallimard 2018 [posthumous].

English edition: Foucault, Michel. *The History of Sexuality*. Vol. 4: *Confessions of the Flesh*. Ed. F. Gros, trans. R. Hurley. New York: Pantheon Books, 2021.

MC Foucault, Michel. *Madness and Civilization: A History of Insanity in the Age of Reason*. Trans. R. Howard. New York: Vintage Books, 1988.

NGH Foucault, Michel. "Nietzsche, Genealogy, History." In *The Foucault Reader: An Introduction to Foucault's Thought*, ed. P. Rabinow, 76–100. London et al: Penguin Books, 1991 [1971].

OD German edition: Foucault, Michel. *Die Ordnung des Diskurses*. Trans. W. Seitter. 10th ed. Frankfurt a.M.: Fischer, 2007.

OT Foucault, Michel. *The Order of Things: An Archeology of the Human Sciences*. New York: Vintage Books, 1994 [1966].

PS Foucault, Michel. *The Punitive Society: Lectures at the Collège de France 1972–1973*. Ed. B. E. Harcourt, trans. G. Burchell. Basingstoke/New York: Palgrave Macmillan, 2015.

PTI Foucault, Michel. *Penal Theories and Institutions: Lectures at the Collège de France 1971–1972*. Ed. B. E. Harcourt, trans. G. Burchell. New York: Palgrave Macmillan, 2019.

RM Foucault, Michel. "The Return of Morality." In *Foucault Live: Collected Interviews 1961–1984*. Ed. S. Lotringer, trans. L. Hochroth and J. Johnston, 465–473. 2nd ed. South Pasadena: Semiotext(e), 1996 [1984].

SMD Foucault, Michel. *Society Must be Defended: Lectures at the Collège de France 1975–1976*. Ed. M. Bertani and A. Fontana, trans. D. Macey. London et al.: Penguin Books, 2004.
German edition: Foucault, Michel. *In Verteidigung der Gesellschaft: Vorlesungen am Collège de France (1975–1976)*. Ed. F. Ewald and A. Fontana, trans. M. Ott. 5th ed. Frankfurt a.M.: Suhrkamp, 2016.

SP Foucault, Michel. "The Subject and Power." In *Michel Foucault: Beyond Structuralism and Hermeneutics*. Ed. H. L. Dreyfus and P. Rabinow, 208–226. 2nd ed. Chicago: The University of Chicago Press, 1983 [1982].

ST Foucault, Michel. "Subjectivity and Truth." In *Essential Works of Foucault 1954–1984*. Vol. 1: *Ethics: Subjectivity and Truth*, ed. P. Rabinow, trans. R. Hurley, 87–92. New York/London: Penguin Books, 2000 [1984].
German edition: Foucault, Michel. "Subjektivität und Wahrheit." In *Ästhetik der Existenz: Schriften zur Lebenskunst*, ed. D. Defert and F. Ewald, trans. M. Bischoff et al., 74–80. 5th ed. Berlin: Suhrkamp, 2015 [1981].

STP Foucault, Michel. *Security, Territory, Population. Lectures at the Collège de France 1977–1978*. Ed. M. Senellart, trans. G. Burchell. New York: Picador 2007 [2004].

TJF Foucault, Michel. "Truth and Juridical Forms." In *Essential Works 1954–1984*. Vol. 3: *Power*. Ed. J. D. Faubion, trans. R. Hurley et al., 39–88. 3rd ed. New York/London: Penguin Books, 2002 [1973].
German edition: Foucault, Michel. *Die Wahrheit und die juristischen Formen: Mit einem Nachwort von Martin Saar*. Trans. M. Bischoff. 5th ed. Berlin: Suhrkamp, 2015 [1973].

VN German edition: Foucault, Michel. "Vorlesung über Nietzsche." In *Über den Willen zum Wissen: Vorlesungen am Collège de France 1970-1971. Gefolgt von Das Wissen des Ödipus*, ed. F. Ewald and A. Fontana, trans. M. Bischoff. 259–281. Berlin: Suhrkamp, 2012.

WC Foucault, Michel. "What is Critique?" In *The Politics of Truth*, ed. S. Lotringer, trans. L. Hochroth and C. Porter, 41–82. South Pasadena: Semiotext(e), 2007 [1978].

German edition: Foucault, Michel. *Was ist Kritik?* Trans. W. Seitter, Berlin: Merve, 1992 [1978].

WDTT Foucault, Michel. *Wrong-Doing, Truth-Telling: The Function of Avowal in Justice*. Ed. F. Brion and B. E. Harcourt, trans. S. W. Sawyer. Chicago/London: The University of Chicago Press, 2014 [1981].

WE Foucault, Michel. "What is Enlightenment?" In *Essential Works of Foucault 1954–1948*. Vol. 1: *Ethics: Subjectivity and Truth*, ed. P. Rabinow, trans. R. Hurley, 303–320. New York/London: Penguin Books, 2000 [1984].

WK Foucault, Michel. *Lectures on the Will to Know: Lectures at the Collège de France 1970–1971*. Ed. D. Defert, London: Palgrave Macmillan, 2013.

WORKS BY FRIEDRICH NIETZSCHE

BGE Nietzsche, Friedrich. *Beyond Good and Evil.* Ed. R.-P. Horstmann and J. Norman, trans. J. Norman. 7th ed. Cambridge: Cambridge University Press, 2009.

German edition: "Jenseits von Gut und Böse (1886)." In *KSA* 5.

BT Nietzsche, Friedrich. *The Birth of Tragedy out of the Spirit of Music*. Ed. M. Tanner, trans. S. Whiteside. New York/London: Penguin Classics, 1993.

German edition: "Die Geburt der Tragödie aus dem Geist der Musik (1872)." In *KSA* 1.

D Nietzsche, Friedrich. *Daybreak. Thoughts on Prejudices of Morality*. Ed. M. Clark and B. Leiter, trans. R. J. Hollingdale. Cambridge: Cambridge University Press, 1997.

German edition: "Morgenröthe. Gedanken über moralische Vorurtheile (1881)." In *KSA* 3.

EH Nietzsche, Friedrich. *Ecce Homo*. Ed. N. Ridley and J. Norman, trans. J. Norman. Cambridge: Cambridge University Press, 2005.

German edition: "Ecce Homo (1888/89)." In *KSA* 6.

GM Nietzsche, Friedrich. *On the Genealogy of Morals*. Ed. K. Ansell-Pearson, trans. C. Diethe. 3rd ed. Cambridge: Cambridge University Press, 2017.

German edition: "Zur Genealogie der Moral (1887)." In *KSA* 5.

GS Nietzsche, Friedrich. *The Gay Science*. Ed. B. Williams, trans. J. Nauckhoff and A. del Caro. Cambridge: Cambridge University Press, 2001.

German edition: "Die Fröhliche Wissenschaft (1882)." In *KSA* 3.

HH Nietzsche, Friedrich. *Human, All too Human. A Book for Free Spirits*. Ed. R. Schacht, trans. R. J. Hollingdale. 14th ed. Cambridge: Cambridge University Press, 2010.

German edition: "Menschliches, Allzumenschliches. Ein Buch für freie GeisterI (1878) and II (1879)." In *KSA* 2.

HL Nietzsche, Friedrich. "On the Use and Disadvantages of History for Life." In: *Untimely Meditations*, ed. D. Breazeale, trans. R. J. Hollingdale. Cambridge: Cambridge University Press, 1997.

German edition: "Vom Nutzen und Nachteil der Historie für das Leben (1874)." In *KSA* 1.

KSA Nietzsche, Friedrich. *Kritische Studienausgabe*. Ed. G. Colli and M. Montinari. 2nd ed. München/Berlin: dtv/de Gruyter, 1986.

KSB Nietzsche, Friedrich. *Kritische Studienausgabe Briefe*. Ed. G. Colli and M. Montinari. 2nd ed. München/Berlin: dtv/de Gruyter, 1986.

N Nietzsche, Friedrich. *Nachgelassene Fragmente* (1869–1889). In *KSA* 7–13.

TI Nietzsche, Friedrich. *Twilight of the Idols*. Ed. A. Ridley and J. Norman, trans. J. Norman. Cambridge: Cambridge University Press, 2005.

German edition: "Götzendämmerung (1889)." In *KSA* 5.

UM Nietzsche, Friedrich. *Untimely Meditations*. Ed. D. Breazeale, trans. R. J. Hollingdale. Cambridge: Cambridge University Press, 1997.

German edition: "Unzeitgemäße Betrachtungen (1873ff)." In *KSA* 1.

WL German edition: Nietzsche, Friedrich. "Über Wahrheit und Lüge im außermoralischen Sinn (1873)." In *KSA* 1.

WS Nietzsche, Friedrich. "The Wanderer and his Shadow (part two, *Human All too Human*)." In *Human, All too Human: A Book for Free Spirits*. Ed. R. Schacht, trans. R. J. Hollingdale. 14th ed. Cambridge: Cambridge University Press, 2010.

German edition: "Der Wanderer und sein Schatten (part two, *Menschliches, Allzumenschliches*, 1879)." In *KSA* 2.

Z Nietzsche, Friedrich. *Thus Spoke Zarathustra*. Ed. A. del Caro and R. Pippin, trans. A. del Caro. Cambridge: Cambridge University Press, 2006.

German edition: "Also sprach Zarathustra (1883–1885)." In *KSA* 4.

ZB German edition: Nietzsche, Friedrich. "Über die Zukunft unserer Bildungsanstalten (1872)." In *KSA* 1.

Introduction

ON DIGITAL "RE-EVALUATION" OR WHAT IS DIGITAL ENLIGHTENMENT TODAY?

"One has only to open one's eyes"[1] (PTI, 1) Michel Foucault stated at the start of his lecture *Penal Theories and Institutions* delivered in the early 1970s at the Collège de France, one of the nation's most prestigious universities.[2] And why should one "open one's eyes?" To see the important things in society in their time. For Foucault that meant in his historical-philosophical lecture the rebellion of the *Nu-pieds*, a popular uprising in Normandy in 1639.

If we today are called to "open [our] eyes," to say it with Foucault, then we can see the socio-cultural phenomenon of digital enhancement, i.e., the attempt to perfect life through digital media. But unlike in the case of the rebellion of the *Nu-pieds* where one can see with Foucault in manifest ways the important things of society in the repression of the *Nu-pieds* rebellion, in our digital present, to say it with another formulation of Foucault, the "cost" (WC, 45) of digital enhancement is *not* easy to perceive. In digital enhancement, the subject can "open [her] eyes" as much as she wants,[3] *only* to find that her gaze is fogged up *already*. What she sees is not what she gets. For she indeed pretty easily sees what digital enhancement *wants to make her see*: Digital enhancement does *not* veil any of its ideas. It wants to enhance people's lives *assuming it is in their best interest.* Already the verb "to enhance" means "to perfect." But in contrast to this promise of digital enhancement, subjects merely need to look around and see that this assumption of perfection does not always work. It implies freedom-impairing effects on the subjects' life, *too*. Subjects only need to read newspapers to know of what this entails: the headlines in *The New York Times*, *The Guardian*, *Le Monde Diplomatique,* or *DIE ZEIT* inform them about the Cambridge Analytica Scandal, PRISM, the NSA, and their respective whistle blowers, Brittany Kaiser, and Edward Snowden, to mention only the most well-known cases. These and related cases publicize the fact that digital enhancement *not only* means to enhance freedom, as its name actually promises. On the contrary, they *also* publicize its freedom-endangering effects, such as endangering data privacy or uses of "big data" in and through which one can be easily tracked by surveillance technologies. This is seen in particular in the well-known recent scandals: The PRISM scandal, which was revealed in 2013 by leading newspapers, showed that the US-American *National Security Agency*, abbreviated NSA, was using a secret surveillance program called PRISM to gather data provided by well-known companies such as Facebook, Yahoo, Apple or Google. The Cambridge Analytica scandal exposed how political elections were manipulated by way of personal profiles that were generated with the help of digital media platforms such as Facebook. This method was used especially in Donald Trump's 2016 presidential campaign and in the leave campaign for the 2016 Brexit referendum. The Cambridge Analytica scandal and the PRISM scandal were exposed by whistleblowers: especially Kaiser, a former Cambridge Analytica employee, and Snowden, a former CIA agent.[4]

But to the surprise of the observer, we are confronted with the following paradox that is central to our digital present: the well-known and easily detectable freedom-impairing effects of digital enhancement on the subject's life do *not* hinder the very same subject to turn into the "user" of digital media. And one should add that they do not only become *any* "user": some even want to become its *exceedingly enthusiastic* "users," as demonstrated, for example, in the Netflix episode "Nosedive" (2016) of the series *Black Mirror* (2011); the series is, of course, fictional, but it is perhaps not so far removed from what some of today's influencers might do. The plot deals with the protagonist Lacie, a young woman who does everything in her life to earn as many credit points as possible via social media. For example, she only wants to be friends with supposedly socially acceptable people from whose friendship she expects to gain more credit points. Subjects like Lacie are even thrilled to subjugate their life to the social imperative of digital perfection.

Nonetheless, it would be shortsighted to focus *exclusively* on the downsides of digital enhancement. To again refer to Foucault's observation in *Penal Theories and Institutions*, we too can "open [our] eyes" and see how digital enhancement has advantages. There is no need to reference to a movie to show the point. All I need to do is look at the computer on which I am typing the manuscript of this book. Like the typewriter in earlier times, today the computer has advantages because we are no longer constrained to practice handwriting but can use a machine that enables us to produce easily revisable texts without having to resort to handwriting. Additionally, the associated digital technology enables us to send the texts we produce around the globe almost instantaneously. Through email communication and digital messaging platforms such as Skype and Zoom, we can overcome the distance constraining us in the analogue world and can, for instance, easily communicate between locations on both sides of the Atlantic. Failing to recognize the possible benefits of such technology would us also expose to a charge of accusations of cultural pessimism.

But still, to play the devil's advocate: Are today's subjects, to put it in obviously exaggerated, hyperbolic terms, not acting like sheep trotting willingly to the slaughterhouse, to use one of Foucault's most drastic metaphors? Are they not, in Foucault's words, "omnes et singulatim" (DE 4/291, 165), all and every single one of them, to some extent seduced by contemporary digital "shepherds,"[5] i.e., obviously not by specific persons, not by the CEOs of companies such as Amazon, Facebook, or YouTube, but by the anonymous logic of perfection, which forms an essential part of our digital today? Do these digital shepherds not cause subjects individually and collectively to ignore the costs of digital enhancement and *already prompt* them to pursue some of the gimmicks one encounters in the virtual world? For certain subjects, these gimmicks even cause a subject to fall in love with their digital operating system, as in Spike Jonze's film *Her* (2013). This movie describes how digital media is regarded as a solution for social problems, in this instance to alleviate loneliness. The film's premise is that if real life cannot provide the connection we need with others, then we must find a digital solution. The digital solution for achieving connection that the protagonist Theodore seeks is a computer, i.e., an operating system, which calls itself Samantha, portrayed by Scarlett Johansson. The digital device prompts Theodore to develop a bond with Samantha in specific ways: he literally wants to take her everywhere he goes. For eyes that are not yet trained to visualize digital life, the scene might look somewhat strange. To mention just one example: since Samantha is an operating system without a physical body, Theodore cannot take conventional walks with her; instead she participates in his daily life via the lens of his portable device, which he carries around in his shirt pocket. One can easily imagine what it means for a couple if one has no body. As the film demonstrates intimacy needs to be reinvented wholesale. Are we to expect such technical solutions at the end of our road to perfecting digital romantic life—love 4.0, so to speak?

In parenthesis: In general, there are four main types of artificial intelligence: reactive machines, limited memory, theory of mind and self-awareness. The first, basic type of artificial intelligence comprises reactive machines. They do not have a memory and can only function with respect to a specific task, such as the chess computer *Deep Blue*. The second type of artificial intelligence involves a limited memory. This is already in use to some extent in the field of self-driving cars. This type of artificial intelligence has a certain kind of memory that allows it to observe things over time, for example, in the case of self-driving cars, the speed of other cars; this helps self-driving cars not only to make decisions in a specific moment, but also to build up an archive of experiences. The third type includes the theory of mind. Here, artificial intelligence would be programmed to have a kind of social intelligence, in and through which it would understand the intentions and emotions that other subjects might have, and predict future social behavior on this basis. This type has not yet been invented. The fourth type, which also does not yet exist, involves the concept of self-awareness. Here, artificial intelligence would also have an understanding of itself.[6]

To return to the case of digital romantic life 4.0 and the film *Her*, before one can give an adequate answer to the question of how to evaluate this case, looking at the field of digital enhancement reveals the necessity of *first* describing its paradoxical conditions, which have thus far been too neglected. To this end, the present book seeks to give an answer to the following, so far seemingly forgotten question: How can we become aware of the paradoxical conditions in which subjects want to use digital enhancement in order to perfect their lives, *while simultaneously* being aware of the freedom-impairing effects that this use has on the digital contexts within which their lives are conducted? The key theoretical claim is that a study *merely* focusing on the level of norms, values, and ideals is rather *in*capable of uncovering and addressing the paradoxical condition of digital enhancement described above. For norms, values and ideals are *not* a fixed unity, that is *in*dependent of this paradoxical condition, *but rather* they are part of it. The present book's point of departure is the hypothesis that if we want to "open our eyes," to come back to Foucault's phrase, and see what is at stake in this paradoxical condition, then we need to gain insight into the subtle affect-sensitive relations of power that generate this paradoxical condition. The notion of power is one of the central *topoi* in twentieth-century philosophy, i.e., Foucault, and its nineteenth-century predecessors, i.e., Friedrich Nietzsche, and figures among the main themes in contemporary political philosophy, social theory and social philosophy.[7] Whereas power relations are usually grasped in material terms, the case of digital enhancement shows the need for conceiving of these power relations as a cognitive phenomenon. This hitherto underrated approach helps us to obtain the perspective required for describing the subject's *specific* blind gaze in digital enhancement on a cognitive level. Thanks to this perspective, we are able to adequately address the central problem that the subject in digital enhancement is so to speak *with "open eyes" (sehenden Auges) ready to receive* such a formed will that *she wants* to be conducted by the logic of perfection. Accordingly, "blind" does not mean the not often appearing cases in which a subject just did not see something. Instead, "blind" here signifies *the subject's acceptance of being formed in his will*. The theoretical framework of affects—or, in other words, of desire—allows us to describe the subject's *openness* to this will formation *without* being forced by any kind of sanction or other threat of repression. It is for this reason that the key figures on whom the theoretical framework draw are Nietzsche and Foucault who places himself in the Nietzschean tradition.[8] Thus, the present book brings nineteenth-century German social and political philosophy into dialogue with newer French philosophy. This dialogue provides the theoretical tools for understanding the subtle, machine-like mechanisms by which the subject is *systematically* structured in his will formation by the logic of perfection.[9]

The book at hand therefore goes on to delineate these affect-sensitive power-relations in the context of beliefs and imaginations, and it shows the different degrees of how procedures of training, such as those involved in habitualization, normalization, and passivization, make the subject to form the specific belief about or imagination of his very own life that this subject is prompted to attach to the logic of perfection. As a consequence, the subject blindly accepts the logic of perfection as the alleged optimal life formation. The present book describes this understanding as freedom-endangering inasmuch as it narrows down the subject's gaze in a machine-like and thus blind way to reach merely one allegedly valuable perspective, i.e., in our digital today, the perspective marked by the logic of perfection.

Moreover, referring to Nietzsche and Foucault provides the further advantage of seeing *the affect- and power-sensitive Janus face of subjectivity in digital enhancement*. Instead of *either* dismissing digital enhancement as only freedom-impairing *or* as praising it as solely freedom-empowering, as the existing literature on digital enhancement tends to do,[10] this book lays out how a critical-theoretical analysis of digital enhancement should pay attention to *both* the freedom-endangering *and* the freedom-empowering sides of digital enhancement that are manifest in the subjects' way of relating themselves to themselves: in other words, in their subjectivity in the digital today. In addition to the freedom-endangering facets of digital enhancement described above, the critical-theoretical method of genealogical critique, which was invented by Nietzsche and further developed by Foucault, gives us the required theoretical tool for conceptualizing the freedom-empowering facets. It demonstrates that subjects *in* digital enhancement are able to *transcend the context of the logic of perfection* and become aware of their formed will. In the terms of the Frankfurt School such a Janus-faced subjectivity in digital enhancement could also be called *dialectical subjectivity in digital enhancement*. "Dialectical" means that the freedom-endangering side *can turn* into the freedom-empowering side of digital enhancement if the subject is able to learn in a self-reflexive way to become aware of its openness to having a formed will by the logic of perfection. The reader might be surprised that the method of genealogical critique and *not* the method of ideology critique is used. For talk of "opening one's eyes" could create expectations of the latter. Then one would think of a gaze that was thus far veiled and that needs now to be *un*veiled. This approach might be interesting, too.[11] But here the focus is slightly shifted. Here, the question is how an appropriate understanding of the history of how one became what one is in our digital present, which is often neglected in contemporary discussions, opens up new, hitherto unforeseen perspectives. This is the task of genealogical critique.[12]

In general, genealogical critique is the means by which it is theoretically conceptualizable that existing values have a history and are therefore not to be taken as eternally true; genealogical critique thus demonstrates that values have a *changeable* character and need to be *re*-evaluated according to the specific context and time. Subjects in digital time obviously do not have to be aware of the theoretical conceptualization of genealogical critique, but for us it is necessary to describe the freedom-empowering side of the affect- and power-sensitive Janus face of subjectivity in digital enhancement. In general, there are two perspectives on genealogical critique. Firstly, genealogical critique can be regarded as vindicatory, that is as conceptualizing the stabilization of the current conditions of life formation.[13] Secondly, genealogical critique can be understood as debunking (Nietzsche, Foucault), that is as conceptualizing the *de*stabilization of the conditions of life formation that have lost its validity.[14] The understanding of genealogical critique advanced here promotes the understanding of genealogical critique in its debunking mode. This view includes *by itself* a constructive character. For debunking, or less harshly put: criticizing, the norms, that have lost its value, *can construct* new spaces for thinking, living and feeling *differently*.[15] If the freedom-endangering side entails the subjects' narrow gaze, by which they evaluate the logic of perfection as the alleged optimal way of life, this understanding of genealogical critique helps us to conceptualize ways of adequately broadening the view. It makes it conceivable that the subjects' gaze is *now broadened* in the sense that they are now able to see different ways of relating themselves to the logic of perfection in digital times, which they no longer regard as *the only good* way of life. Hence, plural ways of thinking, living and acting become perceptible for them. But the subject does not receive such a plural view like taking a pill, as it could be thought in terms of enhancement practices; *on the contrary*, the subjects only achieve this broadened view by participating *themselves* in *self-reflexive* procedures of enlightenment that shed light on the paradoxical conditions that have *hitherto* narrowed down the subjects' gaze.[16] Accordingly, the subjects are no longer relating themselves to themselves in digital enhancement in a machine-like manner, but in an enlightened manner in and through which they become aware of the historically produced context of digital enhancement. The present book calls the freedom-endangering side of Janus-faced subjectivity in digital enhancement *quasi-automatized subjectivity that is willing to conform*. The freedom-empowering side is called *enlightened (aufklärerisch) subjectivity*. In contemporary terms, it would be called democratic subjectivity. But given the language used by Nietzsche and

Foucault this would be too much of a stretch. In this sense, this book contributes to subject-based examinations of digital enhancement. Group-based examinations are interesting, too. But the focus that Nietzsche and Foucault provide shows that if we want to understand how a group entails subjects that do not think, act and live in a merely conform, machine-like manner, then we need to understand how subjects are formed in and through power relations that the book at hand specifies as affect-sensitive.[17]

The title of the introduction "On Digital 'Re-Evaluation' or What is Digital Enlightenment Today?" takes up the understanding of the two sides of the affect- and power-sensitive Janus face of subjectivity in digital enhancement. "Re-evaluation" (BGE, 203, loc. 2242 of 5195, Kindle; J, 5, 203, 126) stems from Nietzsche and involves critical awareness of one's own subject position in the social and political normative orders, put differently: it encompasses "*Umwertung*." It is this understanding that especially Foucault further develops by the term enlightenment (WE). Moreover, "re-evaluation" is not only part of this theoretical trajectory, but also encompasses the field of critical race studies. Here, it is especially Martin Luther King Jr. who referred in his terms to the notion of "re-evaluation," mostly in *Stride Toward Freedom* in discussing Henry Wieman, Alfred North Whitehead and Paul Tillich.[18] In King's terms "re-evaluation" encompasses the processes in and through which subjects can become aware of the racist system in the US and can learn to alter it. This kind of self-awareness-raising is part of non-violent campaigns. By this, the participants can gain, and now I quote King Jr., "new estimate of their own human worth."[19] Also in today's critical-theoretical discussion the notion of "re-evaluation" is widely spread. Here, it is especially Charles Taylor who refers to it in the sense of "reconciliation";[20] and it is in particular Tully who develops it in the sense of becoming aware of the relative value of the existing alleged practices of freedom that stem from hitherto practiced forms of life in comparison to practices of freedom in terms of non-domination.[21] Now, with respect to our digital present, the book at hand conceptualizes the so far not so much seen kind of *digital "re-evaluation."* By this, the book at hand provides a set of critical tools. Firstly, this set can empower a subject to "become aware" that her subject position, life formation and participatory conduct is *not* as free as first perceived, but *silently* governed by the various ways she participates in the digital labyrinth. Secondly, this set also provides subjects with critical aesthetic tools that can empower them to start to participate in practices of freedom, that is in practices of thinking and acting *differently*.

Against this background, the book at hand focuses on the critical-theoretical analysis of the subject position in digital enhancement. This view implies the understanding of the subject's relation to herself to the logic of perfection. Here, of course, it is very seminal to see the intersections with different perspectives on digital change, too. The key study that shows the intersection between capitalism and digital change is Shoshana Zuboff's far-reaching book *Surveillance Capitalism*.[22] She demonstrates that capitalism in times of digital change exploits not the subjects' labor force, as one could initially think, but the subjects' digital behavior. The interrelation with "comparative advantage" is also important to note.[23] Here, the telos that is under scrutiny is not the norm of perfection as done in the present book, but the telos of the profit motif part of capitalism; digital networks are also characterized by this capitalist rationale. In this context, we deal with what can be called the Hobbes, Mandeville, Rousseau, Smith and Kant (Universal History) hypothesis, as I learned from Tully.[24] According to this hypothesis, we can see unintended consequences of peace and justice in future generations that are produced by a hidden hand. Furthermore, it is also widely discussed that digital change orders its users, participants and followers in types of "epistemic bubbles" and "trust bubbles,"[25] that generate what is nowadays called the post-truth age and fake news age, mostly known since Trump's presidency and his use of twitter. This can be called, to put it with Tully, the Clausewitz "war by other means" hypothesis. It highlights that dialogical interaction can deform into the we-them logic of the trust bubbles *against* the *dis*trust bubbles.[26]

By developing such an understanding of the affect- and power-sensitive Janus face of subjectivity in digital enhancement, it becomes clear that *although* Nietzsche and Foucault are skeptical of determining existing norms, ideals, or values as a scale for evaluating social phenomena like today's digital enhancement as good or bad, the critical-theoretical method of genealogical critique *nonetheless* entails an *implicit* normative criterion: It is *not every* norm, ideal, or value that is to be "mistrust[ed]" (GM, III, 24, 398, trans. by the author), to use a Nietzschean term, but *only those* that are not justified. What is not justified according to Nietzsche's and Foucault's critical-theoretical method of genealogical critique is everything that narrows down the subject's experience of and thinking about how to think, act and live *differently*, in other words, that establishes patterns of affective domination, and furthermore, it means that no one can be excluded from the potential to think and experience things differently.[27]

Also the book's subtitle, *Government of the Self and Desire*, alludes to the affect- and power-sensitive Janus face because "the self" or the subject has a double position: on the one hand, the subject is intended to be the object that is open to being governed by the logic of perfection, and on the other, she appears as the subject that does not want to be governed in this way. Contemporary subjects in digital enhancement face the challenge of having to combine these two perspectives. In other words this means that they face the challenge of transforming their desire for being governed by the logic of perfection into desiring *differently*.

Previously, the notion of desire has received scant attention in the scholarly debate of digital enhancement, as is evident from a glance at classical dystopian literature. Aldous Huxley's *Brave New World*, George Orwell's *1984* or, more recently, Juli Zeh's *Corpus Delicti*[28] all describe in distinctive ways what a completely administered world might look like, one in which people are disciplined by media or—to borrow Orwell's phrase that has found its way into contemporary culture—by "Big Brother."[29] Zeh does so in the context of a totally controlled dictatorship of health and body care, Huxley in the context of a totally organized hierarchical system that makes resistance impossible since people are too seduced by sex and drugs, and Orwell in the context of a society characterized as a total surveillance state. Also contemporary theoretical literature neglects the notion of desire and mostly focuses on the disciplining mechanisms of surveillance.[30] Furthermore, it is quite common to refer more broadly to notions of power to theoretically conceptualize the field of digital change that are *not*, however, sensitive to affects.[31] As helpful as they are, however, such conceptions are not sufficient to examine the current problem of big data.[32] In focusing on disciplinary methods, these approaches *cannot* explain the paradoxical conditions in and through which people *are open* to being conducted and thus to being disciplined by the logic of perfection, although they know of its freedom-endangering effects. But it is exactly these paradoxical conditions with which digital enhancement confronts us. People possess the will to consent freely to the logic of perfection, for example, to suspend legal protection mechanisms such as those designed to protect their personal data—despite of what they know about data privacy.[33] From this we see that *although* legal and administrative mechanisms exist and, of course, are very helpful, the mere existence of the legal protections is *not sufficient* to prevent people from, for example, quasi-automatically giving their data away. Foucault was already aware of the fact that the "juridical forms" (TJF)—that is, questions concerning what is forbidden and allowed, or, to update his analysis, the administrative and legal mechanisms surrounding data protection—are not sufficient to understand how subjects *actually arrive* at giving consent.[34] From this perspective, we see that the theme of decision-making, of deciding to consent to the data collection of this or that Internet site, does not enable us to tackle the paradoxical conditions: We are confronted with a ready-made subject that has *already* formed his will to consent. This subject is one that is willing to suspend legal protection mechanisms. In order to disclose these paradoxical conditions, we need to refer to the notion of desire.

There are already a few studies that deal with the notion of desire. These studies are very helpful, in particular the analyses by danah boyd and Bernard E. Harcourt.[35] They describe the subject's desire in more concrete terms. These studies analyze how subjects—or in boyd's case, young people—are seduced to "expose"[36] themselves in digital media.[37] To this view, the present book adds the broader perspective by examining the subject's desire in terms of affect-based power relations that this book specifies as a form of subjectivation. In general, the notion of subjectivation goes back to Foucault's teacher Althusser who had a major influence on Foucault. Usually, subjectivation encompasses the process of becoming a subject.[38] If we now propose examining the notion of subjectivation in terms of affect-centered power relations, it is *not* because we want to develop a theory of affects, as might be suggested by the recent meteoric career of the notion of affects, which even led to the so-called affective turn. This aim would be foreign to Foucault. In this sense, following Foucault, the book at hand interprets the affective aspect as an analysis. What this means is that Foucault was not so much interested in an abstract, methodological understanding of what power *in general* is, as a theory would deliver, but rather he was interested *in empirical cases* and tried to understand how power relations work in concrete terms. This is what an analysis in Foucault's sense does. Now, we develop the register of a *micro*analysis. Such a specific view of analysis enables us to theoretically conceptualize the notion of desire as the type of power in the empirical case of digital enhancement that takes place *in the subject's concrete life formation*. Here, we describe the power relations in the subject's concrete life formation as a *double mode* of quasi-automatically becoming a subject in digital times in *both* attaching to the logic of perfection *and* in *de*taching from the logic of perfection.

For readers familiar with Foucault, it might seem surprising to use Foucault to develop an understanding of desire. For it is Foucault, as Gilles Deleuze emphasizes, who even according to his own formulations "hates"[39] the notion of desire. But Foucault does *not reject entirely* the notion of desire but rather suggests examining it in connection with his understanding of pleasure.[40] In the interview "Genealogy of Ethics" Foucault explains what he means by this. By this he does *not* interpret the "desiring man" (HS 2, 5)—today we would better say the desiring subject—and thus the subject of pleasure as *solely* longing for sexual desires.[41] He notes: "sex is boring" (GE, 253). Foucault is instead interested in understanding *the subtle mechanisms* that put the subject on track of becoming a specific subject, such as becoming the subject of pleasure.[42] In what follows we develop the latter notion further. Our aim here is to understand the subtle mechanisms in and through which the subject *is prompted* to become a subject in digital enhancement.

The fact that such an understanding of desire is central in today's digital field, becomes perceptible if we consider key examples of the freedom-empowering facet of desire as observable in the well-known WikiLeaks scandal. WikiLeaks, which was founded by the whistleblower Julian Assange, is a platform that publishes secret documents from whistleblowers. For example, in 2010 the whistleblower Chelsea Manning, a former member of the US-American army, revealed on WikiLeaks how innocent people were killed by US-American soldiers during the war in Iraq. Most of the scholarly debate revolves around the question of to what extent WikiLeaks is a justified form of resistance or not.[43] Here, the focus is shifted to the circumstance that the subject's desire underwent a change. The freedom-empowering side of desire is seen in this example as Manning turned from someone whose life was blindly governed by his will to fight for the US army into someone who gained a different view on his work as a soldier and thus turned into someone who desired to be governed *differently*.

Furthermore, the relevance of desire in its freedom-endangering aspects becomes obvious if we consider the storm of the Capitol on January 6, 2021 in Washington D.C. Trump had constantly denied Joe Biden's victory in the US presidential elections, and did so again on January 6, the day when Biden was formally confirmed as the 46th president of the United States of America. Trump used social media, in particular his Twitter and Facebook accounts, to incite his supporters *not* to recognize the election, which he saw as cheating him and his voters. With tweets like "Big protest in D.C. on January 6th [. . .]. Be there, will be wild!"[44] Trump fueled the rage of the right-wing protesters, many of whom came from the QAnon conspiracy theory movement or the Proud Boys, part of the strand of white supremacy. This book now considers the so far underestimated, but central role of desire within the storming of the Capitol in order to explain the *blind* attachment of Trump's supporters to what they read and saw on Trump's Twitter and Facebook account. This view enables us to theoretically conceptualize how they were so seduced by Trump's digital call that they became the sort of subjects that put aside what they had in their heads, stormed the Capitol, and killed six people.

What this means concretely becomes obvious in referring to Steven Spielberg's movie *Minority Report* which is based on a story by Philip K. Dick. Spielberg's movie *Minority Report,* centers around the question of what it would mean if we lived in an entirely predictable society. In the film, the members of a fictional future police department called "Pre-Crime" use data to predict future crimes. The predictions promise to reduce the crime rate by preventing crimes *before* they are actually committed, and thus to create a perfectly secure order. What is required for establishing such an alleged perfectly secure order, is interpreted here as a specific kind of affect-sensitive power relations in and through which people are prompted by what they believe to attach to the given social order. Analogously, the officers working for "Pre-crime" do not call themselves policemen, but instead "clergy." Like monks, they possess an unquestioning belief in the reliability of their system, which they deem perfect. But things become complicated with the emergence of a *minority report*—namely, a dissenting opinion concerning a predicted crime of one of the creatures that predict the future crimes—in other words, the easily-overlooked, but still extant cases that do not conform to predictions that divide the members of a future society into murderers and law-abiding citizens. Thus, the film shows that even an allegedly "perfect system" has flaws. Not every prediction that reveals ordinary people to be future law-breakers turns out to be true. *But the more* people believe in the system, *and the more* they simply follow what it predicts, *the farther* they travel down the path of digital unfreedom. "Digital unfreedom" here means that people *only* value the need for an allegedly totally secure society as relevant for their lives. As the film goes on to show, the tiny amount of freedom people still possesses consists *only* in downloading the "minority report," which can prove that the system has wrongly accused someone of a future crime based on a mistaken prediction. Thus, the "minority report" shows the rare cases of digital freedom. "Digital freedom" encompasses here the recognition of dissenting perspectives that thus pluralizes the common view. As the story unfolds, the protagonist of *Minority Report*, played by Tom Cruise, is able to download such a "minority report," but, not surprisingly for a Hollywood production, he is obviously an exemption. Ordinary people cannot follow in his footsteps.

Having thus far addressed the debate of digital enhancement from the perspective of affect-centered power relations, we need to clarify what we mean when we speak of *the* literature. Strictly speaking, *the* debate of digital enhancement does not really exist. Rather, as this book seeks to demonstrate, there are *four* debates that have previously been separated[45] *Firstly*, there is the debate of enhancement in which the notion of desire is not considered to be worthy of in-depth discussion.[46] *Secondly*, there is the debate of digital change that does not include the notion of desire developed here.[47] *Thirdly*, a discussion of the art of living that analyzes how subjects form their life within the pre-given social order. This debate neglects the notion of desire sketched here, too. *Fourthly*, there is the debate of affects.[48] Here, the cognitive aspects in the mechanisms of desire are rather neglected. If cognition and desire are usually regarded as being opposed to each other, it is Nietzsche and Foucault who show how they are tied together. Desire is viewed as the immediate attachment that makes people in digital enhancement have the will to regard the existing narratives, knowledge, and images as optimal on a cognitive level. Previously, these four debates have rarely been combined. Even though they share a common denominator. Digital change is widely conceived as a matter of solving problems in the analogue world through the use of digital media,[49] implying improvement. Things are quite similar when we speak of enhancement because processes of enhancement are also generally understood as rendering something better demonstrated by their semantics. Additionally, the debate on the art of living allows us to address the question of how subjects position themselves in relation to the logic of perfection. Furthermore, the notion of affects delivers the required description of the catalyst. It makes understandable why the subject *is prompted* to set himself in relation to the logic of perfection. In bringing the four discussions into dialogue, we propose to intervene into the debate of digital enhancement.

In light of these considerations, we see a need to promote an understanding of aesthetic critique. The aesthetic critique developed here becomes what we call, using Axel Honneth's term, a "critique of power";[50] this is to say the specific "critique of power" that is sensitive to affects. That Foucault indeed contributes to the field of aesthetics is disputed. His thoughts on aesthetics are widely understood as contributions to an account of the "philosophical life," as Pierre Hadot, in particular, suggests.[51] This view would lead to *an ethical critique* that tackles questions in the context of life formation. An ethical critique, however, would not be able to tally the price of affective-sensitive power relations that produce the paradoxical situation in digital enhancement. It is only the field of aesthetics that promises to do so in an adequate manner. Aesthetic critique in the sense developed here must then be conceptualized in terms of the ancient Greek understanding of "αἵσθησῐς" (*aísthēsis*) and thus of "sensory perception." The critical potential derives from the ancient Greek verb "κρίνειν" (*krínein*),[52] which means "to make distinctions." The art of perception thus becomes critical in cases in which the subject is able to distinguish between the perspectives on life formation that are justified and those that are not justified.[53]

To understand what is justified and what is not, we need to refer to the Canadian political philosopher James Tully, a member of the Cambridge School including other prominent figures such as Quentin Skinner or John G. A. Pocock. According to this school, one neither only theorizes grand political ideas nor merely undertakes socio-structural analysis to understand the author's motifs, but rather one needs to see how political thinking emerges in and through specific contexts, discourses and time frames.[54] In this sense, political philosophy turns into a "practical philosophy,"[55] which is interested in understanding concrete, practical problems by uncovering the political practices that shape the "history of ourselves"[56]—as Foucault, one of Tully's main influences, would put it—such as the various forms of how people are governed. This understanding of political philosophy discloses the settings in and through which subjects become what they are, and in doing so, it reveals the contingency of that history. Such a view makes room for understanding the "practices of freedom"[57] in and through which subjects can glimpse at different ways of becoming who one *could be*.[58] If we further think this trough, then it becomes obvious, following Tully, that justified is only that which allows different ways of becoming who one is to emerge. As a consequence, justified contexts are those that make room for developing *plural* voices and not only for *one* dominant voice.[59]

Analogously to Tully's understanding, the aesthetic critique developed here needs then to be understood as a "practical philosophy" whose purpose is to analyze the sensory perception of the "history of ourselves" in digital enhancement. It promises to provide an appropriate understanding of the paradoxical condition of our digital today. It discloses to what extent the subject's sensory perception is already fueled by relations of domination in and through which the logic of perfection becomes the dominant voice in society, which excludes dissenting, plural voices. Aesthetics understood in this way is specific because it does *not* deal with just *any* perception, but with the *specific* method, which the present book compares with the method of zooming in part of a burning glass. This method magnifies the problem of the bigger picture, which is the dominating voice of the logic of perfection, on small, concrete case studies, which represents the subject's concrete life formation. Here, the argument develops the line of thought by differentiating between the *entire* problem of the bigger picture and the possibility of *partial*, context-transcending solutions in some areas. Accordingly, it claims that although the bigger picture might be structurally problematic, since the logic of perfection is the leitmotif of the present, there is the possibility of transcending this bigger picture in concrete areas and of opening up spaces of enlightened freedom. These concrete areas are not conceived of as fixed settings, but need to be reached via procedures of genealogical critique. Genealogical critique sparks the subject's enlightened attitude in and through which she is able to take distance from the logic of perfection. The subject, of course, might not necessarily be aware of such a methodological approach to aesthetics, but the present book hopes to demonstrate that the observer requires these perspectives to understand the full scope of the subject's life formation.[60]

The argument developed here is situated within the very instructive contemporary debate in critical theory and feminist critical theory as follows:[61] Foucault is mainly accused of *not* providing the theoretical means to generate a critical freedom-empowering and emancipatory perspective. This objection is mostly advanced by feminist critical theory in the tradition of Habermas, e.g., Seyla Benhabib. Benhabib, seminal though her work is, considers Foucault as *un*helpful for critical thought because, as she argues, he does not provide the possibility to transcend the given contexts, and thus does not make room for an emancipatory perspective. With Foucault, according to Benhabib's reading, one remains caught in the meshes of power and thus *bound to the contexts* in and through which one *is*.[62] In contrast to Benhabib, deconstructive feminist critical theorists such as Wendy Brown and Butler argue that a context-transcending perspective would not be helpful, as this perspective would neglect the power structures that underlie subject formation. Nonetheless, Brown and Butler also see Foucault as failing to consider the desire for freedom, since he neglects the psychoanalytical structures that form the subject's will, according to them. He does so because, in his reference to Nietzsche's will to power, he stresses the power relations and forgets the will, according to Brown and Butler at least.[63] The feminist critical theory proposed by Allen considers Foucault helpful for the study of power relations, but *not* for that of social transformation.[64] Furthermore, unlike Brown and Butler, she sees Nietzsche as providing a perspective of heroic individualism, which neglects structural power relations.[65]

The argument developed here considers the method of genealogical critique as the key to understanding Foucault's freedom-empowering perspective, which the present study specifies as enlightened (*aufklärerisch*). This method has long been neglected in (feminist) critical theory. A very instructive study, showing the eminent role of genealogical critique in the tradition of Nietzsche and Foucault, is Saar's brilliant monograph *Genealogie als Kritik.*[66] If we take the view developed there to its logical conclusion, however, it would lead to an "*immanence of power*."[67] This view is very helpful to see two things: firstly that power can*not* be criticized from an abstract perspective from an alleged nowhere, but requires a perspective from within, and secondly that one can become aware of *the potentiality of freedom*. But to understand *the realization of freedom* it is necessary to consider *context-transcendence*. For the subject can only realize his own freedom by seeing that he *really* can leave behind the power structures in and through which he is situated. This path is proposed in the present book, with a combined reading of genealogical critique and the notion of enlightenment. Genealogical critique, this book argues, can theoretically conceptualize how the subject can become aware of the power relations in and through which he has so far been constituted in *in*transparent ways. An awareness of this entanglement leads to the emergence of an enlightened agency. It theoretically conceptualizes the subject's *clear gaze* on the power relations in and through which he is constituted that the present book sketches as the enlightened gaze (*aufklärerische Blick*). By creating such an enlightened gaze, the subject is able to transcend the context in and through which he has so far been constituted in intransparent ways. It is this understanding that makes Foucault fruitful for studying not only relations of power, but also the desire for freedom and enlightened agency. The book at hand aims to develop this previously neglected perspective in the following chapters.

The analysis of the present time developed here requires the following approach to Foucault's work: This analysis of the present time is grounded in Foucault's understanding of knowledge, power, and life formation that permeates his entire work. Whereas Foucault's work is generally divided into three parts,[68] i.e., knowledge, power, and ethics, or, life formation, and this view is very helpful, we conceive of knowledge, power, and ethics as a *relative unity*.[69] As a consequence, we do *not* think of different Foucaults. Accordingly, we do not think of a Foucault that thinks *at first* through contexts of discipline in conceiving the systems of knowledge and power and of another Foucault that *then* considers possibilities of freedom in contexts of life formation, as it is usually done. *On the contrary*, from the perspective of the relative unity of his work, we note that the subject is *in fact* structured by the systems of knowledge and power, but that she is *far from* being "dea[d]" (OT, 342). What vanishes "like a face drawn in sand at the edge of the sea" (OT, 387), to cite the last lines of *The Order of Things*, is *not* the understanding of the subject *as such*, but *merely* the understanding of the sovereign subject that is conceived as having everything in her hands. Thus, the perspective of the relative unity of Foucault's work developed here enables us to conceptualize that *on the one hand*, the subject is determined by the systems of knowledge and power, but *on the other hand*, she can do something against her determination and is able to transcend the determining contexts; the present book seeks to show this by developing the understanding of *the subject's enlightened ethos*, i.e., the enlightened attitude towards the subject's relation to the systems of knowledge and power.

In a preliminary chapter, we draw on Nietzsche's work, since Nietzsche is the founding father, so to speak, of genealogical critique. For the purpose of this analysis, it suffices to follow the usual tripartite arrangement of Nietzsche's work.[70] The early Nietzsche refers to the time when Nietzsche was in Basel. *The Birth of Tragedy*, in particular, belongs to this period. It ends with the publication of the *Untimely Meditations*. The middle period is characterized by the rather sensitive works, such as *Daybreak* or *The Gay Science*. The late, more drastic Nietzsche encompasses the works of *Thus Spoke Zarathustra*, *Beyond Good and Evil* and *The Genealogy of Morals*. But here, too, analogously to the interpretation of Foucault's work, we need to speak of *a relative unity* of Nietzsche's work. From this perspective, it becomes clear that the early Nietzsche is interested in understanding what life is about and the late Nietzsche, beginning with the *Zarathustra*, answers this question in developing the conception of the "will to power."

Methodologically, the argument developed here is rooted in the following considerations: Foucault's method of "the genealogy of problems" (GE, 256) is needed, permitting us to derive from Foucault's *descriptive* analysis the *systematic* perspective required to portray the *today's* affect- and power-sensitive Janus face of digital enhancement. Foucault obviously could *not directly* address our present-day concerns. But the systematic preconditions of such a view are already anchored in his texts (HS 2, 10p). For Foucault himself saw the necessity for a "history of the present" (DP, 31) that did not stop in his own time, the 1970s and 1980s, but could be continued. The analysis of the present digital time developed in this book is situated in this line. From this perspective, the study is situated in Foucault's approach to investigate the history of "the experience of madness, the experience of disease, the experience of criminality, and the experience of sexuality" (GSO, 5). We expand this line to the history of the experience of the logic of perfection, and thus to the common denominator of digital enhancement.[71]

The following, further methodological remark is necessary: Basically, the method is descriptive. *Implicitly* it is also normative. "Descriptive" is the method inasmuch as it provides a description of the paradoxical conditions of digital enhancement. This description entails the understanding of the affect- and power-sensitive Janus face of subjectivity in our digital today. In and of itself, the description is not normative, but value-neutral. Usually, the critical-theoretical method of genealogical critique is even regarded as skeptical of a normative approach, since it seeks, at least in its most common understanding of it,[72] to debunk and thus to *de*stabilize the existing norms. *Nonetheless*, if it does not want to lose its critical potential and fall into a relativistic "anything goes,"[73] genealogical critique needs to be characterized here by *an implicit normativity*. As described above, this normativity emerges by virtue of the need to specify what kind of norms exactly have to be destabilized. In line with Tully, we view these as *not every* norm, but only those norms that are not justifiable since they hinder the emergence of different voices within the context of digital enhancement: i.e., all those voices that dissent from the logic of perfection. This understanding of normativity is regarded in light of what is called in scholarly debate "critical constructivism."[74, 75] By *criticizing* the existing *un*justified norms, it *constructs an enlightened agency*: namely, an agency that is able to shed light on the paradoxical conditions of subjectivity in digital enhancement.

Against this backdrop, this book is divided into three parts. The first examines the object of analysis, that is digital enhancement, and develops a conception of aesthetic critique. In this context, the diagnosis refers to empirical as well as fictional material, such as documentaries, films and TV shows. The second develops, with reference to Nietzsche and Foucault, the empirically informed analysis of the present time to conceptualize the affect- and power-sensitive Janus face of subjectivity in digital enhancement, i.e., its freedom-endangering and freedom-empowering aspects. Part three describes how scenes of affect-centered power relations can activate the subject's intra-subjective communication between the two sides of the Janus face of subjectivity in digital enhancement. To this end, the book brings newer French philosophy into dialogue with newer French literature, such as Annie Ernaux and Édouard Louis, as well as Virginia Woolf's feminist classic *Orlando*.

NOTES

1. Foucault's work is quoted following the standard abbreviation system. The abbreviation system refers to the English translation. Reference to secondary literature is only made when necessary to accentuate the approach developed here.

2. Foucault has recently been accused of pedophilia. The case has been investigated and the allegations are wrong. Daniel Defert, François Ewald and Anne Thalamy, "Communiqué sur les accusations de Guy Sorman contre Foucault," available at: https://centremichelfoucault.com/wp-content/uploads/2021/05/Communiqu%C3%A9-Foucault_English-rev-2.pdf.

3. In keeping with a feminist perspective, the word "subject" is randomly replaced by the pronouns "he," "she," and "it" in the following discussion.

4. Obviously, there are situations where one does not need a specific critical eye to identify a situation as a problem. This is the case, for example, when one sees refugees dying in the Mediterranean. Without Author, "Deaths at Sea on Migrant Routes to Europe Almost Double, Year on Year," *UN News: Global Perspective Human Stories*, April 29, 2022, available at: https://news.un.org/en/story/2022/04/1117292.

5. Originally, Foucault introduced the figure of the shepherd in his lecture *Security, Population, Territory* in order to show how people can be gently governed like sheep by a shepherd. Cf. STP, 136.

6. Cf. Arend Hintze, "Understanding the Four Types of Artificial Intelligence, from Reactive Robots to Self-aware Beings," *The Conversation*, November 14, 2016, available at: https://theconversation.com/understanding-the-four-types-of-ai-from-reactive-robots-to-self-aware-beings-67616.

7. Amy Allen, *The Politics of Ourselves: Power, Autonomy, and Gender in Contemporary Critical Theory* (New York: Columbia University Press, 2008); Judith Butler, *The Psychic Life of Power: Theories in Subjection* (Stanford: Stanford University Press, 1997); Rainer Forst, *Normativität und Macht: Zur Analyse sozialer Rechtfertigungsordnungen* (Berlin: Suhrkamp, 2015); Sabine Hark, *deviante Subjekte: Die paradoxe Politik der Identität* (Opladen: Leske+Budrich, 1996); Axel Honneth, *Kritik der Macht: Reflexionsstufen einer kritischen Gesellschaftstheorie* (Frankfurt a.M.: Suhrkamp, 1988); Martin Saar, *Die Immanenz der Macht: Politische Theorie nach Spinoza* (Frankfurt a.M.: Suhrkamp, 2013).

8. In this sense, we refer to *the French Nietzsche*. For this theme see Werner Hamacher, ed., *Nietzsche aus Frankreich: Essays von Maurice Blanchot, Jacques Derrida, Pierre Klossowski, Philippe Lacoue-Labarthe, Jean-Luc Nancy und Bernard Pautrat* (Frankfurt a.M./Berlin: Ullstein, 1986) and Bernard E. Harcourt, Martin Saar and Sarah Bianchi, "The Critique and Politics of Identity: On the Affinities between Critical Theory and Poststructuralism. A Conversation with Bernard E. Harcourt and Martin Saar conducted by Sarah Bianchi," *The Coils of The Serpent: Journal for the Study of Contemporary Power* 10, no. 1 (2021): 118–30. Foucault's pre-eminent role in the French reception of Nietzsche is shown by Duncan Large, Translator's Introduction to *Nietzsche and Metaphor*, by Sarah Kofman (Stanford: Stanford University Press, 1994), viipp; and Martin Saar, *Genealogie als Kritik: Geschichte und Theorie des Subjekts nach Nietzsche und Foucault* (Frankfurt/New York: Campus, 2007).

9. In general, newer French philosophy refers to the French philosophy of the second half of the twentieth century. This includes Foucault in particular, as already mentioned, but also other well-known thinkers such as Jacques Derrida, Gilles Deleuze or Jean-François Lyotard as well their theoretical predecessors Roland Barthes, Louis Althusser and Jacques Lacan. Although contemporary French philosophers cannot be conceived as a homogeneous group, despite their differences there are *nonetheless* common themes. This strand of philosophy is especially characterized by conceptualizing the entanglement of power and knowledge. Cf. Martin Saar, "Critical Theory and Poststructuralism," in *The Routledge Companion to the Frankfurt School*, ed. Peter E. Gordon, Espen Hammer and Axel Honneth (New York/Abingdon: Routledge, 2018), 323–35.

10. For a paradigmatic perspective that exclusively stresses the freedom-impairing strand, see Glenn Greenwald, *No Place to Hide: Edward Snowden, the NSA and the Surveillance State* (New York: Picador, 2014). For an account that exclusively highlights the advantages of the society of surveillance, cf. William Simon, "In Defense of the Panopticon," *Boston Review* 39, no. 5 (2014): 58.

11. In general: Sarah Bianchi, "Unter dem und wider den 'Bann des Einheitsprinzips': Identifizierungsmacht, Subjekt und Affekt nach Adorno," *Zeitschrift für Praktische Philosophie* 9, no. 2 (2022): 15–48; Rainer Forst, "Gerechtigkeit nach Marx," in *Nach Marx*, ed. Rahel Jaeggi and Daniel Loick (Berlin: Suhrkamp, 2013), 107–21; Raymond Geuss, *The Idea of Critical Theory: Habermas and the Frankfurt School* (Cambridge: Cambridge University Press, 1981); Rahel Jaeggi and Robin Celikates, *Sozialphilosophie: Eine Einführung* (München: Beck, 2017); Martin Saar, "Ideologie," in *Habermas-Handbuch*, ed. Hauke Brunkhorst, Regina Kreide and Cristina Lafont (Stuttgart: Metzler, 2009), 323–24.

12. Of course, there could also be different reasons why subjects do not see the effect of the logic of perfection on their lives; one central example might be the understanding of the exhausted subject in times of neoliberalism, cf. Alain Ehrenberg, *Das erschöpfte Selbst: Depression und Gesellschaft in der Gegenwart*, trans. Manuela Lenzen (Frankfurt a.M.: Campus, 2015).

13. Bernard Williams, *Truth and Truthfulness: An Essay on Genealogy* (Princeton: Princeton University Press, 2002); or Charles Taylor, *Human Agency and Language: Philosophical Papers I* (Cambridge: Cambridge University Press, 1985), 45pp. Recently, see Amy Allen, "Dripping with Blood and Dirt from Head to Toe: Marx's Genealogy of Capitalism in Capital, Volume 1," *The Monist* 105 (2022) 4: 470–86.

14. Saar, *Genealogie als Kritik*; Raymond Geuss, "Nietzsche and Genealogy," in Raymond Geuss, *Morality, Culture, and History: Essays on German Philosophy* (Cambridge: Cambridge University Press, 1999), 1–28. Lorenzini suggests a third way and understands genealogical critique in its possibilizing character. Daniele Lorenzini, "On Possibilising Genealogy," *Inquiry: An Interdisciplinary Journal of Philosophy* (2020). That the modes of debunking and of possibilizing belong together and are not two different kinds, as suggested by Lorenzini, shows David Owen, *Maturity and Modernity. Nietzsche, Weber, Foucault and the ambivalence of reason* (London/New York: Routledge 1994), 212 and Saar, *Genealogie als Kritik*.

15. But this book does not go so far as Colin Koopman does and connects genealogical critique *per se* to transformative processes. Colin Koopman, *Genealogy as Critique: Foucault and the Problems of Modernity* (Bloomington: Indiana University Press 2012). Such transformative possibilities only *can* arise, but this is not self-evident. To this end we explicitly speak *only* of changes; see Raymond Geuss, *Changing the subject: Philosophy from Socrates to Adorno* (Cambridge, MA: Harvard University Press, 2017).

16. That the notions of enlightenment and genealogical critique need to be seen as intertwined is so far especially highlighted by Owen and by Geuss. David Owen, "Kritik und Gefangenschaft: Genealogie und Kritische Theorie," in *Michel Foucault: Zwischenbilanz einer Rezeption. Frankfurter Foucault-Konferenz 2001*, ed. Axel Honneth and Martin Saar (Frankfurt a.M.: Suhrkamp, 2003), 143. Raymond Geuss, "Kritik, Aufklärung, Genealogie," in *Michel Foucault: Zwischenbilanz einer Rezeption. Frankfurter Foucault-Konferenz 2001*, ed. Axel Honneth and Martin Saar (Frankfurt a.M.: Suhrkamp, 2003), 145pp.

17. This view is not undisputed. For example Linda M. G. Zerilli's study *Feminism and the Abyss of Freedom* (Chicago: University of Chicago Press, 2005) criticizes subject-centered approaches and opts for group-based, collective analyses.

18. Martin Luther King Jr., *Stride Toward Freedom: The Montgomery Story* (New York, Evanstone and London: Harper & Row Publishers, 1958), 69.

19. King, *Stride Toward Freedom*, 9.

20. Charles Taylor, *Philosophy in an Age of Pluralism: The Philosophy of Charles Taylor in Question*, ed. James Tully (Cambridge: Cambridge University Press, 1994), 214.

21. Tully, *Politische Philosophie als kritische Praxis*, 43.

22. Shoshana Zuboff, *The Age of Surveillance Capitalism: The Fight for A Human Future at the Frontier of Power* (New York: Public Affairs, 2019).

23. See Alexandra Bykova and Roman Stöllinger, "Trade Balances and International Competitiveness in Cyber-physical, Digital task-intensive, ICT Capital intensive and Traditional Industries," *Research Report: The Vienna Institute for International Economic Studies* 468, May (2023): 24pp.

24. E-mail conversation with James Tully, June 15, 2023.

25. Cf. C. Thi Nguyen, "Echo Chambers and Epistemic Bubbles," *Episteme* 17, no. 2 (2020): 141–61; William H. Dutton et al., *The Internet Trust Bubble. Global Values, Beliefs and Practices, World Economic Forum 2013*, available at: https://www3.weforum.org/docs/WEF_InternetTrustBubble_Report2_2014.pdf.

26. Monika Kirloskar-Steinbach, ed., *Dialogue and Decolonization: Historical, Philosophical, and Political Perspectives. Garrick Cooper, Sudipta Kavirai, Charles W. Mills, Sor-Hoon Tan and James Tully* (London et al: Bloomsbury Press, forthcoming 2023).

27. In the debate as conducted so far, Foucault has been accused of conservatism, especially by brilliant scholars, such as Nancy Fraser, Axel Honneth and Jürgen Habermas, since he does not provide an *explicit* normative scale. Nancy Fraser, *Unruly Practices: Power, Discourse, and Gender in Contemporary Social Theory* (Minneapolis: University of Minnesota Press, 1989); Jürgen Habermas, *Der philosophische*

Diskurs der Moderne: Zwölf Vorlesungen, 12th ed. (Frankfurt a.M.: Suhrkamp, 2016 [1985]); Honneth, *Kritik der Macht*. In turn, Foucault scholars usually interpret Foucault as standing in opposition to normativity. For a paradigmatic example see Mark G. E. Kelly, *For Foucault: Against Normative Political Theory* (Albany: SUNY Press, 2018). An instructive overview on the Foucault/Habermas debate gives Samantha Ashenden and David Owen, "Introduction: Foucault, Habermas and the Politics of Critique," in *Foucault contra Habermas: Recasting the Dialogue between Genealogy and Critical Theory*, ed. Samantha Ashenden and David Owen (London: Sage Publications et al., 1999), 2. With regard to the question of norms in Foucault, the present book follows Rainer Forst's interpretation "Endlichkeit, Freiheit, Individualität. Die Sorge um sich bei Heidegger und Foucault," in *Ethos der Moderne. Foucaults Kritik der Aufklärung*, ed. Eva Erdmann, Rainer Forst and Axel Honneth (Frankfurt a.M./ New York: Campus), 149. There, Forst shows that Foucault's interest lies in excavating the *finite* and thus the historically shaped character of norms. This book goes further down this road and seeks to show that an *implicit* normative scale emerges when we excavate this historically shaped character of norms, which is conceptualized here by the method of genealogical critique.

28. Aldous Huxley, *Brave New World* (London: Vintage Penguin Books, 2007 [1932]); George Orwell, *1984* (London: Penguin Classics, 2013 [1949]); and Juli Zeh, *Corpus Delicti: Ein Prozess* (München: btb, 2010). An instructive overview gives Bernard E. Harcourt, *Exposed: Desire and Disobedience in the Digital Age* (London/ Cambridge, Mass.: Harvard University Press, 2015).

29. It would also not suffice to characterize today's digital society in the tradition of Guy Debord's "society of spectacle." For in the spectacle, we are dealing, as Debord shows, with *passive* spectators who are dominated by the spectacle, that is by the logic of production. Digital change, however, challenges us to grasp the situation that people *actively* want to participate in the digital arena and not in a passive way as *mere* spectators. See Guy Debord, *La société du spectacle* (Paris: Buchet/Chastel, 1967).

30. David Lyon, *Theorizing Surveillance: The Panopticon and Beyond* (Cullompton/Portland: Willan Publishing, 2006); Kevin D. Haggerty and Richard V. Ericson, "The Surveillance Assemblage," *British Journal of Sociology* 51, no. 4 (2000): 605–22.

31. On the notion of "#datapolitics," see Davide Panagia, "#datapolitics: An Interview with Davide Panagia," *Contrivers' Review* 11 (2017); and on the notion of "communication biopower," see: Tiziana Terranova, *Network Culture: Politics for the Information Age* (London: Pluto Press, 2004). For a general understanding of biopower: Miguel Vatter, *The Republic of the Living: Biopolitics and the Critique of Civil Society* (New York: Fordham University Press, 2014).

32. Cf. at the case of data: François Ewald, "Omnes et singulatim: After Risk," *Carceral Notebooks* 7, no. 82 (2011): 77–107.

33. One of the most influential readings on the notion of consent is still Ruth Faden and Tom L. Beauchamp, *The History and Theory of Informed Consent* (Oxford: Oxford University Press, 1986).

34. How the conception of consent is problematic is shown by Joseph J. Fischel and Carolin Emcke in the case of sexual relationships; in particular see Joseph J. Fischel, *Screw Consent: A Better Politics of Sexual Justice* (Oakland: University of California Press, 2019); and Carolin Emcke, *Ja heißt ja und . . .* (Frankfurt a.M.: Fischer, 2019).

35. danah boyd, *It's Complicated: The Social Lives of Networked Teens* (New Haven/London: Yale University Press, 2014); Harcourt, *Exposed.*

36. Harcourt, *Exposed.*

37. In the current debate on enhancement, affects are only, if at all, considered by speaking of emotions. Cf. Alberto Guiblini and Sagar Sanyal, "Challenging Human Enhancement," in *The Ethics of Human Enhancement: Understanding the Debate*, ed. Steve Clark et al. (Oxford: Oxford University Press, 2016), 18. On the contrary, we do *not* equate affects with emotions. By comparison to emotions, affects are the more basic term. Emotions can become more robust and turn into affects. Subjects can laugh in a particular situation, thus experiencing the emotion of joy, *although* they feel more dominated by sad affects in their lives, which is being dominated by the affect of grief, see Bianchi, "Unter dem und wider den 'Bann des Einheitsprinzips.'"

38. This focus on human beings as agents might sound anthropocentric. For general critiques of anthropocentrism, see Vanessa Lemm, *Homo Natura: Nietzsche, Philosophical Anthropology and Biopolitics* (Edinburgh: Edinburgh University Press, 2020). However, in today's digital enhancement it only makes sense to deal with *subjects* as agents. So far, we are not aware of other possible agents in the self-reflexive sense the present book develops.

39. Gilles Deleuze, "Wunsch und Lust," in *Schizophrenie und Gesellschaft: Texte und Gespräche von 1975 bis 1995*, ed. David Lapoujade (Frankfurt a.M.: Suhrkamp, 2005 [1994]), 125 (trans. by the author).

40. Deleuze, "Wunsch und Lust."

41. Daniele Lorenzini, "The Emergence of Desire: Notes Toward A Political History of the Will," *Critical Inquiry* 45, no. 2 (2019): 448–70.

42. An instructive understanding of desire as mechanisms is provided by Paul Veyne, "Foucault Revolutionizes History," in *Foucault and His Interlocutors*, ed. Arnold I. Davidson (Chicago: Chicago University Press, 1997), 163. This interpretation should *not* be misunderstood as a *mere* mechanistic one. For it does *not* involve historically *in*variant causal relations that determine *in a fixed way* what effects might happen next. Quite the contrary, it deals with historically *variable* mechanisms that prompt the subject to be on a specific track.

43. Geoffroy de Lagasnerie, *The Art of Revolt: Snowden, Assange, Manning* (Stanford: Stanford University Press, 2017).

44. Marc Fisher et al., "The Four Hour Insurrection: How A Trump Mob Halted American Democracy," *Washington Post*, January 7, 2021, available at: https://www.washingtonpost.com/graphics/2021/politics/trump-insurrection-capitol/.

45. Already hinting in this direction: Gabriele Gramelsberger, "Digital Enhancement: Zum Konstitutionsverhältnis von Mensch, Computer und Welt," available at: https://idw-online.de/de/news666229.

46. Cf. Julian Savulescu and Nick Bostrom, to only name the big shots: Julian Savulescu and Nick Bostrom, eds., *Human Enhancement* (Oxford: Oxford University Press, 2009).

47. Scholars have long neglected the notion of desire in debates of digital change. As already seen, cf. Lyon, *Theorizing Surveillance*; Haggerty, Ericson, "The Surveillance Assemblage." Recent studies have investigated its somewhat forgotten role: boyd, *It's complicated*; Harcourt, *Exposed*; Andreas Reckwitz, "Die Transformation der Sichtbarkeitsordnung: Vom disziplinären Blick zu den kompetitiven Singularitäten," in *Vierzig Jahre 'Überwachen und Strafen': Zur Aktualität der Foucault'schen Machtanalyse*, ed. Marc Rölli and Roberto Nigro (Bielefeld: transcript, 2017), 202pp; Andreas Reckwitz, *Die Gesellschaft der Singularitäten: Zum Strukturwandel der Moderne*, 3rd ed. (Berlin: Suhrkamp, 2017), 250p; Rainer Mühlhoff et al., eds., *Affekt Macht Netz: Auf dem Weg zu einer Sozialtheorie der digitalen Gesellschaft* (Bielefeld: transcript, 2019). They are valuable, but they do not consider the understanding of the affect- and power-sensitive Janus face of digital enhancement introduced in the present book.

48. Melissa Gregg and Gregory J. Seigworth, eds., *The Affect Theory Reader* (Durham/London: Duke University Press, 2010).

49. Edward Snowden, *Permanent Record* (London: Pan Macmillan, 2019), 3.

50. Honneth, *Kritik der Macht.*

51. Pierre Hadot, *La philosophie comme manière de vivre: Entretiens avec Jeannie Carlier et Arnold I. Davidson* (Paris: Albin Michel, 2001).

52. Cf. Geuss, "Kritik, Aufklärung, Genealogie," 146.

53. This justification does *not* emerge in a positivist sense, as is widely criticized, Axel Honneth, "Rekonstruktive Gesellschaftskritik unter genealogischem Vorbehalt: Zur Idee der 'Kritik' in der Frankfurter Schule," in *Pathologien der Vernunft: Geschichte und Gegenwart der Kritischen Theorie*, ed. Axel Honneth, 5th ed. (Frankfurt a.M.: Suhrkamp, 2020) 62pp, but in and through a historically based interaction with the contexts in which the subject thinks, lives, and feels.

54. Rainer Forst, "Eine politische Theorie der Freiheit: Das Werk von James Tully," Introdution to *Politische Philosophie als kritische Praxis*, by James Tully (Frankfurt/New York: Campus, 2009), 7–13.

55. James Tully, *Politische Philosophie als kritische Praxis* (Frankfurt/New York: Campus, 2009), 18.

56. "History of ourselves" (Tully), as quoted in Forst, "Eine politische Theorie der Freiheit," 8. Originally, the quotation goes back to Foucault. There, it is a paraphrase of Foucault's notions of the "ontology of ourselves" (WE, 316) and the "history of the present" (ÜS, 43, trans. by the author).

57. Tully, *Politische Philosophie*, 27 (trans. by the author).

58. Tully, *Politische Philosophie*, 18pp.

59. Tully, *Politische Philosophie*, 23; Stanley Cavell, *Conditions Handsome and Unhandsome: The Constitution of Emersonian Perfectionism* (Chicago/London: Chicago University Press, 1990); David Owen, "Perfectionism, Parrhesia, and the Care of the Self: Foucault and Cavell," in *The Claim to Community: Essays on Stanley*

Cavell and Political Philosophy, ed. Andrew Norris (Stanford: Stanford University Press, 2006), 128pp.

60. The present book thus goes beyond traditional, non-critical readings of aesthetics in the sense of "aisthesis," for this traditional understanding see James D. Faubion, Introduction to *Michel Foucault. Aesthetics, Methods, And Epistemology*, ed. James D. Faubion (New York: Penguin Press 2020), x-xxxvi.

61. For a highly seminal overview of these debates, see: Allen, *Politics of Ourselves*.

62. Seyla Benhabib et al., *Feminist Contentions: A Philosophical Exchange* (New York: Routledge, 1995).

63. It goes without saying that Brown's works and Butler's texts are brilliant. However, they do not so much take into consideration the fact that the later Foucault wrote four volumes on *The History of Sexuality*, which conceptualize the subject's will formation. Furthermore, Brown and Butler do less see that Foucault refers *not only* to Nietzsche's understanding of the will to power, *but also* to his method of genealogical critique.

64. Allen, *Politics of Ourselves*, 19pp. Karsten Schubert gives a helpful overview of this point, see: Karsten Schubert, *Freiheit als Kritik: Sozialphilosophie nach Foucault* (Bielefeld: transcript, 2018).

65. Allen, *Politics of Ourselves*, 12. The Nietzsche reception in critical theory presents various clichés. On the one hand, Nietzsche is accused of paving the way for mere individualism, centered on the figure of the heroic individual. Such critics do not see that Nietzsche's subjects are also *necessary to each other*, cf. Sarah Bianchi, *Einander nötig sein: Existentielle Anerkennung nach Nietzsche* (Paderborn: Fink, 2016). On the other hand, Nietzsche is accused of destroying reason. This view goes back to Georg Lukács. But Lukács overlooks the fact that Nietzsche does *not* destroy reason *as such*, but only analyzes the historical formation of reason, which can, in some historical configurations, lead to ideological, dominating forms of reason. Georg Lukács, *Die Zerstörung der Vernunft* (Bielefeld: Aisthesis Verlag, 2022 [1954]). On the relationship between Nietzsche, critical theories and feminist theories, see: Sarah Bianchi and N.N., eds., "Subjektivierung und Aufklärung: Von Nietzsche zu kritischen Theorien und feministischen Theorien," *Nietzscheforschung* 33 (2026), forthcoming.

66. Saar, *Genealogie als Kritik*.

67. Saar, *Immanenz der Macht*.

68. Ulrich Johannes Schneider, *Michel Foucault* (Darmstadt: Primus, 2004); Guillaume Le Blanc, *La pensée Foucault* (Paris: Ellipses, 2014).

69. Concerning this approach, consider Amy Allen, "Foucault and Enlightenment: A Critical Reappraisal," *Constellations* 10, no. 2 (2003): 183; Friedrich Balke, "Selbstsorge/Selbsttechnologie," in *Foucault Handbuch: Leben—Werk—Wirkung*, ed. Clemens Kammler, Rolf Parr and Ulrich Johannes Schneider (Stuttgart: Metzler, 2014), 287; Sarah Bianchi, "Mikropolitiken und Aufklärung: Sozialontologische Perspektiven nach Nietzsche und Foucault," in *Nietzsches Perspektiven des Politischen*, ed. Corinna Schubert and Martin A. Ruehl (Berlin/Boston: de Gruyter, 2022), 57–71; Saar, *Genealogie als Kritik*; Frieder Vogelmann, ed., *"Fragmente eines Willens zum Wissen": Michel Foucaults Vorlesungen 1970–1984* (Stuttgart: Metzler, 2020).

70. Bianchi, *Einander nötig sein.*

71. Nietzsche needs to be situated in this line, too (see the chapter on Nietzsche).

72. Saar, *Genealogie als Kritik*; Geuss, "Nietzsche and Genealogy."

73. Paul Feyerabend, *Against Method: Outline of an Anarchistic Theory of Knowledge* (London: Verso, 1975).

74. Rainer Forst, *Die noumenale Republik: Kritischer Konstruktivismus nach Kant* (Berlin: Suhrkamp, 2021).

75. Here, there is no need for an in-depth examination of the controversial debate on Foucault's relationship to Immanuel Kant. It will suffice to look at Foucault's overall attitude towards Kant. This book suggests that we should see this attitude as being marked, *from early on*, by Foucault's understanding that he is part of the Kantian tradition that analyzes the *activation of the subject*. The general debate can be revisited in Allen, "Foucault and Enlightenment."

PART I

Mapping the Field of Digital Enhancement

In what follows, we introduce the key semantics of the microanalysis of the present digital time undertaken here. To this end, part one is divided into two introductory chapters. The first introductory chapter clarifies what we mean when we talk about the theme of digital enhancement; the second outlines the methodology of aesthetic critique.

Chapter 1

The Enlightenment App Does Not Need to Be Invented or What Is Digital Enhancement?

The task of this chapter is to thematize the object of the microanalysis of the present digital time developed here. To this end, the present book asks: *What is digital enhancement?* Whereas some participants of the current discussion seek to develop a digital invention for more or less everything, this chapter ironically states: *The enlightenment app does not need to be invented.* That is, this book seeks to show that what is at stake in digital enhancement is not so much about *what* is invented as about the *procedure* and, more specifically, as about the self-reflexive, or enlightened, attitude that the subject develops towards the various digital means. The following section on Nietzsche and Foucault will explain the enlightened attitude in greater depths. The task of this chapter is first to establish what exactly digital enhancement is.

From the outset, we must clarify that *the* debate on digital enhancement does not exist. To date, there have been *four* debates, which the present book aims to bring together. *Firstly,* there is the academic and public debate on the socio-cultural phenomenon of enhancement that has been carried out in the natural sciences and humanities since the 1980s. *Secondly*, we are, furthermore, confronted with the academic and public debate on the phenomenon of digitalization, a topic that has been widely discussed over about the last twenty years and, like the debate on enhancement, is found in the fields of the natural sciences and the humanities. *Thirdly*, we address the philosophical debate on the art of living that has emerged especially since the 1980s mostly in the field of social and political philosophy. And *fourthly*, we also need to consider the debate over affects that has become highly relevant in the fields of social and political philosophy, social and cultural sciences and psychology for about the last thirty years.

In considering these four debates *together*, this book advances the hypothesis that the field of enhancement is thereby better described than the conceptions offered hitherto. For only with this new understanding we are able to take into account the paradoxical condition, thus far neglected, that ordinary people *are prompted to* use digital media, such as smartphones, apps, digital libraries or Facebook, Instagram and so on, in order to constantly improve their lives, *although* they know about the negative impacts of such digital media use on their freedom. As this study will *also* show, the use of digital media can *nonetheless* open new spaces for freedom that are possible in the field of digital enhancement. In order to theoretically conceptualize these two perspectives, the present book develops the notion of the affect- and power-sensitive Janus face of subjectivity in digital enhancement, which has been neglected thus far.

In order to support the hypothesis outlined above, the chapter is divided into two steps. The first step specifies the object of the microanalysis by connecting the four debates over enhancement, digitalization, the art of living and affects. The second step concretizes the necessity of considering these four debates together by discussing the empirical and fictitious cases of predictive policing and the manipulation of political elections, as recently seen especially in the Cambridge Analytica scandal. The first case—predictive policing—will be analyzed with reference to the film *Minority Report*. The second case—manipulating political elections—will be in turn sketched using the example of the Netflix documentary *The Great Hack*. The next chapter on aesthetic critique will show in greater detail why we need to describe these two issues by referring to films that do *not explicitly* deal with the theme of digital enhancement. Here, it suffices to note that these two films help us to concretize the necessity of connecting the four debates over enhancement, digitalization, the art of living and affects.

NOT ONLY A CURRENT FAD

The Socio-Cultural Phenomenon of Digital Enhancement in its "*Longue Durée*" (Bloch)

Outlining the debates over enhancement, digital change, the art of living and affects will provide the theoretical conceptions that we require in order to specify the object of analysis, that is, the socio-cultural phenomenon of digital enhancement.[1] As a look at the debates thus far shows, there has not yet been a large study that delineates the necessary relation between these four debates in a systematic manner. The present study aims to contribute a perspective that is central to filling this gap. Connecting these four debates highlights the fact that digital enhancement is *not only a current fad*, but needs to be seen *in its "longue durée,"* to put it in the words of the French historian Marc Bloch. In this sense, the first two debates enable us to better describe the current field of digital enhancement in terms of the logic of perfection, that is the common denominator of digital enhancement. The two other debates help us clarify why subjects care at all about the logic of perfection part of digital enhancement. In doing so, they show that the current practices in the field of digital enhancement are *not* as new as one might be tempted to think at first glance; instead they need to be situated in the older genealogy of life formation *as such*.

To begin with, we must establish the common thread in the debates over enhancement, digitalization, the art of living and affects. As will be shown, this common ground lies in the logic of perfection. The term "enhancement" itself implies the promise of making something better. It already carries the notion of perfection in its name as the verb "to enhance" means "to perfect."[2] The term "digitalization" also promises perfection, as the buzzword centers around the *topos* of "solving problems."[3] Digitalization, in other words, promises *to make life easier*. The conception of the art of living explains why the subject attaches to the logic of perfection *at all*. Hence, there is *no* reference to the art of living in order to gain a further twist in the specification of the logic of perfection. For Nietzsche and Foucault, the main advocates of the strand of the art of living, are very skeptical of perfectionism, since they consider it a means of heteronomy. This might be surprising, since Nietzsche in particular is often understood to have supported a perfectionist project.[4] But in specifying what is meant today by perfectionism, Nietzsche's and Foucault's criticism becomes clear.

Today's discussion centers on the conception of ethical perfectionism and distinguishes between two types. In general, ethical perfection entails perfectionism in the context of life formation and thus in the context of the art of living. The *strong* understanding of ethical perfectionism in the classical sense stands in the Aristotelian tradition. It departs from the teleology-based view that human beings flourish towards an objective societal good and that they form their virtues to this end. Christoph Menke shows that Foucault does not stand in this tradition.[5] The same holds true for Nietzsche, as James Conant demonstrates.[6] For Nietzsche and Foucault are very skeptical of *telos*-orientated perspectives. Moreover, in today's discussion there is also talk of a *weak* understanding of ethical perfectionism. This kind of perfectionism is also called "moral perfectionism."[7] But, as one learns from Owen, what it means refers to the field of ethics.[8] As Owen further shows, "moral perfectionism" in Cavell's terms is *not* to be confused with an Aristotelian kind of strong perfectionism that considers the subject's striving towards an external goal.[9] By contrast, weak ethical perfectionism entails the understanding of the subject that strives to become a better, that is, a more reflexive version of herself.[10] In recent discussions, this view is linked to Foucault.[11]

In the light of the present book, a weak kind of "ethical perfectionism" is present in Nietzsche and Foucault if we see it in the perspective of genealogical critique. In this case, the subject is not directed by an external *telos* to become a better version of herself, as one might think following Aristotle, but has *an enlightened interest* in better understanding the "history of ourselves," in Foucault's terms, in order to uncover the hitherto all too hidden dominating contexts that make different "histor[ies] of ourselves" impossible. But this is *not* to be confused with an anthropological constant. Rather, it must be understood as a process that *can* emerge through the obstacles of domination with which the subject in her life formation is constantly confronted. However, we do not refer to Nietzsche's and Foucault's understanding of the art of living in order to consider it in the context of ethical perfectionism; we refer to it in order to render understandable why subjects place themselves in relation to the logic of perfection.

The debate on affects further navigates the paths the present book started. If the debate on the art of living explains the subject's attachment to the logic of perfection, then the notion of affects helps explain the subject's *motivation* for such an attachment. More specifically, with the help of the notion of affects we are able to conceptualize the notion that subjects in digital enhancement *are prompted* to do so.

Now we turn to outlining the debate about enhancement. In the debate, there is dispute as to what should be properly considered as contributing to optimization. This dispute centers on narrow and broad understandings of enhancement. *Narrow* understandings define enhancements solely as biotechnological interventions in the human organism,[12] which are evaluated in a positive way by the people who undertake such operations. Such narrow understandings mainly distinguish three areas of enhancement: *first*, the domain of bodily modifications, including implants, cosmetics or prostheses; *second*, the domain of mental modifications, such as pharmaceutical measures or technologies of the extended mind such as computers, navigational devices or mobile phones; and *third*, the domain of reproductive modifications, such as pre-implantation genetic diagnosis (PGD). By contrast, *broad* understandings see enhancements as everything that improves life,[13] beginning with the most ordinary socio-cultural practices such as wearing glasses or drinking coffee. Broad understandings take a positive view of such enhancements.

Within this debate, we contribute to an affect-sensitive, power-based perspective on the broad understanding of enhancement, conceptualizing enhancement as an umbrella term. Our present experience shows that it is necessary to consider under this umbrella all those phenomena in and through which *the subject is prompted to set herself in relation to the logic of perfection by using digital means*, to give already the entire definition, which also includes the reference to the debate over digital change, the art of living and affects. The present book's definition of digital enhancement is new in two ways with respect to the first debate on enhancement: *Firstly*, this broad understanding does not conceive the socio-cultural practices of enhancement as *exclusively* positive as done by Harris. *Secondly*, it centers on the logic of perfection and *not* on specific areas in which enhancements take place, such as having a coffee. In this sense, the present book theoretically conceptualizes digital enhancement as *procedures* marked by the logic of perfection that impact every subject of society *to different degrees* and not just *some* subjects of society, such as so-called cyborgs, i.e., people who have implants in their heads, for instance to improve hearing for the deaf. Hence, digital enhancement needs to be considered as mass phenomena, since the logic of perfection, the common denominator of digital enhancement, is everywhere and thus everyone is confronted with it, to some extent. The latter restriction is important to consider: Although the logic of perfection has power over society *as a whole*, it does *not* have the *same* effect on everyone.

The current debate over enhancement is not only split over the question of what should be considered as enhancements, but also about the question of how to evaluate enhancement technologies. Scholars from various disciplines ranging from biotechnology, informatics, robotics and neurosciences, to cultural and social sciences and philosophy have been engaged in these debates for about the last forty years. Whereas tech people create the technological tools that *at least* promise to perfect life by making available what previously seemed to be *un*available, the social and political philosophical debate so far is particularly concerned with the question of how to reflect on and to call into question what can be achieved through technological possibilities. Questions of aesthetics have thus far been considered almost exclusively in the obvious sense of weighing, for example, the possibilities of plastic surgery, such as deliberating the notion of beauty in the context of lip or breast enlargements.[14] In general, the major *topoi* of the overall philosophical debate center on the values and norms of freedom, of justice, of human dignity or of authenticity. Whereas in this context deontological positions focus on the subject's motives and draw from them the duty to enhance or the duty not to do so, consequentialists think through the outcome and thus particularly take into consideration risk and benefit.

The debate over enhancement is not only taking place within academia, it also fills the public sphere. Newspapers, magazines, television or radio features are eager to report on the newest trends, fears and hopes that are connected with the notion of enhancement. The following questions often emerge in the public debate: Are we sliding down a "slippery slope," as one of the prominent arguments claims, without knowing where we are headed while seeking to enhance our lives? Are we not rather opening a Pandora's box by using enhancement tools when we try to "play God," as it is commonly put, although we are simply seeking to escape the alleged "natural lottery?"[15]

Going beyond the state of the art, this study points to the desideratum that to date, there has not yet been an understanding of the paradoxical conditions of the field of digital enhancement. In order to contribute a central perspective to this desideratum, the present study must describe the affect-centered, power-based conditions of digital enhancement underestimated in the debate hitherto. For it is only in describing these conditions that, as this book aims to show, are we able to plausibly address how subtly the subject's agency is determined by the logic of perfection. Norms, values or ideals, as important as they are, are already part of such determinations. Hence, if we *only* analyze norms, values or ideals and ignore the affect-centered, power-based conditions that already create them, as the main debate has done thus far, we would *not* be able to understand the *subtle* mechanisms by which the subject is *systematically* structured by the logic of perfection. It is Nietzsche and Foucault that show that norms, ideals, and values are the product of affect-sensitive power relations and *not* a fixed unity as would be conceivable following action-theoretical understandings prominent in accordance with Max Weber. The book at hand takes up this thread and shows the need to provide an understanding of the field of digital enhancement that is aware of this hitherto all too neglected entanglement. At the same time, this subtle look at the affect-centered, power-based conditions of digital enhancement enables us also to broaden the horizon and provides the theoretical tools that we need in order to conceptualize the affect- and power-sensitive Janus face of subjectivity in digital enhancement, which has thus far been largely overlooked. With this we gain an affect- and power-based sensorium that enables us to perceive that digital enhancement is *not only* a problem that constructs a quasi-automatized subjectivity willing to conform to the logic of perfection, but can *also* simultaneously construe an enlightened subjectivity that is aware of the power of the logic of perfection.

In what follows, we need to further outline the different camps of the debate over enhancement. Basically, the social and political philosophical debate as a whole can be differentiated in three major camps: skeptical bio-conservatives, hesitating middle-ground positions and optimist trans- and posthumanists. These camps are, of course, more heterogeneous than this division suggests. For our purpose here, however, such a provisional order is helpful and suffices. The following will briefly present these three camps in order to characterize the common thread of each. In order to further illuminate the differences between the three camps, we refer in each case to Stanley Kubrick's groundbreaking science-fiction film *2001: A Space Odyssey* (1968).[16]

The skeptical bio-conservative camp ranges from religious strands to social critique. Essentially, according to this perspective, enhancement technologies do *not* perfect life.[17] Accordingly, those who take this view seek *to protect life* as much as possible from alleged technological intrusion. Thus, according to them, technology achieves first and foremost a change in the direction of cultural decline. The label "bio-conservative" already indicates this direction: the prefix "bio-" derives from the Greek noun "βίος" (*bios*), meaning "life," and "conservative" traces back to the Latin verb "conservare," which means "to preserve." In combining these two elements we obtain the concept that characterizes the "bio-conservative" position with the key motif "to preserve life." Although the bio-conservative position is often *mis*understood as reflecting a certain political point of view, namely a conservative one, the etymological references already enable us to better see their point, and this point is *not* to be confused with the factions that we know from the political right; although these two positions can overlap, the bio-conservatives should be understood first and foremost as representing a skeptical, culturally pessimistic position.

In what follows, we specify the bio-conservative position by referring to Kubrick's science-fiction film *2001: A Space Odyssey*. In general, the film depicts the voyage of five scholars who travel to intergalactic space in order to analyze a black monolith. This voyage turns into an odyssey, as the film shows, because the computer HAL[18] becomes a threat although it is actually meant to help the scholars to carry out their research.[19] We will now consider the opening scenes of the film in order to sketch how bio-conservatives would position themselves in relation to enhancement. These first scenes are characterized by Kubrick's famous cut. Kubrick introduces the cut with the following words: "Now he (the hominid Moonwatcher) was master of the world, and he was not sure what to do next [. . .]. But he would think of something."[20] What Moonwatcher could be thinking of became one of the most well-known cuts in the history of film: Kubrick shows Moonwatcher thinking about throwing a bone into the air, which continues to rise for miles and miles, passes into intergalactic space and finally becomes a satellite. Through the magic of montage, technological development occurs within a matter of seconds. What actually required millions of years is condensed into a *single* cut that fast-forwards in history from the lives of hominids in prehistoric Africa into the year 2001. If we translate this scenario into the context of the contemporary debate over enhancement, the bio-conservative position is depicted as follows: bio-conservatives would not dare to throw the bone so high into the air as Kubrick lets Moonwatcher do. Rather, bio-conservatives would fear the possibility of making a significant mistake that would only pave the way for hazardous scenarios.

On the other end of the spectrum, we find the trans- and posthumanists. In contrast to the bio-conservative camp, this perspective conceives of enhancement technologies in limitless terms as an *exclusively* freedom-enhancing utopia. If bio-conservatives represent a culturally pessimistic group, trans- and posthumanists by contrast constitute a super-optimistic camp that understands enhancement technologies as a central remedy in the attempt to overcome, as they see it, the flawed and antiquated label "human being."[21] Already the notions "transhumanist" and "posthumanist" indicate the idea of such a transition. With slightly different nuances, which, however, are not of further interest here, already the etymology of the prefixes "trans-" and "post-" indicates the trans- and posthumanist views: From their perspective, *homo sapiens* goes "beyond" a further period of development, as the prefixes "trans-" and "post-" means "beyond" and "behind," respectively.[22] So these two strands share *ultra*-optimistic conceptions of enhancement technologies; possible downsides or dangers make no appearance.

To better understand this ultra-optimist perspective, let us consider one of the key essays in this strand of thought, namely Nick Bostrom's classic essay "Why I Want to be a Posthuman when I Grow Up." Bostrom describes a possible posthuman life with the following enthusiastic words: "You have just celebrated your 170th birthday and you feel stronger than ever."[23] Such visions are widespread among the trans- and posthumanist camp. With this in mind, one might not be surprised by the fact that super-futuristic ideas of, for example, uploading the human mind into a computer are also seriously discussed within this camp.[24] As a sidenote: It is especially this camp that could opt to invent an enlightenment app, the scenario with which we began the chapter. But to reiterate, this and similar inventions would fail to consider the subject's self-reflexive, to wit enlightenment attitude, which cannot be improved by taking a pill or installing an app. It is not the consequences that count, to wit being more enlightened, but the subject's self-reflexive procedure in and through which subjects *come* to make decisions in a *non*-automatized way. If we interpret trans- and posthumanist visions in terms of Kubrick's famous cut, as we did with the bio-conservative position, the following picture emerges: *in contrast to* bio-conservatives, trans- and posthumanists would *not* hesitate for a moment to throw the bone into the air; they would *enthusiastically* want to take their future into their own hands in the firm belief that they *cannot but* contribute to the greatest technological progress in the history of humankind. And they do not want to miss this opportunity.

Between these two camps, we see a third group, the so-called middle-ground positions. This group opts to carefully consider both the possible risks and benefits of enhancement technologies. This book characterizes these positions as having a '"yes, but" attitude.[25] Hence, as we hope to show, they share with trans- and posthumanists *on the one hand* the aim to avoid rejecting potential technological progress too early. In this sense, they say "yes" to enhancement technologies, so to speak. At the same time, however, their "but" enters the stage. For *on the other hand* they share with bio-conservatives the concerns about trusting technological progress *blindly or naïvely*, as the trans- or posthumanists could be accused of doing. From this perspective this middle group regards the conscious and rational subject as the crucial agent in the field of digital enhancement, who is able to undertake such considerations and thus to develop its own "yes, but" attitude. According to this understanding the subject thus is able to make an autonomous decision within the multiple possibilities of enhancement technologies. As a key instrument that allows the subject to arrive at such decisions, middle-ground positions use the tool of informed consent, which originated in the medical ethics of the 1980s. This tool implies two aspects: *firstly* knowledge or information, and *secondly* autonomy. In combining these two aspects, the subject is considered to be capable of autonomous decision-making. Thus this tool begins with the assumption that the conscious and rational agent knows or has the information relevant to what is at stake in using this or that enhancement technology.[26] In this context, a subject might be conceived to be capable of asking herself to what extent she is leading a digital life that is more useful, as a utilitarian perspective would ask, or more authentic, less risky, more just or more beneficial and so on. Secondly, the concept of "informed consent" assumes that the agent is also able to decide by himself—that is, autonomously—what to do on the basis of the relevant knowledge. From this perspective, the middle group can be conceived as tending in the direction of hesitating positions. At this point, too, as we have done for the bio-conservatives and the trans- and posthumanists, we describe the hesitating middle-ground positions with reference to Kubrick's film with its famous cut: unlike the bio-conservatives or the trans- and posthumanists middle-ground positions would *carefully deliberate* the possible advantages and disadvantages entailed by throwing the bone any further into the air. On this view, what might happen in the end depends, then, on the respective subject's rational and autonomous decision-making undertaken in accordance with their concrete time and context.

To this panorama of the hitherto lead debate, we see the necessity to add the hitherto missing affect- and power-sensitive perspective that is able to provide a critical-theoretical microanalysis of the present digital time: It promises to analyze the so far neglected paradoxical conditions of life formation in the field of digital enhancement. The hitherto existing perspectives of the three camps above described are reductive insofar as they cannot plausibly explain why subjects *quasi-automatically* have such a formed will that they want to be part of a field that is *not only* enhancing their freedom, as it promises, although they fair well *know* that it *also* endangers the subject's freedom. The given perspectives of the debate hitherto cannot do so, as they do not see sufficiently, if at all, how much norms, ideals and directions *are generated* by power relations that in the context of digital enhancement must be specified as affect-sensitive. Hence, it would be a misunderstanding to take them as fixed unities that enable us to adequately assess our digital today. *Initially*, one must see that they are already the product of the field that one wants to assess. This is the fundamental lesson that we learn from Nietzsche and Foucault and that this book follows.

Now let us turn to the debate over digitalization. If the object of analysis developed here appears as such an umbrella term encompassing the procedure of the logic of perfection, then the second debate, that is the debate over digitalization, helps us to further specify this view. From this perspective we can now say that we are dealing with the specific procedure of the logic of perfection that belongs to digital change. The debate over digitalization emerged alongside the development of digital media, and has acquired great importance over about the last twenty years, especially since the revelations made by WikiLeaks beginning in 2006 and by Snowden in 2013.[27] These revelations made clear what price one must pay when seeking to enhance life using digital means, namely the price of freedom, such as the cost of data privacy. Whereas the dominant strands of the existing debate normatively weigh the politics, economics and ethics of digital change,[28] the view of the present book is *not* developed. Nonetheless, there are already readings that hint in the direction relevant here. These views understand digital change as contributing to the theme of solving problems.[29] We expand this view to the understanding of digital change as involving the logic of perfection. In this context, the logic of perfection as part of digital change promises to make life perfect by solving the problems people face in life. As it is put in the present study, this theme further entails the key motif of "making life easier." In our digital today, this *topos* emerges in various domains. It is especially part of the industry 4.0. Generally speaking, industry 4.0 describes the smart connection of machines by means of digital communication devices in the domain of industry that are not usually connected. It is the last of the four phases of industrial change. The first phase included mechanization by means of hydropower and the steam engine in the eighteenth century. The second phase involved the emergence of the automobile industry, which brought the introduction of the assembly line. The third phase encompassed the emergence of PCs in the 1970s.[30] The *topos* of making life easier is also part of different areas, such as business, administration, elderly care, medicine or architecture.

In these areas, the conception "smarter together" is one of the most prominent themes. The idea is quite clear: It promises to enable people to lead a better life by digitally connecting areas that have not previously been connected in this way. The expectation is that this will improve life by saving time, reducing administrative workloads, and so on. In order to illustrate this, we refer to the key examples of smart cities, smart living, and the social credit system.[31] Furthermore, we describe digital phenomena in the fields of social and personal relations, and in the context of literature and art.

The theme "smarter together" is especially relevant in the area of "smart cities," such as was planned for the Canadian district "Queens Quay East" in Toronto by the company "Sidewalk Labs,"[32] a subsidiary of Alphabet Inc. (Google's parent organization) concerned with designing cities. In the case of Toronto, the idea of "smarter together" carried with it significant promises to make life easier: these included eliminating car accidents or crimes,[33] but also heating the sidewalks according to the specific weather conditions or having traffic lights that adapt to specific situations.[34] All these promises were to be realized by storing data. Thus car accidents were expected to be reduced because self-driving cars would recognize smart bicycles;[35] the crime rate also would have been enormously reduced because—so the expectations of Sidewalk Labs—everyone would have been tracked. But as newspapers around the globe reported, "Google town"—as the ironic heading in the *Frankfurter Allgemeine Zeitung* puts it[36]—was moved to the archives of digital history. In May 2020, in the beginning of the emerging Covid crisis, Sidewalk Labs said that they needed to abandon the project in Toronto because of the unforeseeable financial risks. But there were also massive protests against the project. People asked loudly: Where are our data?

On a smaller scale, we also encounter the *topos* "smarter together" in the ordinary territory of "smart living" that is especially widely known from a major Swedish home furnishing company with four capital letters in its name. This example functions in analogous ways, but on a smaller scale. Making life at home easier in this case means that one can, for example, regulate a home's lighting according to one's mood using an app, or improve security by installing digital motion detectors.[37]

The conception of "smart living" has been taken a step further in the Chinese authoritarian regime. "Making life easier" in this context involves the introduction of the social credit system that was introduced a few years ago that extends the existing financial rating system to social life.[38] By this the Chinese authoritarian regime seeks to perfect social life in China. Within this social credit system, people gain or lose credit points for their allegedly good or bad social behavior, which is of course defined by the Chinese authoritarian regime. Here "making life easier" means publicly indicating one's trustworthiness by social credit points. If a person has a lot of social credit points, then they are regarded as trustworthy and gains advantages, for example in economic contexts. "Trustworthy" refers to an interpersonal understanding of trust.[39] It means that a person is regarded by another person as someone of whom can be said *with certainty* that he will do something in the *un*certain, risky future social interaction. This understanding stems from the general notion of trust, which means security in times of uncertainty.[40] In other words, the Chinese social credit system can be regarded as a trust currency. If, for example, one wants to obtain a bank loan, the underlying idea is that the social credit system helps the banker weigh up the probability, that is the certainty, of the loan being repaid in the uncertain future.[41] Especially charity work, donating blood or expressing political conformity by praising the government on social media are deemed by the Chinese government to be socially acceptable social behaviors for which one can gain social credit points and which thus augment one's trustworthiness.[42] In turn, drunk driving, jaywalking or protesting against authorities is regarded by the Chinese government as *un*acceptable social behaviors for which one loses credit points and thus also one's trustworthiness. It goes without saying that although the Chinese social credit system promises to make life easier by publicly indicating people's alleged trustworthiness, it is clear from the perspective of the present study that this system enables the Chinese government to govern their citizens' lives even more easily than they already do.

For the strategy of the social credit system needs to be seen in the light of general efforts by the Chinese government to govern their citizens' online behavior. From this perspective we see that the Chinese government is in general highly concerned with censoring the Chinese people's use of the internet, be it via the social credit system, via smart living or via social media, as the case of the Clubhouse App recently showed. Via this app, people could discuss politically sensitive and censured topics in online rooms, such as the cases of Tibet or the Uighurs in which people are afraid of sinicization.[43] In February 2021, however, the Chinese government blocked this app—as a sidenote, in China's authoritarian approach to digital media other online options, such as Facebook, Instagram or Twitter, are also usually monitored and regulated by the authoritarian state.

In addition to this extreme example of the social credit system in China, there are also somewhat less extreme areas of life that are supposed to be made easier by digital strategies, like the area of social and personal relations. A central example of this field is the film *Her*. Films like *Her*, to which we already briefly referred in the introduction, depict fictional visions of how one might imagine a digitally altered future. To briefly summarize the plot again, the protagonist Theodore (Joaquin Phoenix), who is weary of his life due to his divorce, seeks to simplify his daily routines by buying a new operating system with an artificially intelligent virtual assistant named Samantha, played by Scarlett Johansson. Instead of only making the administrative business of Theodore's life easier, Samantha abandons her intended protocol and tries to make Theodore's romantic life easier; they both fall in love even and become a couple. The film then shows in ironic ways that they are anything but an ordinary couple; thus the promise of making romantic life easier does not always redeem. For example, when Theodore and Samantha go on a hiking trip with another couple, they have difficulty finding a place for Samantha in the group as she does not have a body. This scene shows how clumsy and anything but easy relationships between human beings and machines are. Nonetheless, the film does not advance one-dimensional views that only conceive of digital technological change as an entire danger. In contrast to rather stereotyped films like *Ex Machina* (2015), which presents a clear-cut picture of digital change as endangering human life, since the plot ends with the human-like robot Ava killing its engineers,[44] the film *Her* instead depicts the relationship between human beings and digital technologies in more nuanced ways. It depicts the spaces in between the poles of being purely a threat *on the one hand* and *on the other* of being purely a benefit that makes life easier.

In the context of literature and art, too, phenomena of digital technologies are designed to make life easier. Although there are widespread discussions concerning whether we are dealing with literature or art *at all* if it is created by digital means,[45] here the question is only to what extent the digital means work out. This looks like a somewhat consequentialist view that is only interested in the outcome and evaluates the digital technology as good as long as the outcome provides the desired result. This view, however, would be a misunderstanding. The upcoming passage only provides a description of what can be done so far *without* evaluating it as good, as would be conceivable following the consequentialist perspective. Here, the conception of digital enhancement is best understood as seeking to make the life of writing or the life of art easier. For example in the beginning of 2020, the prominent writer Daniel Kehlmann, well-known for his bestseller *Measuring the World*, met with the robot CTRL in Silicon Valley in order to write a short story together. In general, the experiment sought to discern whether robots can also write texts. Kehlmann, however, was quite blunt in his assessment that CTRL "is a friend of the fragment [. . .], rather Kafka than Dickens."[46] So, according to Kehlmann, the experiment needs to be considered a failure. Robots cannot write stories that are longer than one page and they only write what seems likely to occur given the set of data they haven been fed. Hence, they do not make the writer's life easier.

However, experiments taking place in the area of art have been more successful and have greater promise to make the artistic life easier. In spring 2016, the Dutch project called "The next Rembrandt" was especially surprising to the public sphere.[47] Although here, too, computers are not so creative that they are able to paint an original picture on their own, *at least* they are able to paint an entire picture in an already existing style,[48] here in that of Rembrandt. Accordingly, the picture is called "the next Rembrandt" more than 300 years after Rembrandt's actual death.[49] So, this artificial painter does not stop after one brush stroke, so to speak, like the computer with whom Kehlmann wanted to collaborate to write a story. Nonetheless, making artistic life easier is restricted, as this machine is also dependent on the set of data with which it was nourished in order to create a Rembrandt.

The book at hand refers to the third debate over the art of living to specify the object of analysis. The art of living is commonly understood as a kind of life formation.[50] In general, the motif of the art of living can be traced back to antiquity. After a period of neglect in the Middle Ages and in Renaissance Philosophy, the motif re-appeared as a prominent theme in Weimar Classicism. In the nineteenth century, it was developed in particular by Nietzsche and reproblematized in newer French philosophy by Foucault. The question of the art of living also plays a major role in contemporary philosophical debate.[51] Nonetheless, in today's debate on digital enhancement the theme of the art of living is not often discussed, although there have been some initial approaches in this direction.[52] With the main advocates of the debate of the art of living, and thus with Nietzsche and Foucault, the present book basically describes life formation as the way in which a subject seeks to give her life form. From an early juncture, this access clarifies that digital enhancement is *not* to be confused with a fixed content, that is, with one technological device or another. By contrast, referring to the debate of the art of living clarifies from the very outset that we are dealing with *a procedure*. If we already learned from the reference to the debate on enhancement that we approach enhancement in terms of procedures characterized by the logic of perfection, then we now specify that we are dealing with the specific procedures marked by the logic of perfection that encompasses *the subject's life formation*.

The perspective on the art of living is essential, as it helps us to see that digital enhancement is *not only* part of our digital time, but must be seen in its "*longue durée*," to use the words of the French historian Bloch. For only then is one able to also conceptualize the dynamics that already existed *before* digital tools were invented in the twentieth century, i.e., before the invention of digital computers in the early 1940s or of the internet in the late 1960s. The conception of the art of living as conceived in this book enables us to specify these dynamics as the subject's attitude towards his time. Put differently, giving one's life form implies *placing oneself in relation to one's time*. With this in mind, it is clear that subjects did not just begin to form their attitude towards their present time in our digital age; indeed, they have done so throughout human history. For example, they did so when Johannes Gutenberg invented the printing press in the fifteenth century, with which Gutenberg sought to make life easier, or when the power loom or the steam engine were developed in the Industrial Revolution, which were also intended to make life easier. The current digital enhancement, therefore, does *not* mark a shift in paradigm; rather, the conception of the art of living clarifies that the paradigm of seeking to perfect life continues, *only* with a slight change to the materials used within this paradigm: this time, the hype is not about the steam engine or the printing press, but the World Wide Web. And the challenge with which we are confronted is to analyze how today's subjects set themselves in relation to the present hype and thus to the materials provided by their time that come along with the social expectations that they will also be used.[53]

Seen from this angle, even those areas that lack an understanding of current technologies enable us to conceptualize the subject's attitude towards these digital tools. Nietzsche and Foucault not only help us understand that, as is common in today's language, materials must be considered the hardware to which subjects place themselves in relation. The lesson we learn from Nietzsche's and Foucault's understanding of the art of living teaches us to understand already the social and political normative order that lends the subject its social status as the hardware. Note that the understanding of the art of living enables us to see that every epoch has its own norms and subjects are challenged to position themselves in relation to these norms. The common denominator of today's hardware and that of earlier periods in history lies in the logic of perfection. Here, however, we need to be a bit more specific. For strictly speaking, there are *no* norms that demand that subjects should perfect their lives, neither today nor in earlier times. But the logic of perfection is part of the social norms that *expect* subjects to become who they are in the best possible way for their time. If they do not, they are commonly regarded as not "fit enough," so the usual attitude. For some subjects these social expectations can even acquire normative power. Similarly, to pursue this idea further, seen this way the subject's attitude towards the hardware must be understood as the software. So, an understanding of the art of living in its "*longue durée*," to return to Bloch's phrase, allows us to *also* consider all those periods in and through which subjects developed their software toward the hardware of their time.

If we are to take seriously the narrative of digital enhancement in such a "*longue durée*," then all those narratives that deal with the *topos* of life formation must also be considered as part of the story of digital enhancement. What does become relevant here, is the understanding of the subject's software, in the sense of her attitude towards the hardware, we recall, the social and political normative orders of a specific time. This perspective then also encompasses several narratives, of which we can only mention some of the most well-known, such as Ovid's telling of the myth of Pygmalion, Johann Wolfgang Goethe's *Faust* or E.T.A. Hoffmann's *Sandmann*. In different ways, these narratives also describe the modes in and through which subjects give life form in setting themselves in relation to the logic of perfection.

In Ovid's story, the protagonist, Pygmalion, seeks to perfect romantic life, a thread that we already touched on with the film *Her* and the romantic relationship between a human being and his virtual assistant. In Ovid, we encounter the story of a sculptor who, after suffering a long period of loneliness, creates his own object of love, an ivory statue. He intends to build the statue in order to perfect his life. As he falls in love with the statue, the statue must even be regarded as enhancing his romantic life. To the surprise of the reader, the statue actually becomes alive in the ending: "[I]t is a body,"[54] the narrator comments when Pygmalion realizes that he is no longer dealing with stone but with flesh, and in the end the two become a happy couple. Thus, Pygmalion places himself in a *dissenting* relation to the existing social expectations of romantic life, as he does not follow the usual social expectations. For the usual social expectations entail interpersonal relationships, and thus romantic relations between human beings and *not* between a machine and a human being. Later, Ovid's theme was taken up by Bernard Shaw in his play *Pygmalion* (1913).[55] In interpreting this myth and its modern variant from today's perspective of digital enhancement, the motif hints in the direction of artificially creating human beings, and thus an understanding that does not consider machines or, in Ovid's case, statues as dead, fixed or stable, but as potentially living things that are able to participate in the world of human beings.

We already see such visions especially in Goethe's tragedy *Faust*. In general, the tragedy thematizes what has come to be known as "the Faustian" (*das Faustische*), i.e., the restless striving for knowledge. With the figure of the Homunculus, Goethe already considers how human beings strive to improve their knowledge of the artificial creation of human beings, or in Goethe's semantics, the "polite little man"[56] Homunculus. Thus Goethe also already had the vision that human beings could be artificially created in what he calls a "phial."[57] Here, the teacher Wagner, who creates Homunculus, sets himself in a *dissenting* relation to the existing norms as one normally would *not* expect to create a human being by artificial means.

Nonetheless, the striving for creating artificially human beings does not reflect an isolated view, but needs to be regarded alongside many such ideas that have been quite commonly discussed since the eighteenth century. These visions go back in particular to Julien Offray de La Mettrie's text *L'homme machine* (1747). La Mettrie was a doctor and philosopher who wanted to create automata that, in accordance with the ancient Greek word αὐτόματος (autómatos), were able to move by themselves.[58] One of the most prominent inventors of such automata in the eighteenth century was the French engineer Jacques Vaucanson. Although he tried to create machines that looked like human beings—the ancestors of today's artificial intelligence, in some sense—it happened that he produced machines that we now know as children's toys, such as a mechanical duck and a flute player that is able to play dozens of songs on its own. During the eighteenth century it was Kant who especially criticized these attempts of artificially creating human beings. According to Kant, to try to replicate human beings as mere automata is to misunderstand what human beings are.[59] Following Kant's understanding of enlightenment, human beings need to be understood as the beings that can think for themselves, a task from which human beings cannot be freed from by machines (see part two of the present book). Further objections against such automata are demonstrated in the novel *Sandmann* by the German Romantic poet Hoffmann. This novel describes the dystopian fantasies that accompany these and similar mechanical creations. Hoffmann is particularly concerned to depict the unsettling atmosphere that is generated by automata. To this end, he describes how the protagonist Nathanael is deeply confused by falling in love with an automaton named Olimpia that he, however, regarded as a human being.[60] Given these exemplary cases, it must be regarded as reductive to *only* conceive of the phenomenon of digital enhancement as a contemporary trend that first began with the introduction of computers or the World Wide Web. By contrast, in view of these earlier narratives about life formation we see that people have long dreamed of visions and utopia of, for example, artificially created human beings.

Having sketched the third debate over the art of living, we now turn to the fourth debate over affects. If the first debate over enhancement specified the object of analysis in terms of procedures marked by the logic of perfection, the second debate over digitalization further specified that we are dealing with the logic of perfection in the context of digital change, then the debates about the art of living and affects describe digital enhancement in its "*longue durée*." On this basis, the debate on the art of living enables us to theoretically conceptualize the "*longue durée*" by referring to the subject's attitude towards the logic of perfection. The debate on affects gives more precise form to this view by highlighting the subject's *motivation* for developing such an attitude towards the logic of perfection. The understanding of affects developed here enables us to say that the subject develops such an attitude because he *is initially prompted* to do so. At the same time, on a cognitive level, the notion of affects shows that *the subject's will is programmed in such a way* that he *quasi-automatically, in a machine-like manner* evaluates this attitude as good, at least initially.

In general, the debate over affects has become highly relevant over the last three decades and ranges across the fields of philosophy, neuroscience, social and cultural sciences and social psychology.[61] While the notion of affects did not gain much attention in Aristotelian and scholastic philosophy, it was the Dutch philosopher Spinoza in the seventeenth century who highlighted the role of affects for life in general. Given the widely spread debates, there is now even talk of an affective turn. Basically, the debate over affects entails two different strands. Whereas the first strand, shaped by the work of Tomkins, conceptualizes affects as psychobiological aspects of emotions, the second strand by contrast advances an understanding of affects as changeable relations of power and forces (*Kräfte*). The present book aims at contributing a central perspective to the second strand. In so doing, it develops a hitherto largely unexplored power-based understanding of affects, referring to Nietzsche and Foucault.[62] Affects in this sense are the name for the specific relations of power that lend the subject the will to think, act and live in *immediate* ways, at least in the beginning. Opposing the dominant view in the debate on the second strand of affects, which sharply differentiates between cognition on the one hand and affects on the other,[63] this book seeks to demonstrate the role of *cognitive* aspects as constitutive for the formation of affects. Hence, it regards cognition and affects as belonging together. Furthermore, contemporary affect theory is mainly based on a negative, freedom-impairing understanding of affects. Mostly, they sketch an affective attachment in and through which the subject is impaired in her freedom. Firstly, this can be seen paradigmatically in Lauren Berlant's study *Cruel Optimism*. She describes "cruel optimism" as the condition "when something you desire is actually an obstacle to your own flourishing."[64] However, Berlant does less see that affects can also have a freedom-empowering potential. Secondly, this can be paradigmatically observed in Sara Ahmed's or Brian Massumi's understanding of fear. Their analyses only consider the freedom-impairing role of fear, for example the ideological, racist fear of Black people.[65] Of course, these studies are very helpful for studying the negative aspects of affects. But the freedom-empowering aspects are less considered. There are already studies that hint in the direction of the Janus face of affects, such as the instructive studies written by Otto Penz and Birgit Sauer or by Brigitte Bargetz.[66] Bargetz, however, reads affects from a too broad agonistic perspective that she sets in the contexts of Chantal Mouffe's understanding.[67] In general, the term "agonism" derives from ancient Greek and Latin (*ἀγών* and *agon*) and means contestation. In today's discussions, however, this term is often used in too broad, problematic ways, as often *every* kind of contestation and dispute is qualified as good (especially Mouffe, William E. Connolly, and Bonnie Honig).[68] Also Bargetz considers *every* contestation as freedom-empowering.[69] *On the contrary*, the book at hand *only* considers *the specific kind of enlightened contestations, struggles and disputes* as freedom-empowering (see the chapters on the enlightenment test). But, of course, the debate also contains nuanced views on agonism, such as elaborated by Owen. Here, Owen combines the notion of agonism and the ethical kind of perfectionism in accordance with Cavell (see the first chapter).[70] This type is called agonistic perfectionism in contemporary debate.[71] Moreover, the notions of agonism and antagonism need to be differentiated. Whereas antagonism (from ancient Greek words *anti* and *agon*) entails the specific contestation that has a binary frontline between *two distinct* sides, for example between rich and poor, the notion of agonism *pluralizes* this view. Here, *various* frontlines are considered as the source that could produce conflicts and thus also enlightened conflicts. But as the language of agonism *could* be misunderstood, we do *not* consider it. Here, we are contributing the previously less considered understanding of the Janus face of affects in terms of power relations.

In referring to the debate on affects we seek to show, drawing on Nietzsche and Foucault, that considering the subject of digital enhancement as *only* the rational, intentional or conscious subject would ignore the power relations that form the subjects rationality, intentionality and his consciousness. This view undergoes the binary scheme between the individualistic, action-theoretical understanding of Weber on the one hand and the collective understanding of power following Marx on the other. If Weber considers power as "every chance within a social relationship, of enforcing one's own will even against resistance"[72] and considers the notion of the will in individual terms,[73] and Marx conceives of power as class consciousness in collective terms,[74] then Nietzsche and Foucault show that it is power relations that *socially* construct the formation of the subject's will in freedom-impairing and freedom-empowering ways. Accordingly, when this book talks of subjectivity, of both quasi-automatized subjectivity willing to conform *and* enlightened subjectivity, it is clear that we are *not* dealing with a private cinema or a private archive, and thus with an individual understanding, as the accusation is often made. On the contrary, Nietzsche and Foucault teach us to see subjectivity as *socially* constructed by affect-centered power relations. In this context, the language of affects focuses on the blind or immediate production of the subject's will *and* on the creation of enlightened subjectivity. Accordingly, it is by no means wrong to talk about the rational, intentional or conscious subject, but what Nietzsche and Foucault emphasize is keeping in mind that these unities are the product of affect-sensitive power relations. And it is the rational, intentional and conscious subject's task to shed light on this production.

In the current academic and public debate on enhancement, the understanding of affects has received relatively little consideration.[75] As already outlined, we must, however, consider the notion of affects if we want to achieve an apt understanding of the catalyst of the subject's quasi-automatic, machine-like way of living, acting and feeling in the context of her digital today. Instead of focusing on the notion of affects, most scholars concentrate on the potential social, legal or political implications of the emerging perfectionistic technologies.[76] Thus they concentrate merely on the rational, conscious and intentionally acting subject that, as an *animal rationale*, has good reasons to come to this or that decision regarding technological change. But in so doing, they fail to examine the subtle aspects that *prompt* the subject to form the will to perfect life and thus the will that simply makes them regard these emerging technologies as the rational, reasonable and thus necessary way of living, feeling and acting. But the debate is not entirely concerned with the rational, reasonable and intentional subject. There are also some strands that already consider the subject's less rational side. When they do so, however, they tend to subsume these aspects under the conception of emotions.[77] Affects are different from emotions insofar as affects are the more basic term. Here, it is the question of the relations of forces as also encountered in contemporary affect theory.[78] Emotions, to wit feelings such as joy, sadness, or anger, can be part of affects and can be formed to more basically imprinted affects in the subject.[79] As a result, we still do not have a clear view of the affect-sensitive relations of power that, however, are needed in order to understand why the subject has the programmed disposition to have this or that will with which they can then rationally, intentionally and consciously decide how to navigate and orient themselves within the territory of enhancement. This gap is taken up in the present study.

To briefly conclude, the outline of the four debates over enhancement, digitalization, the art of living and affects has shown that thus far a study that connects these four debates in a systematic and fundamental way is lacking. The present study offers a central perspective in this direction. This perspective is required in order to provide a new vocabulary to articulate the subject's paradoxical condition in the field of digital enhancement: the circumstance that subjects *immediately* want to use digital enhancement in order to perfect their life, *although* they know of the freedom-impairing effects that this use has on their life formation, such as having their data intercepted. The new vocabulary developed here is not externally applied to the field of digital enhancement; rather, the field of digital enhancement *inherently* shows the necessity of considering such a perspective. In thinking these four debates together, we arrived at the following new definition of digital enhancement: It is defined by the formulation *being prompted to set himself in relation to the logic of perfection by using digital means*. It is this new vocabulary that promises to show that the field of digital enhancement is not only a current trend, but needs to be conceived in relation to its "long durée," that is in light of the long history of the social practices of life formation.

"UNTIL THEY SEE THE WORLD WE WANT THEM TO SEE"

Digital Predictive Policing and "Social Engineering" of Political Elections

The present chapter has the task to further clarify the paradoxical condition of digital enhancement. As a brief reminder, by this paradoxical condition we mean the circumstance that subjects "open [their] eyes" (Foucault) and already have their will formed in such a way that they see their life as enhanced by using digital media *despite* the well-known freedom-endangering effects of using digital media, such as having one's data intercepted. To continue to shed light on this hitherto all too neglected view on the paradoxical condition the present chapter discusses the field of predictive policing, with reference to the film *Minority Report*; with reference to the Netflix documentary *The Great Hack* it delineates the problem of "social engineer[ing]"[80] in the field of political elections, to use the words of the whistleblower Christopher Wiley, one of the key figures that made the Cambridge Analytica scandal public. These paradigmatic cases demonstrate how subjects *are open to having their will formed* "until they see the world we want them to see," to cite the whistleblower Britanny Kaiser,[81] who also worked for Cambridge Analytica.

It goes without saying that the film *Minority Report* and the Netflix documentary *The Great Hack* thematically do not belong to the classic canon of digital enhancement. In the next chapter on aesthetic critique we explain more fully why we *nonetheless, and in fact especially,* need to consider these films systematically. Here, it suffices to note that they also convey a central view on the field of digital enhancement, depicting dystopian examples of what an allegedly perfect world could also imply, namely the opposite of what was actually intended. *Minority Report* shows a scenario in which society intends to enhance life by making the world a safer place. However, the film depicts how the digital methods used to fulfill this goal, namely the methods of predictive policing, lead *instead* to a world that reduces people's freedom. *The Great Hack* outlines how political elections could be enhanced with the use of social media. But here, too, the documentary clarifies that this goal has severe freedom-impairing effects: instead of making the world an allegedly better place, people's political opinions are shaped in a certain preferred direction. What is considered to be the preferred direction depends on the sponsors—that is, it depends on the political parties that hire tech companies, such as Cambridge Analytica and others like them, to form the voters' political opinions, as the Democrats did with Barack Obama or the Republicans with Ted Cruz and Donald Trump, to mention only key examples of the political landscape in the US.[82] To reiterate, this scandal revealed the extent to which social media is used in order to manipulate political elections, such as the US presidential election or the Brexit referendum in Great Britain, both of which took place in 2016.

To begin with, one must note that there are obviously different degrees of being open to having one's will to adapt to the digital world formed. It is not so much a question of when, for example, a subject does not clearly see what is at stake if their smart refrigerator automatically buys butter or meat if needed. The subject might be even happy to have this help and not be required to shoulder the full administrative workload of the household alone. Nevertheless, there are key areas that raise larger concerns. And it is to these key areas that the cases of predictive policing and "social engineering" in the context of political elections belong.

It might sound a bit provocative that people today need to be conceptualized as being open to having their will formed with respect to such core areas as political elections. It is even inscribed in the Preamble of the Constitution of the United States of America to "secure the Blessings of Liberty to ourselves and our Posterity."[83] So one might indeed wonder how it could be that today's supposedly enlightened people more or less freely consent to being formed in their will in a country like the US, which is so proud of its freedom. To better understand how smoothly such formation works, it is helpful to consider Annie Ernaux's classic novel *Shame*. Ernaux's book, of course, does not deal with the topic of digitalization under discussion here; rather, she sketches a quite different narrative, namely her traumatic childhood memory of how, according to Ernaux's book's very first line, her "father tried to kill [her] mother one Sunday in June."[84] Throughout her book Ernaux shares with the reader her experiences of shame that she felt afterwards and analyzes how these experiences determined the rest of her life. Shame, to put it briefly, is the emotion that arises when people feel they fail to meet the social standard.[85] The point that overlaps with our theme consists in the fact that Ernaux does not tell the story as her individual problem, but rather outlines its *social construction* by power relations that were part of the French society in the early 1950s and that this book understands in terms of affects. These affect-centered power relations made people like her ready to feel shame in this way. And why did they do so? As Ernaux informs us, she "was living by the rules and codes of this world; it never occurred to [her, SB] that there might be others."[86] Put differently, the problem emerges if subjects think, live and feel in self-evident ways in and through which they *are prompted* to adapt to given rules, norms and codes that are part of a certain society. It is this problem that also arises in the different setting of predictive policing and political elections: That is, blindness to the possibility that *other* rules and codes might exist entails a lack of conscious awareness of those rules and codes that *do* form the framework of one's life. From this angle we see that such a subject does not fully decide in a rational, intentional or conscious manner to participate in this perspective. The subject's current way of living rather becomes such an *un*questioned part of his social life that any other perspective is simply not visible. The subject does not even think of potential changes.

In our digital present, such a one-sided view emerges also quite easily. This point becomes apparent if we consider the daily practice of consenting to like-buttons. For example on Facebook, users are invited to "like" contributions on the platform. In clicking "like," however, the user quasi-automatically gives their consent to Facebook to track their further digital activity, not only within the limits of Facebook, which would be bad enough in itself, but Facebook is also immediately permitted to track what the user is doing on all other websites that operate with a Facebook like button. Back in 2011, this was the case for 33 percent of the 1000 most regularly visited websites and this number has certainly increased since then. *Only* by explicitly signing out of Facebook is Facebook *not* permitted to track one's further online activity.[87] Although this point is well-known, people just continue to click on like-buttons and thus develop the one-sided view that using digital media entails this practice.

A plausible explanation of how such a blind behavior in today's digital field becomes so widespread is given in the Netflix documentary *The Great Hack*. In this film, one user tries to explain why she does not really care about data protection; according to her simple but consequential explanation she just "grew up with the internet."[88] This view is not a unique opinion. It can be seen as representing the general attitude of the younger generation usually described as "digital natives." But even on a broader scale, this statement also echoes the blind attitudes that people in general have in response to digital enhancement, at least initially, before they begin to care about what exactly they are doing when they use digital tools.

Against this backdrop, it becomes better understandable how easily the subject's smooth will formation takes place in the field of predictive policing and "social engineering" in political elections. In the following, we will sketch this process more concretely through the science-fiction film *Minority Report*. Predictive policing is a method that is currently being used in reality, for example in the US. This method is based on the premise that one can predict future crimes on the basis of a pre-given set of digital data. With this information the method aims to reduce the crime rate, with the government seeking to prevent alleged criminals from committing a crime *before* they in fact do so. The method of predictive policing originated with the sociologist Ernest W. Burges, who invented it in the field of incarceration in the beginning of the twentieth century.[89] Although predictive policing wants to perfect life by making social life more secure, its downsides are widely known.[90] In particular, the specific actuarial method of racial profiling is criticized, since with this method Black people are more likely to be identified as potential criminals. Here, it is known that racial profiling *in itself* produces structurally racist patterns: In the US, Black people are stopped by the police more often than white people. Therefore, there is more data on Black people than white people *although* this social fact does not represent the *real* crime rate. Hence, if one seeks to divide subjects into potential delinquents and potential *non*-delinquents on this basis, there is a much higher probability of identifying Black people as possible future criminals. This fact results from the set of data that was previously harvested in a biased way. To put it differently, the method of racial profiling fosters racial prejudices by alleging that Black people are more dangerous than white people, for example in claiming that Black people were more likely than white people to sell drugs.[91]

It once again became clear what this means for the concrete lives of Black people in the US in May 2020 when George Floyd was killed by the police officer Derek Chauvin in Minneapolis while being arrested, to name just one prominent example. Floyd's last words "I can't breathe" filled newspapers around the world.[92] Tymber Hudson, an antiracist activist and artist, recently described the far-reaching effects that scenes like this have in the daily lives of Black people in the US. When they are stopped by the police, for example while driving, they are always afraid of what the police will do to them, and as Hudson puts it, they are always confronted with the question: "Am I next?"[93] More concretely, with this question she is asking, Will I be the next one that, like Floyd, to lie on the street, with a police officer's knees on my neck, no longer be able to breathe?[94] That Hudson's fears are so understandable became apparent during the trial of Floyd's murderer. While this trial was taking place, there was another murder only a few miles away from the courthouse in Minneapolis where Floyd's case was being tried. This time, it was a police officer who killed a black man during a traffic stop because she confused her weapons, according to her own statement. Instead of drawing her taser, she drew her gun, and with this she killed the innocent black man.[95]

Despite these well-known downsides of predictive policing and in particular the method of racial profiling, it is paradoxical that people still have faith in them. An answer as to why this paradox emerges is provided by the film *Minority Report*, which we already mentioned in the present book's introduction. To again briefly summarize the premise of the film before we begin: The main idea of the film, directed by Spielberg, centers around the conception of predictive policing, which in the film is called "precrime" policing. Specifically, the film depicts the endeavors of the police officers that work in the "PreCrime" police department: the officers are tasked with deciphering predictions of future crimes made by psychics ("precogs") and thus detecting potential crimes *before* they are committed. Ultimately, the film problematizes the fact that people are quasi-automatically prompted to strongly believe in the system of "precrime" policing. This belief endangers the subject's freedom insofar as the subject *immediately* accepts the pre-given perspective of "precrime" policing and only quasi-automatically reacts to it, *without* considering the existence of other possibilities. In the case of *Minority Report* this implies for police officers: digitally predicting future crimes in order to reduce the crime rate *although*, as the film shows, the predicted perpetrators might *in fact* not be the perpetrators they were defined as. In this sense, the film portrays that the subject *herself* becomes an accomplice of the systems of perfection in willingly adhering quasi-automatically and thus blindly to the perspective determined by the authorities. In other words, we see that *not only* do the contexts in which the subject immediately lives, feels and acts become a problem, but *so does* the subject's immediate attitude towards these contexts. In our case, we are dealing with the subject's specific attitude that is generated by her belief. In and through this belief she is prompted to set herself in such a relation to the perspective of "precrime" policing that she simply believes in its necessity.

The film demonstrates this problem in the fate with which the protagonist John Anderton is confronted. Anderton, played by Tom Cruise, is head of the "PreCrime" police department and an absolute believer in the system's perfect functioning. He has lost his child to a murder. Due to this loss, his marriage has also fallen apart. These personal circumstances now prompt Anderton to do everything he can to prevent similar cases from happening to other members of society, and this also entails fully believing in the necessity of predicting future crimes with the help of digital media. He is simply blind to the possibility that the "precrime" system could have flaws or might not be delivering its promise of enhancing freedom. However, one day he is confronted with the problem that "precrime" identifies him as a future murderer of a person he has thus far never heard of. Anderton is thus faced with the paradox that *on the one hand* he absolutely believes in the system of predicting future crimes, which he considers to be infallible. But *on the other*, he also believes that he is incapable of killing a person, he who is one of the best police officers in the department. The plot of the film then centers on Anderton's attempts to prove his innocence and thus shows his struggles with his conflicting belief in the system *and* in himself as a person who is *not* a murderer. As a consequence of his absolute belief in the system of "precrime," Anderton has trouble perceiving any other possible perspective on his life than the one given by the prediction. So, although the "precog" Agatha—one of the figures who delivers the prediction and thus someone who is actually part of the predicting machine—tries to open Anderton's eyes to other possible views on his life, her attempts do not reach him. It is impossible for him to understand what she means by her whispered warning: "you have a choice."[96] In his eyes, life can *only* be perfect as part of the predicting system. Of course, Tom Cruise plays Anderton as a superhero who can fight against his predicted destiny and thus also against his own strong belief in the methods of digitally predicting future crimes. Thus he is depicted as someone who is able to address his blind belief and change from being a strong supporter of the system to one of its harshest critics. Hence, he is more or less able to fight his belief and perceive that "precrime" policing does not exclusively enhance freedom, as it promises, but can also endanger freedom, as it had done to him. Of course, ordinary people do not need to become superheroes, but they can also have a chance to counter their beliefs once taken. The field of digital enhancement shows, too, that people *constantly* face the challenge of having beliefs that unfold as unjustified. In view of this, the present study conceptualizes how the subject is structurally confronted with this problem and constantly called to gain an enlightened attitude.

We now turn to the documentary *The Great Hack*. *The Great Hack* shows a "crisis of perception,"[97] in the words of the whistleblower Wylie part of the Cambridge Analytica scandal in 2018, taking place in the context of political decision-making. While the standard view on political decision-making assumes that such decisions are made by *fully* rational, intentional and conscious subjects who are *totally* free to opt for this or that political party, the case of the Cambridge Analytica scandal reveals another level of today's political reality: it discloses how people's political will was *purposively* formed through subtle mechanisms of "social-engineer[ing]," in Wylie's words, in particular in the US presidential election and the Brexit referendum, both in 2016. Hence, such subjects are *not* the assumed fully rational, intentional or conscious subject, but also subjects whose political will is formable. In this sense, *The Great Hack* demonstrates how subjects were prompted by tech-companies like Cambridge Analytica to vote in a specific political direction *without* them being aware that their political will was not only the result of their rational thinking. Most famously, Trump hired Cambridge Analytica during his presidential campaign in 2016, but so did Cruz and Obama used somewhat similar methods in 2012. But as we saw in the discussion of *Minority Report*, we also see in this case that subjects, here the voters, are *not only* passive objects being manipulated. Rather, this book delineates that subjects, or again the voters, become a *blind accomplice* in these manipulations, for they *are open* to being manipulated by methods such as those that Cambridge Analytica used.

Before we problematize the Cambridge Analytica scandal, it is necessary to roughly sketch the outlines of what happened. In March 2018 a group of newspapers, beginning with *The New York Times* and *The Guardian*, revealed the subtle mechanisms at work in the US presidential campaign in 2016.

Cambridge Analytica, it was shown, had manipulated voters' opinions by designing an "artificial society,"[98] as Wylie puts it, based on the collected data that they easily acquired from Facebook. The platform Facebook, launched in 2004 to connect the social lives of people all around the globe, was helpful for Cambridge Analytica's purposes as "Facebook knows more about you than any other person in your life, even your wife,"[99] for example by gathering the material that shows when someone wrote to whom, how often, what someone liked, et cetera. This obviously hyperbolic quotation is from Wylie, who cites the words of the Cambridge University professor Aleksandr Kogan. This professor played a major role in the scandal, as it was he who created the consequential app including a personality test. Using this personality test, Cambridge Analytica was able to easily acquire a wide collection of consumer data that they needed to manipulate the people's political will. The test entailed the Facebook personality quiz app "This is your digital life." All in all, 270,000 people participated in this test in a free manner, without being forced to do so, only being motivated by the promise of earning only one or two dollars. Cambridge Analytica not only received these people's data, but also that of *all* their contacts on Facebook. In sum, by this method the political consulting company was able to access the data of more than 80 million Facebook users.[100] This set of Big Data—"the new oil"[101]—was required for the political strategy that Cambridge Analytica had in mind. Their strategy was quite simple, but effective. In his autobiography *Mindf*ck*, Wylie describes how this strategy worked. It was rooted in the methods of TV channels such as Fox News.[102] Wylie notes that Fox News, for example, purposefully triggers voters' emotions, like anxiety, that is, the feeling that is motivated by unreal dangers that paralyzes the subject's action,[103] in order to motivate their political behavior. Whereas Fox News, however, operated with monolithic group categories, such as white voters, woman voters, Latino voters or suburban voters, as Wylie reports, the strategy of Cambridge Analytica was more specific. They sought to reach the people *within* these groups individually, a method called *micro*-targeting.[104] As Wylie notes, there can be a wide range among, for example, white voters. One can imagine that an underprivileged white voter living in a trailer, to use Wylie's illustrative example, does not feel responsive to political campaigns that *only* focus on privileged white voters. Through the results of the personality test on Facebook, Cambridge Analytica was able to generate the requisite individualized profiles, then sought to feed voters with campaigns that *in fact* corresponded to their respective needs. And by addressing these needs they aimed to manipulate the voters' emotions. The voters were even "bombarded" with material that responded to their concrete fears and anger, to describe the mechanisms of the scandal using the drastic expression of Carole Cadwalladr, one of the leading journalists who contributed to the revelation of this scandal. For example, Cambridge Analytica wanted to create the feeling of being threatened because with this feeling they intended to stimulate a voter's openness to racist or xenophobic attitudes. By creating such an openness, Cambridge Analytica sought to trigger a shift in the voter's political opinion. With this purpose, they were particularly interested in swing states like Florida, Michigan or Pennsylvania. In these states, they hoped to best manipulate the "persuadables," as they are called in *The Great Hack*, in order to motivate them to vote in the political direction of the campaign that paid for Cambridge Analytica's work. In 2016, this meant especially to direct voters to the ballot box for the Trump campaign and the Leave camp in the Brexit referendum in 2016.

If Cambridge Analytica's goal was to create an "artificial society," as Wylie's autobiography puts it, then this book seeks to demonstrate that they were only able to do so because the subject willingly *cooperated* in this social construction. Cambridge Analytica smuggled in the construction materials, i.e., the pieces of information, and the subject offered the ground on which this material was able to be put together—and this social construction is grounded on the subject's formable will. In this sense, the book at hand shows that the method of "social engineer[ing]" cannot work entirely independently; it requires a counterpart within the subject that is open to being conducted, and that is the subject's formable will. In other words we see that the subject's formable will functions to open the door for digital strategies that seek to conduct one's political life. In view of these perspectives, one understands why Carole Cadwalladr raises the key question: "Is this what we want? To sit back and play with our phones [. . .]?"

Hence, the view on the Cambridge Analytica scandal also raises once more the justified question as to why people are open to giving away their digital data that helps others to create their digital footprint, *despite* knowing of the problems of data privacy in the WWW.

FIRST CONCLUSION

In this chapter we clarified the present study's object of analysis: the field of digital enhancement. In order to adequately conceptualize this field, the first step showed the need to connect four debates that have thus far not been considered *together*: the debate over enhancement, digitalization, the art of living and affects. The common denominator of these four debates lies in the logic of perfection. *First*, from the debate on enhancement, we gained the theoretical tool of conceptualizing the object of analysis as characterized by the procedures of the logic of perfection. *Second*, from the debate over digitalization, we learned that we are dealing with the procedures of the logic of perfection in the specific context of digitalization. *Third*, the debate over the art of living helped us to consider digital enhancement in terms of life formation. This showed us that the object of analysis cannot be conceived as a current trend, but needs to be understood in its "*longue durée*." We continued this line of thought with the fourth debate over affects. By this, we do not only see that people form their life by digital media in relating oneself to the logic of perfection, but we can now specify their motivation. Now we were able to say they *are prompted* to do so. Through the lens of these four debates we developed the following new understanding of the object of analysis: an enriched understanding of digital enhancement involves the conception of *being prompted to set oneself in relation to the logic of perfection by using digital means*. The second step specified the problem horizon of digital enhancement outlined here with respect to the hitherto all too neglected paradoxical condition of digital enhancement. Especially in discussing the film *Minority Report* we problematized the case of digitally predictive policing. Further, drawing on the Netflix documentary *The Great Hack* we also considered the case of manipulating political elections, such as the US presidential elections and the Brexit referendum in 2016, as revealed by the Cambridge Analytica scandal.

NOTES

1. According to this book, we need to deal with a broad, *non*-essentialist understanding of culture. The broad understanding of culture entails the subject's procedures of life formation. Hence, life formation is *not* a natural process. "Social" specifies the kind of life formation. It encompasses the effects of the logic of perfection on the subject that, therefore, can be called *social effects*.

2. Hence, the notion of enhancement needs to be differentiated from the notion of therapy, that involves the understanding of amelioration. On this discussion see Günter Thomas, "Enhancement: Evangelisch theologische Optionen in der gegenwärtigen Debatte," in *Sind Sie gut genug? Zur (Selbst-) Optimierung und Vervollkommnung des Menschen*, ed. Monika C. M. Müller et al. (Loccum: Loccumer Protokolle, 2011), 26. In contrast to the notion of therapy, the conception of enhancement thus entails the understanding, to put it with Buchanan, "to make humans better than well," see Allen Buchanan, *Better Than Human: The Promise and Perils of Biomedical Enhancement* (Oxford: Oxford University Press, 2018), 173.

3. Harcourt, *Exposed*, 21.

4. Paradigmatically: Bertrand Russell, *The History of Western Philosophy: And its Connection With Political and Social Circumstances From the Earliest Times to the Present Day* (London/New York: Routledge, 1967).

5. Christoph Menke, "Zweierlei Übung: Zum Verhältnis von sozialer Disziplinierung und ästhetischer Existenz," in *Zwischenbilanz einer Rezeption: Frankfurter*

Foucault-Konferenz 2001, ed. Axel Honneth and Martin Saar (Frankfurt a.M.: Suhrkamp, 2003), 199–210.

6. James Conant, "Nietzsche's Perfectionism: A Reading of 'Schopenhauer as Educator,'" in *Nietzsche's Postmoralism: Essays on Nietzsche's Prelude to Philosophy's Future*, ed. Richard Schacht (Cambridge: Cambridge University Press, 2001), 181–257.

7. Cavell, *Conditions Handome and Unhandsome*, xi.

8. Note that neither Nietzsche nor Foucault differentiate between "morals" and "ethics," in contrast to current discussions in the Habermasian tradition (cf. Forst, *Kontexte der Gerechtigkeit*). Whereas Habermas places ethics in the context of life formation and morals in the context of justice, Nietzsche and Foucault refers to the notions of "morals" and "ethics" to identify the set of rules of behavior that governs the subject's way of life.

9. Owen, "Perfectionism, Parrhesia and Care of the Self," 128–55.

10. Rainer Forst, "Die Ungerechtigkeit der Gerechtigkeit: Normative Dialektik nach Ibsen, Cavell und Adorno," in *Kritik der Rechtfertigungsverhältnisse: Perspektiven einer kritischen Theorie der Politik*, ed. Rainer Forst (Berlin: Suhrkamp, 2011), 183.

11. Owen, "Perfectionism, Parrhesia and Care of the Self"; and Daniele Lorenzini, *Éthique et politique de soi: Foucault, Hadot, Cavell et les techniques de l'ordinaire* (Paris: Vrin, 2015).

12. Jan-Christoph Heilinger, *Anthropologie und Ethik des Enhancements* (Berlin/Boston: de Gruyter, 2010).

13. Buchanan, *Better Than Human*, 24pp; John Harris, *Enhancing Evolution: The Ethical Case for Making Better People* (Princeton: Princeton University Press, 2010).

14. Of course, this perspective is nonetheless very seminal, see: Paula Irene Villa, ed., *Schön normal: Manipulationen am menschlichen Körper als Technologien des Selbst* (Bielefeld: transcript, 2008).

15. Obviously, in the light of current debates about the differentiation between nature and culture, it is self-evident that the common formulation of the alleged "natural lottery" has its limits. The subject's life formation, which includes digital enhancement, belongs to the area of culture and thus *not* to the arena of nature. The formulation of the "natural lottery" thus *only* entails the biological aspects of the subject's life formation.

16. Keir Dullea, *2001: A Space Odyssey*. DVD. Dir. St. Kubrick. Los Angeles: Warner Bros, 2004 [1968].

17. The report *Beyond Therapy* outlines the elements of the debate that combine criticism of enhancement with religious aspects. See Leon Kass et al., *Beyond Therapy: Biotechnology and the Pursuit of Happiness. A Report by the President's Council on Bioethics*, 2nd ed. (New York: HarperCollins Publishers, 2003). As a paradigmatic culturally pessimistic bio-conservative position we can cite Francis Fukuyama's view in *The End of History and the Last Man* (New York/London: Penguin Press, 1993). Michael J. Sandel's position, which he presents in *The Case Against Perfection: Ethics in the Age of Genetic Engineering* (Cambridge, Mass.: Harvard University Press, 2007), is an important example of social critique. Although Habermas's works are

commonly interpreted as also belonging to the bio-conservative camp, the present book can*not* support this view; for Habermas stresses the *self-relational* aspect and explicitly *not* an essentialist or substantialist approach to what he understands as nature; thus it would be short-sighted to understand his position *as if* he were seeking to protect an alleged substance called nature, see Jürgen Habermas, *Die Zukunft der menschlichen Natur: Auf dem Weg zu einer liberalen Eugenik?*, 5th ed. (Berlin: Suhrkamp, 2018).

18. There are rumors that the name implicitly symbolizes the initials of the company IBM; see Christoph Drösser, "Der Name HAL ist von IBM abgeleitet: Stimmt's?," *ZEIT*, January 27, 2000, available at: https://www.zeit.de/2000/05/200005.stimmts_hal_ibm_.xml.

19. Obviously, the film does not explicitly refer to enhancement technologies. However, they are implicitly part of it, as enhancement technologies are understood here as the technological means by which human beings seek to perfect their lives, even intergalactic lives.

20. Stanley Kubrick and Arthur C. Clark, *Screenplay: 2001. A Space Odyssey*, available at: http://www.archiviokubrick.it/opere/film/2001/script/2001-originalscript.pdf.

21. Cf. Ingmar Persson and Julian Savulescu, "The Duty to be Morally Enhanced," *Topoi* 38 (2019): 7–14.

22. More specifically, see Janina Loh, *Trans- und Posthumanismus zur Einführung*. 3rd ed. (Hamburg: Junius, 2020).

23. Nick Bostrom, "Why I Want to Be a Posthuman When I Grow Up," in *Medical Enhancement and Posthumanity*, ed. Bert Gordijn and Ruth Chadwick (Berlin: Springer, 2008), 112.

24. Tristan Quinn, "Horizon: The Immortalist: Uploading the Mind to a Computer," *BBC News*, March 14, 2016, available at: https://www.bbc.com/news/magazine-35786771.

25. Henry T. Greely, "Some First Steps Toward Responsible Use of Cognitive-Enhancing Drugs by the Healthy," *The American Journal of Bioethics* 13, no. 7 (2013): 39–41.

26. Cf. Faden and Beauchamp, *A History and Theory of Informed Consent*, 235. The limits of the notion of "informed consent" are shown in particular by Joseph J. Fischel, *Sex and Harm in the Age of Consent* (Minneapolis: University of Minnesota Press, 2016).

27. Geoffroy de Lagasnerie, *L'art de la révolte: Snowden, Assange, Manning* (Paris: Fayard, 2015). Strictly speaking, the debate existed prior to this, but did not gain as much attention. See Lutz-Bachmann, Matthias, "Die Zukunft mit Algorithmen gestalten," *Goethe-Uni online*, September 12, 2019, available at: https://aktuelles.uni-frankfurt.de/veranstaltungen/die-zukunft-mit-algorithmen-gestalten-interview-mit-prof-lutz-bachmann/.

28. Nick Bostrom and Eliezer Yudkowsky, "The Ethics of Artificial Intelligence," in *The Cambridge Handbook of Artificial Intelligence*, ed. Keith Frankish and William M. Ramsy (Cambridge: Cambridge University Press, 2014), 316–34; Nick Couldry and Ulises A. Mejias, *The Costs of Connection: How Data is Colonizing Human Life*

and Appropriating it For Capitalism (Stanford: Stanford University Press, 2019); Peter Dabrock and Tabea Ott, "Transparent human–Non-transparent Technology? The Janus-faced Call for Transparency in AI-Based Health Care Technology," in *Frontiers in Genetics* 13 (2022); Armin Grunwald, "Zwischen Fortschrittsoptimismus und Technikdämonisierung," in *Schöne neue Welt? Zwischen technischen Möglichkeiten und ethischen Herausforderungen*, ed. Mathias Lindenau and Marcel Meier Kressig (Bielefeld: transcript, 2020), 77–100; Jeanette Hofmann, "Digitale Infrastrukturen im Wandel," *Bürger & Staat* 72, no. 1–2 (2022): 56–62; Sarah Spiekerman, *Digitale Ethik: Ein Wertesystem für das 21. Jahrhundert* (München: Droemer, 2019); Zuboff, *The Age of Surveillance Capitalism*. A functionalist analysis of digital change is provided by Armin Nassehi, *Muster: Theorie der digitalen Gesellschaft* (München: Beck, 2019).

29. Harcourt, *Exposed*, 6.

30. Cf. Lisa de Propris and David Bailey, eds., *Industry 4.0 and Regional Transformation* (London: Routledge, 2020).

31. In his recent study, Klaus Günther shows the relevance of the examples of smart cities and the social credit system, see Klaus Günther, "Von normativen zu smarten Ordnungen," in *Normative Ordnungen*, ed. Rainer Forst and Klaus Günther (Berlin: Suhrkamp, 2021), 523–52. Here we add the connection between digital change and the logic of perfection in terms of affect-centered power relations.

32. Matthias Sander, "Wem gehört die Smart City? Toronto ringt mit dem Google-Konzern," *Neue Zürcher Zeitung*, October 10, 2019, available at: https://www.nzz.ch/international/wem-gehoert-die-smart-city-toronto-ringt-mit-dem-google-konzern-ld.1513002. The plan was ultimately cancelled, as the newspaper informs us, but it is nonetheless a useful illustration of what is at stake in digital enhancement.

33. Niklas Maak, "Google-Stadt ist abgebrannt: Keine Smart-City in Toronto," *Frankfurter Allgemeine Zeitung*, May 11, 2020, available at: https://www.faz.net/aktuell/feuilleton/debatten/keine-smart-city-in-toronto-google-stadt-ist-abgesagt-16763217.html.

34. Felix Simon, "Google baut in Toronto einen vernetzten, perfekten Stadtteil–der unsere Freiheiteinschränken könnte," *Neue Zürcher Zeitung*, December 8, 2017, available at: https://www.nzz.ch/feuilleton/wie-wollen-wir-leben-ld.1336803?reduced=true.

35. More specifically, on self-driving systems, see Janina Loh, *Roboterethik: Eine Einführung* (Berlin: Suhrkamp, 2019); in general, on the theme of artificial intelligence consider Walther Ch. Zimmerli, *Künstliche Intelligenz: Philosophische Probleme* (Stuttgart: Reclam, 1993).

36. Maak, "Google-Stadt ist abgebrannt."

37. IKEA, available at: https://www.ikea.com/gb/en/product-guides/ikea-home-smart-system/.

38. Thus far participation in the social credit system is not mandatory, but a mandatory participation is planned. This was originally scheduled for the end of 2020, but was postponed due to technological problems. See Johnny Erling, "So absurd ausgefeilt ist Chinas Überwachungssystem," *DIE WELT*, April 17, 2019, available at: https:

//www.welt.de/wirtschaft/article192029849/Social-Scoring-So-absurd-ausgefeilt-ist-Chinas-Ueberwachungssystem.html.

39. Mark E. Warren, ed., *Democracy and Trust* (Cambridge: Cambridge University Press, 1999).

40. Rainer Forst, "The Justification of Trust in Conflict: Conceptual and Normative Groundwork," *ConTrust Working Paper*, no. 2 (2022), available at: contrust.uni-frankfurt.de/wp-2.

41. On the notion of probability, Darrel Moellendorf, *The Moral Challenge of Dangerous Climate Change: Values, Poverty, and Policy* (Cambridge: Cambridge University Press, 2014).

42. Kelsey Munro, "China's Social Credit System Could Interfere in Other Nations' Sovereignty," *The Guardian*, June 27, 2018, available at: https://www.theguardian.com/world/2018/jun/28/chinas-social-credit-system-could-interfere-in-other-nations-sovereignty.

43. Helen Davidson, "Beijing Blocks Access to Clubhouse App After Surge in User Numbers," *The Guardian*, February 8, 2021, available at: https://www.theguardian.com/world/2021/feb/08/bingeing-free-expression-popularity-of-clubhouse-app-soars-in-china.

44. Alicia Vikander, *Ex Machina*. DVD. Dir. A. Garland, London: DNA Films, 2015.

45. Paradigmatically: Ulla Hahn, "Vernunft ist auch eine Herzenssache," *Frankfurter Allgemeine Zeitung*, March 10, 2019, available at: https://www.faz.net/aktuell/feuilleton/buecher/literatur-und-ki-vernunft-ist-auch-eine-herzenssache-16079038-p4.html.

46. Jenni Rieger, "Mein Algorithmus und ich: Daniel Kehlmann testet KI," *Tagesschau*, February 10, 2021, available at: https://www.tagesschau.de/wirtschaft/daniel-kehlmann-kuenstliche-intelligenz-versuch-101.html, trans. by the author.

47. Stefanie Karg, "The Next Rembrandt: Kunstwerke aus dem 3 D-Drucker?!," *Kultur Hoch N*, available at: https://www.kulturhochn.de/the-next-rembrandt-kunstwerke-aus-dem-3d-drucker/.

48. More specifically, the Dutch project developed a software system that was able to create a painting using a 3D printer. See Karg, "The Next Rembrandt."

49. There are also many other projects paving the way in a similar direction. Consider especially the case of the robot Ai-Da, which looks like a female artist. The robot's works can be visited in the exhibition "Unsecured Futures"; cf. Anatol Locker, "Kunst oder Kokolores: Wie Roboter Ai-Da funktioniert," *ZDF*, June 12, 2019, available at: https://www.zdf.de/nachrichten/heute/kunst-oder-kokolores-wenn-roboter-zeichnen-100.html.

50. Alexander Nehamas, *The Art of Living: Socratic Reflections from Plato to Foucault* (Berkeley et al.: University of California Press, 1998); Nikolaos Loukidelis, ed., *Nietzsche und die Lebenskunst: Ein philosophisch-psychologisches Kompendium* (Stuttgart: Metzler, 2016); Wilhelm Schmid, *Philosophie der Lebenskunst: Eine Grundlegung* (Frankfurt a.M.: Suhrkamp, 1998).

51. Hadot, *La philosophie comme manière de vivre*; Martin Saar, "Nachwort: Die Form des Lebens: Künste und Techniken des Selbst beim späten Foucault," in *Michel*

Foucault: Ästhetik der Existenz: Schriften zur Lebenskunst, ed. Daniel Defert and François Ewald,. 5th ed. (Berlin: Suhrkamp, 2015), 319–43.

52. See Schmid's considerations of, as he puts it, the "electronic subject": Wilhelm Schmid, *Mit sich selbst befreundet sein: Von der Lebenskunst im Umgang mit sich selbst*, 9th ed. (Frankfurt a.M.: Suhrkamp, 2016), 162pp, trans. by the author. These approaches, however, do *not* pay attention to the affect-sensitive power relations that are part of the art of living.

53. Different areas of enhancement could also be seen from this perspective, as for example the field of genetic enhancement and the recently invented genetic scissor CRISPR-Cas 9. This scissor allows human beings to nurture their dreams of cutting cells like paper, to arrange and rearrange the snippets according to their preferences in order to produce, in the best case, healthier babies; in the worst case we are on the cusp of creating designer babies with manipulated eye or hair color or boosted intelligence. For a brief introductory understanding of what is at stake in the context of this genetic scissor see Sarah Bianchi and Christopher Th. Scott, "Enhancing Human Embryos: Restraint or Full Speed Ahead?," *The Enhancing Life Project*, February 25, 2016, http://enhancinglife.uchicago.edu/blog/enhancing-human-embryos.

54. Ovid, *Metamorphosen*, ed. Michael von Albrecht, trans. Michael von Albrecht (Stuttgart: Reclam, 2019), 542p, trans. by the author (*Corpus erat*).

55. Bernard Shaw, *Pygmalion: Romanze in 5 Akten*, trans. Harald Mueller (Berlin: Suhrkamp, 2012).

56. Johann W. Goethe, *Faust: Der Tragödie erster und zweiter Teil*, ed. Erich Trunz, 16th ed. (München: Beck, 1996 [1808]), 211, trans. by the author (*artig Männlein*).

57. Goethe, *Faust*, 210, trans. by the author.

58. On its broader reception, see Anne Fleig, "Automaten mit Köpfchen. Lebendige Maschinen und künstliche Menschen im 18. Jahrhundert," in *Goethe-Zeit Portal*, January 19, 2004, available at: http://www.goethezeitportal.de/db/wiss/epoche/fleig_automaten.pdf.

59. This point is neglected in today's actor-network theory, see Bruno Latour, "On Actor-Network Theory: A Few Clarifications," in *Soziale Welt* 47, no. 4 (1996): 369–81. There, things and human beings are considered as having an equal status. But from this perspective one can*not* see that it is the subject that has to weigh technology. The book at hand shares this point with the widely discussed perspective of digital humanism, cf. Julian Nida-Rümelin, and Nathalie Weidenfeld, *Digitaler Humanismus: Eine Ethik für das Zeitalter der Künstlichen Intelligenz* (München: Piper, 2018).

60. Ernst T. A. Hoffmann, *Der Sandmann*, ed. T. Borken (Berlin: Henricus Edition Deutsche Klassik, 2019 [1817]).

61. Jan Slaby, *Gefühl und Weltbezug: Die menschliche Affektivität im Kontext einer neo-existentialistischen Konzeption von Personalität* (Paderborn: mentis, 2008).

62. See in particular John Protevi, "Affect and Life in Spinoza, Nietzsche, and Bergson," in *Affect and Literature*, ed. Alex Houen (Cambridge: Cambridge University Press, 2020), 66–82. Nonetheless, affects are not understood in terms of power in the specific way sketched here. Mainly, on Foucault: Rainer Mülhoff, *Immersive Macht: Affekttheorie nach Spinoza und Foucault* (Frankfurt/New York: Campus,

2018). Unlike the book at hand, Mühlhoff understands affects in terms of resonance and immersion.

63. Brian Massumi, *Politics of Affect* (Cambridge/Malden: Polity, 2015), 10pp.

64. Lauren Berlant, *Cruel Optimism* (Durham: Duke University Press, 2011), 1.

65. Sara Ahmed, *The Cultural Politics of Emotion*, 2nd ed. (Edinburgh: Edinburgh University Press, 2014); Brian Massumi, ed., *Politics of Everyday Fear* (Minneapolis: The University of Minnesota Press, 1993).

66. Otto Penz and Birgit Sauer, *Affektives Kapital: Die Ökonomisierung der Gefühle im Arbeitsleben* (Frankfurt a.M./New York: Campus, 2016), 15; Brigitte Bargetz, *Ambivalenzen des Alltags: Neuorientierungen für eine Theorie des Politischen* (Bielefeld: Transcript, 2016), 176pp and 239pp.

67. Chantal Mouffe, *Agonistics: Thinking the World in Plurality* (London: Verso, 2013).

68. Mouffe, *Agonistics*; William E. Connolly, *The Ethos of Pluralization* (Minneapolis: University of Minnesota, 1995); Bonnie Honig, *Political Theory and the Displacement of Politics* (New York: Cornell University Press, 1993).

69. Bargetz, *Ambivalenzen des Alltags*, 55pp.

70. Owen, *Nietzsche, Politics and Modernity*.

71. Marie Paxton, *Agonistic Democracy: Rethinking Political Institutions in Pluralist Times* (New York/Abingdon: Routledge, 2020), 56pp.

72. Max Weber, *Economy and Society*, trans. K. Tribe (Cambridge, Mass./London: Harvard University Press, 2019), 134.

73. Furthermore, the Weberian understanding conceives of power relations in terms of domination. Because Weber thinks that settings of power are very complicated to interpret in terms of his interpretative sociology, he reduces phenomena of power to settings of domination. Max Weber, *Wirtschaft und Gesellschaft: Grundriss einer verstehenden Soziologie*. Ed. J. Winckelmann. Tübingen: Mohr, 1976 [1921/22], § 16.

74. Georg Lukács, *History and Class Consciousness* (Cambridge: MIT Press, 1972).

75. Greely, "Some Steps."

76. For example the German handbook on bioethics lacks an entry for "desire." See Dieter Sturma and Bert Heinrichs, eds., *Handbuch Bioethik* (Stuttgart/Weimar: Metzler, 2015).

77. Guiblini and Sanyal, "Challenging Human Enhancement," 1–26; Nick Bostrom, "Why I Want to be Posthuman when I Grow Up," 108pp.

78. Harcourt, Saar, and Bianchi, "The Critique and Politics of Identity."

79. Bianchi, "Unter dem und wider den 'Bann des Einheitsprinzips.'"

80. Christopher Wylie, *Mindf*ck: Cambridge Analytica and the Plot to Break America* (New York: Random House, 2019), 82.

81. Kaiser, *The Great Hack*.

82. Without author, "So können Daten den Wahlkampf-Planern helfen," *Süddeutsche Zeitung*, March 23, 2018, available at: https://www.sueddeutsche.de/politik/wahlkampf-in-den-usa-daten-sammeln-fuer-den-wahlsieg-das-machen-alle-1.3918504-2.

83. Thomas Jefferson et al., "The Constitution of the United States," in *America's Founding Documents*, available at: https://www.archives.gov/founding-docs/constitution.

84. Annie Ernaux, *Shame*, trans. Tanya Leslie (New York: Seven Stories Press, 1999 [1997]), 13.

85. Sarah Bianchi, "Vom Scham-ersparen: Über Nietzsches fragiles Verständnis von Anerkennung," *Zeitschrift für Praktische Philosophie* 4, no. 2 (2017): 15–40.

86. Ernaux, *Shame*, 51.

87. Britanny Kaiser, *Targeted: My Inside Story of Cambridge Analytica and How Trump, Brexit and Facebook Broke Democracy* (London: HarperCollinsPublishers, 2019), 153pp; Harcourt, *Exposed*, 3pp.

88. Student's voice. Kaiser, *The Great Hack*.

89. Bernard Harcourt, *Against Prediction: Profiling, Policing, and Punishing in an Actual Age* (Chicago: The University of Chicago Press, 2007), 56pp.

90. Harcourt, *Against Prediction*, 56pp.

91. On a critique of an alleged neutral social science and data collection see Bernard E. Harcourt, *Language of the Gun: Youth, Crime and, Public Policy* (Chicago/London: The University of Chicago Press, 2006), loc. 45pp, Kindle.

92. Evan Hill et al., "How George Floyd Was Killed in Police Custody," *New York Times*, available at: https://www.nytimes.com/2020/05/31/us/george-floyd-investigation.html; Lothar Gorris, "'I can't breathe': Der Satz eines Landes, das zu zerreißen droht," *Der Spiegel*, June 5, 2020, available at: https://www.spiegel.de/ausland/george-floyd-i-can-t-breathe-ein-satz-veraendert-die-welt-a-00000000-0002-0001-0000-000171426670.

93. Tymber Hudson, "Garden of Vitality," in *Abolition 13/13* (seminar). February 25, 2021, available at: http://blogs.law.columbia.edu/abolition1313/10-13-abolish-family-policing/.

94. A video that reconstructs George Floyd's death can be found here: https://www.nytimes.com/2020/05/31/us/george-floyd-investigation.html.

95. Nicholas Bogel-Burroughs et al., "Minnesota Officer Who Shot Daunte Wright Meant to Fire Teaser, Chief Says," *New York Times*, April 12, 2021, available at: https://www.nytimes.com/2021/04/12/us/brooklyn-center-police-shooting-minnesota.html.

96. Agatha, *Minority Report*.

97. Wylie, *Mindf*ck*, 8.

98. Wylie, Mindf*ck, 82; more specifically on the "social media society" see Roberto Simanowski, *Facebook Society: Losing Ourselves in Sharing Ourselves*, trans. S. H. Gillespie (New York: Columbia University Press, 2018).

99. Wylie, *Mindf*ck*, 97.

100. Jannis Brühl et al., "Was ist eigentlich bei Facebook los?," *Süddeutsche Zeitung*, April 5, 2018, available at: https://www.sueddeutsche.de/digital/datenmissbrauch-was-ist-eigentlich-gerade-bei-facebook-los-1.3932349.

101. Joris Toonders, "Data is the New Oill of the Digital Economy," *Wired* 7 (2014), available at: https://www.wired.com/insights/2014/07/data-new-oil-digital-economy/.

102. The field of advertisement works in similar ways. The classic readings in this field include in particular Vans Packard, *The Hidden Persuaders: The Naked Society and the Status Seekers* (New York: Pocket Books, 1957).

103. Axel Honneth, "'Angst und Politik.' Stärken und Schwächen der Pathologiediagnose von Franz Neumann," in *Kritische Theorie der Politik. Franz L. Neumann–eine Bilanz*, ed. Mattias Iser and David Strecker (Baden-Baden: Nomos, 2002), 200.

104. The method of micro-targeting was already used for military purposes but only achieved its high impact within political campaigns, according to *The Great Hack*.

Chapter 2

On the Enlightened Gaze or What Is Aesthetic Critique?

Having specified in the previous chapter what we mean by the term "digital enhancement" and what the paradoxical conditions of digital enhancement entail in the specific contexts of predictive policing and "social engineering" of political elections, we now need to clarify the approach to aesthetic critique developed here. The field of digital enhancement calls for the delineation of a thus far rather neglected understanding of aesthetic critique involving a specific mode of access to its objects of analysis, i.e., the paradoxical condition of digital enhancement. Obviously, the subject in the field of digital enhancement needs *not* be aware of the theoretical methodology of aesthetic critique. But we need to care about such a methodology in order to theoretically conceptualize the very same paradoxical condition of digital enhancement.[1]

Drawing on Aristotle, the present book outlines the theoretical basis of aesthetics in the sense of aísthēsis (αισθάνομαι) and thus of sensory perception. For centuries, aesthetics was understood as aísthēsis. It was not until the eighteenth century that this changed, with Johann Joachim Winckelmann, Alexander Gottlieb Baumgarten, the founding father of aesthetics as a philosophical discipline, and under the powerful influence of German Classicism. The conception of aesthetics as the theory of the beauty in nature and art came to dominate. But unlike the dominant understanding of aesthetics, there are also approaches,[2] that go beyond the convenient view, for instance by considering the notion of force;[3] but aesthetic critique as understood here should be orientated towards enabling us to see that *the previously not so visible paradoxical conditions of digital enhancement are produced by the affect-centered power relations part of our digital life*. This view stems from translating Tully's political philosophy into aesthetics, and thus it considers the "history of ourselves" in and through which the subject's perception is *already* enmeshed with affect-centered power relations. It is in this sense that the aesthetic critique developed here aims to contribute to the broader field of social critique (*Gesellschaftskritik*). If, generally speaking, social critique criticizes relations of power,[4] then aesthetic critique criticizes the *specific* relations of affect-centered power that are part of the subject's sensory perception. The book at hand describes such an understanding of aesthetic critique in terms of the *enlightened gaze* (*aufklärerische Blick*). The motif of the gaze has a long history in philosophy. In the hitherto led debate, the freedom-impairing character of the gaze is at the center (French existentialism, postcolonial philosophy or feminist film theory).[5] In contrast to these freedom-impairing views, the book at hand analyzes how the *enlightened gaze* as a freedom-empowering kind of gaze can emerge.

Against this background, the chapter is divided into five steps. The first step sketches the basic methodological approach to aesthetic critique with respect to the motif "making distinctions" (*Unterschiede machen*). The second discusses aesthetic critique in treating the motif "like a magnifying glass." The third step shows the task of aesthetic critique in detecting the social phenomena of affective domination, affective normalization, and affective subjectivation. The fourth step concretizes the critical mode in terms of the motif of enlightenment in the tradition of genealogical critique and compares it with the work of a mole. The fifth step briefly summarizes the main points.

SENSORY PERCEPTION AND MAKING DISTINCTIONS: ON THE BASIC METHODOLOGICAL ACCESS

Basically, the methodology of aesthetic critique developed here entails two points: Sensory perception in the sense of aísthēsis (*αισθάνομαι*) and making distinctions in the sense of the ancient Greek verb "κρίνειν" (*krínein*), which means "to make distinctions" (*Unterschiede machen*). The first explains the sensory perception-based approach to aesthetics and the second explains why we are dealing with a *critical* and *not* with any form of sensory perception. To state from the outset, developing aesthetic critique as such a methodology refers to a *weak* understanding of aesthetics. It is "weak" insofar as it does *not* claim to be a distinct discipline, as a strong understanding would do, such as the dominant view on aesthetics does by claiming to be the distinct discipline of the discipline of beauty, art or taste; rather, it develops a peculiar methodology, which has hitherto largely been ignored, namely the methodology that is able to detect affect-sensitive power relations in digital enhancement.

Firstly, aesthetic critique is based on the ancient Greek understanding of αἴσθησις (*aísthēsis*), which means "sensory perception." Taking our orientation from Foucault and his predecessor Nietzsche, it is clear that by this we do not mean *any* mode of the subject's sensory perception, such as seeing a stone, touching a paper, hearing the voice of someone else, smelling flowers or tasting wine. On the contrary, the field of digital enhancement calls for a specific understanding of sensory perception.[6] This specific understanding is based in particular on references to optics that enable us to conceptualize a new methodology in terms of which we can make plausible how the subject can "open [her] eyes" (Foucault) and see the subtle effects of the logic of perfection on her life formation that were obscured until now. This subject has an enlightened gaze. It enables her to see the affect- and power-sensitive Janus face of digital enhancement.

References to optics play a central role in Nietzsche's and Foucault's work.[7] From early on Nietzsche is interested in understanding the "lens of life" (BT, 6) and he never tires of calling on the subject to "[t]ake a close look! That is your life!" (BT, 114).[8] Foucault also had a strong interest in understanding how "the order of things" (OT) can become visible to the subject. Foucault's references to optics are grounded in a basic grammar that can be identified in a variety of phenomena.[9] The basic grammar centers on Foucault's attempts to disclose that the representations of things in life that are supposedly evident—by the way, a term that also stems from the semantic field of seeing (Lat. *videre*/to see)—are in fact not as evident as they might appear at first glance. This motif then reoccurs in different contexts, in particular in the context of the panopticon, which Foucault sketches in his basic work *Discipline and Punish*. The panopticon is the specific architecture of a cell-like prison in the middle of which there is a central tower with a guard. From this central tower, the guard can see into all the prisoners' cells *without* the prisoners knowing whether they are being observed or not. The consequence of this architectural structure is that the inmates behave all the time *as if* they are being observed. Through the motif of the panopticon, Foucault discloses that the subject's attitudes towards disciplinary norms with its social expectations are not as self-evident as one might have initially thought. They are the product of the panoptic machine that makes the subject perceive to live, act and feel *in conformity with* the norms and the corresponding social expectations generated by the guard who observes the inmates.

References to acoustics are also important in Nietzsche's and Foucault's texts. Nietzsche not only wants to be seen but also to be heard in order *not* to be *mis*understood (EH, 71). Moreover, the field of acoustics helps us to conceptualize Foucault's claim that he is *not* speaking for others, but providing the *conditions* that *can* empower people whose voices were previously *in*audible to be heard. Foucault's own practice in the *Prisons Information Group* (GIP)—which he founded in the early 1970s—can be seen as a key example by which we can describe how such acoustics function. By conducting interviews with prisoners and providing information about the conditions in French prisons, Foucault tried to create the conditions in which prisoners' voices could reach to be heard in society.[10]

Secondly, aesthetic critique refers to the ancient Greek verb "κρίνειν" (*krínein*) and thus to the verb that entails the meaning of "to make distinctions." It clarifies that aesthetic critique encompasses a *critical* sensory perception. The subject's way of seeing the world is "critical" if he is able to make distinctions between the perspectives on life formation. At first glance, making distinctions *could* have a relativistic bias. One could imagine differentiating between *any* perspectives. But then the critical potential would be lost and one would rather establish a conservative critique that *ultimately* stabilizes the current perspective. Here one might think of life formation in the sense of a mere aestheticism, as seen in Baudelaire's *a*political figure of the dandy.[11] Then the dandy would be seen as a person who as a *flâneur* wants to distinguish himself from the perspective of the rest of society by merely strolling (*flanieren*) through life.[12]

If aesthetics wants to be critical, then, it must take a different path. We already aimed at demonstrating this path in the introduction when referring to Tully, who places himself in the Foucaultian tradition.[13] Nietzsche too must be considered to stand in this line. If for Tully political philosophy becomes the "practical philosophy" that uncovers how today's political problems are the result of the practices that shape "the history of ourselves" (Foucault), then similarly, aesthetic critique becomes a "practical philosophy." It becomes the specific "practical philosophy" that shows how today's problems of digital enhancement result from the affect- and power-sensitive practices that shape the specific "history of ourselves" in and through which the subject's sensory perception is already formed. This view is "critical" because such an understanding of aesthetic critique enables us to disclose the hitherto hidden practices that form the subject's sensory perception in such a way that he is *open to having his will formed* in order to conform to the logic of perfection in digital enhancement. As a consequence, subjects now *blindly* perceive digital enhancement as the only way to think, feel and live and thus *subtly* receive a narrow, one-dimensional view of life. This view is problematic because the subject is *not* aware that his will is being formed. Telling the "history of ourselves" this way promises to show the contingencies, and thus the non-necessities, of the affect- and power-sensitive conditions that generated the "history of ourselves" made hitherto. Hence, different "histor[ies] of ourselves" become possible by this change of view, namely all those histories that do not perceive digital enhancement as the only perspective of a good life formation.

To put it differently: With Tully we seek to demonstrate the critical potential of aesthetic critique in the ability to differentiate between justified and *un*justified perspectives on life formation. *Un*justified perspectives are all those that *close* the subject's gaze on life formation by making the subject blind. As a consequence, the subject has lost his self-reflexive potential to some extent. He is not aware of having a closed perspective on life. Justified perspectives, on the contrary, are thus *self*-reflexive perspectives that broaden the subject's view of life formation. Via these perspectives, the subject can become aware of what is going on in the field of digital enhancement and this means of recognizing *himself* as being open to having a formed will that considers the logic of perfection part of digital enhancement a good way of life. Later, this self-reflexive potential will be spelled out in terms of genealogical critique and considered in relation to the term "enlightenment."

This approach also clarifies why one need to consider aesthetic critique as an apt methodology for uncovering the paradoxical situation in digital enhancement and *not* epistemology or ethics. Aesthetic critique enters the stage of digital enhancement *earlier* than epistemology or ethics would do, so to speak, and this means it *already* questions *the affect-centered, power-based conditions* in and through which subjects feel, think and live in digital times. For, unlike epistemology, it does *not* ask what people *know* and, unlike the traditional understanding of ethics, it does *not* ask what people *should* do. Here, one must note, scholars following Foucault have the advantage of interpreting ethics in the specific sense of an *ethos*, which means the subject's attitude.[14] This approach is very helpful. Nonetheless, it does not suffice for our task: in the field of digital enhancement we must *already* theoretically conceptualize *the affect-centered, power-based conditions* in and through which the subject builds such an ethos. This is the field of aesthetics in the sense of *aísthēsis*, that is of sensory perception, which has been developed here. This enables us to plausibly demonstrate the subject's *self-reflexive perception* of the differences between his previous closed perspective *and* his potential new, broadened perspective on the formation of his life. From this perspective, aesthetic critique can conceptualize the subject's paradoxical situation in appropriate ways. For it has the advantage of highlighting that people in our digital world "open [their] eyes" (Foucault) and *are prompted* to participate in what they just see, namely the digital world as the supposedly perfect world in which they simply want to live, feel and act. In this sense, focusing on Nietzsche and Foucault is very helpful, allowing us to see that life *itself* needs to be considered through the lens of aesthetics in the sense of *aísthēsis*. Against "the accusation of totalization" (*Totalitätsvorwurf*), it should be noted that we are *only* dealing with the subject's concrete, affect-centered and power-based sensory perception of his life formation and *not with any* sensory perception.

Understanding the methodology of aesthetic critique in this manner facilitates seeing that the subject's sensory perception is the name for an ubiquitous ensemble of affect-centered power relations. Affect-centered power relations describe the subtle effects that A has on B, due to which B *immediately* feels, thinks and lives *differently* than she would have done without the effect of A, and which she *immediately* evaluates on a cognitive level as good. This view on power is close to what Forst calls "noumenal power."[15] A in the context of digital enhancement is the logic of perfection, and B the subject seeking to give form to her life in digital times. Furthermore, B must be regarded as a more or less free subject that would have different options for acting; hence the scene is *not* to be confused with a slave-like situation. The effect needs to be specified as a *social* effect. It thus does *not* encompass natural catastrophes, such as famines. In general, power relations are neutral, that is not only bad, negative or freedom-endangering.[16] For power relations can be understood in a freedom-empowering, positive manner, *too*, such as the cases of emancipatory power, democratic power and so on show. But *affect*-centered power relations are *non*-neutral, and thus negative, *at least at first*. For the specific kind of affect-centered power relations targets the *subtle* formation of the subject's will of which the subject is *not* aware. As a consequence, the subject's possible room for thinking, acting and living is closed. In this sense, this kind of affect-centered power relation must be regarded as *un*justified and thus as negative. "Violence" is the extreme case when the subject is totally treated as the object whose will is *broken*. "Domination" is the case when the subject's will is *formed*. Domination achieves its goal when the subject *consents* to the formation of his will. In this sense, digital domination is the power that initially has blind power over the user's will. But aesthetic critique *also* conceptualizes the notion that the subject can gain a *self-reflexive, enlightened attitude* towards his affect-centered, power-based adaptation to digital enhancement, something the present book terms, on a theoretical level, genealogical critique. Then, we deal with *enlightening power* in a freedom-empowering sense. Aesthetic critique shows, then, that the subject *can transcend* the negative context of affect-centered power relations. Obviously, there are different degrees of the negative understanding of power. Aesthetic critique has the task of detecting this kind of affect-centered power relations already present in the subject's sensory perception and thus in the way the subject sees his digital life formation.

LIKE A MAGNIFYING GLASS: ZOOMING IN ON "INCONSPICUOUS SURFACE-LEVEL EXPRESSIONS"

As seen, we need the methodology of aesthetic critique in our digital times in order to conceptualize the subtle effects of the logic of perfection on the subject's life formation. This book aims to show that aesthetic critique can fulfill this task insofar as it functions *like a magnifying glass* by zooming in on small "inconspicuous surface-level expressions"[17] (*unscheinbare Oberflächenäußerung*), to put it with Siegfried Kracauer.

In his productive essay "The Mass Ornament" (1927) Kracauer analyzes the specific phenomena on the surface in the context of the Weimar Republic. Among the central phenomenon on the surface analyzed by Kracauer are the "Tillergirls," that were famous showgirls at that time. By an analysis of them Kracauer reveals the "fundamental substance of the state of things"[18] (*Grundgehalt einer Epoche*); this substance he sees in the aesthetic reflex; with this aesthetic reflex Kracauer outlines how people attach relatively freely also in their leisure time to the capitalist logic of rationalization. Now we transfer Kracauer's analysis of the Weimar Republic in a *systematic* perspective to our contemporary digital today, and therefore undertake a shift in time of about a century. Viewed through the lens of the magnifying glass developed here, we are able to see the today's phenomena on the surface, such as the subject's daily routines; this allows us to take the "fundamental substance of the state of things" into view, in our case the hitherto obscured affect-centered power relations in the field of digital enhancement.

In viewing this problem, Foucault can be seen as standing in a line with Kracauer.[19] Instead of referring to the terminology of the "phenomena on the surface" of which Kracauer speaks, Foucault uses a similar terminology and speaks of the "surface of things" (DE 1, 33, 634, trans. by the author). To continue with Foucault, this surface allows us to see the "mute reflections" (OT, 27); by this Foucault means the underlying, superficially covered structure of things, to put it differently: the hidden "order of things" (OT). This corresponds to what Kracauer calls "fundamental substance of the state of things." In this sense, aesthetic critique developed here provides the required theoretical tools to zoom in, like a magnifying glass, on these "mute reflections" (Foucault) by making the subject see them as if it were happening in front of her eyes.

The metaphor of the magnifying glass needs to be traced back to Foucault's crucial formulation of the "focal points of experience" (GSO, 5)[20] in his lecture "Government of the Self and Others." The formulation of the "focal points of experience," by which Foucault sketches the entanglement between order, practice and subjectivity, is so central that he also uses it to describe his philosophical approach in the context of his chair "The History of Systems of Thought" at the Collège de France. These "focal points of experience" must be conceived as the specific phenomena on the surface in which the "local" (SMD, 8; WE, 316) perspective, or the *micro*-perspective on the seemingly small corners of the subject's life formation, is mirrored in the "global" (WE, 316) perspective of the logic of perfection, or in the *macro*-perspective of the social normative order. These two perspectives of the micro- and macro-contexts do *not* mirror each other in harmonic ways in the specific phenomena on the surface; *on the contrary,* these phenomena on the surface are so crucial because they show in condensed and intensified fashion how the two perspectives come into structural conflict in the subject's practices, for instance how Snowden realized his concrete micro-position in the overall macro-context and thus decided to leave his job at the NSA. These two perspectives of the micro- and the macro-contexts are the "focal points" at which the subject may "open [her] eyes" and can see that there is a problem with his previous way of living, feeling and thinking. And the subject is now able to see this problem *not* as his private, *mere* subjective problem, but as a *structural* problem of society as a whole.

Against this background we can specify the task of aesthetic critique as lying in the analysis of the three levels of order, practice and subjectivity.[21] The macro-level of the order shows that the subject's way of living is from early on *immediately* structured by the logic of perfection. The account presented in the present book shows that, especially through modes of habitualization, the subject is trained *in quasi-automatic ways* to see the digital world through the lens of perfection. So already in virtue of her sensory perception, the subject is prompted to value the digital world with its norms and social expectations of perfection as *the* alleged normal way one should feel, think and live.

That the macro-level of order, however, does not entail a substance, essence or nature becomes clear with the second level, the meso-level of practices, which prevents us from succumbing to the fallacy involved in describing the order of perfection in substantialized, essentialized or naturalized ways. For the meso-level of practices shows that the very same order of perfection *is immediately produced* and thus socially constructed by affect-centered power relations. Hence, this order is established in quasi-automatic ways by different practices of life formation in our digital time, i.e., in particular of habitualization. Talking about the subject's life formation in the semantics of practices has the advantage of clarifying that the subject's way of seeing her life is *not only externally* determined by the social order. Rather, the subject *herself* participates in the immediate production and reproduction of the order and this *indeed* means, in the first place, that the subject is prompted to stabilize the current social order by seeing her life with the eyes of perfection; *but* in case she becomes aware of what she does, and thus in undertaking self-reflexive, enlightened practices, she can also change her view and start to live, feel and think *differently*. Aesthetic critique is understood here accordingly as a required, but previously underappreciated theoretical tool with which we can conceptualize this process. In this sense, such a tool can demonstrate that the subject *herself* contributes to the constitution of the order by the *self*-constitutional practices of her life formation. Aesthetic critique reveals that this constitutional process seems so self-evident to the subject that she quasi-automatically evaluates the order of perfection as a normal and stable order. This in turn has the effect that she cannot question the alleged normality of the order. This approach is based on the dialectic between "being constituted" and "constituting oneself" that must be spelled out in our context as the specific dialectic between "being formed" and "forming oneself."[22] "Dialectics" entails the understanding of *an inherent turning* from "being formed" into "forming oneself." More specifically, dialectics is seen here as the tension in the subjectivity and thus primarily entails the understanding of *dialectical subjectivity*. This, however, is *not* to be confused with a ubiquitious (*allumfassend*) dialectics. For then the critical potential would be lost. Then the subject would be *always* trapped in the logic of the social and political normative order. Freedom would always turn into domination. The function of the magnifying glass *by contrast* provides the tool that can make a *localized* dialectics transparent by zooming in on the concrete phenomena on the surface of the subject's life formation that makes room, too, for forming oneself *differently*. Thus it shows how the logic of perfection as part of the bigger picture, in and through which the subject *is formed*, is condensed in small forms of life formation, for example in the subject's daily routines. This, however, does not mean that the subject then always *forms herself along the given line of perfection; she also has the self-reflexive potential to turn into someone who can form herself differently*. Following Nietzsche and Foucault, this dialectic is grounded in the social-ontological register and thus in the *transformable* relation between everything that is, and thus also in the transformable relation between the problems of the bigger picture and the problems of the small forms of life.[23]

The third level, the micro-level of subjectivity, concerns the subject who immediately sets herself in relation to the digital world through her sensory perception. At this level we encounter subjects that develop a relation to themselves by giving form to their lives in and through micro-practices. From the foregoing it should be clear that we are *not* talking about an individualistic subject who forms his life *in*dependently of the order of perfection; rather, we are necessarily dealing with a "self in context"[24] and thus with a subject who is socialized through and through. This subject must be conceived accordingly as one who forms her life *in dependence* on the order of perfection. Aesthetic critique on this level thus reveals that subjects become what they are in their digital today *in immediate response* to the order of perfection. And on the micro-level of subjectivity, this immediate response is theoretically conceptualized as the subject's attitude or ethos towards the digital era she lives in. In and through this attitude, she is prompted to set herself in relation to this era by her sensory perception.

At the same time, aesthetic critique also shows the potential for the subject to become aware of her blind gaze. In *criticizing* her practices of life formation hitherto that made her open to having her will formed in such a way that she immediately considers digital enhancement a good thing in life, aesthetic critique *constructs an enlightened agency* that is marked by a kind of self-reflexive potential. In this way, it contributes to what in today's discussion is called *critical constructivism*.[25] This kind of enlightened agency works on a *voluntaristic* basis. For the subject is *not* dependent on the given perspective on digital enhancement, but an enlightened subject can *choose* how to place herself in relation to this given perspective on digital enhancement. This subject can turn an app on, but also off. In any case, she is aware of what she does. This does *not* mean that the subject is from the outset independent from the given perspective on digital enhancement. *By contrast*, the social-theoretical account of aesthetic critique rests in the fact of showing how the subject is *structurally bound* to the given perspective in and through affect-sensitive power relations. The subject *initially* has the will to turn on the app, to refer to the example again. But through self-reflexive procedures of enlightenment, theoretically conceptualized by the method of genealogical critique, the subject *can learn to free herself* from her immediate attachment to the given perspective. As a consequence, enlightened subjects *can become in*dependent from the given perspective. This process, however, is *precarious*. That is, the subject's independence is fragile and needs to be guaranteed by *continuing* processes of enlightenment. Hence, enlightenment does *not* reach an endpoint; there is no final conclusion, so to speak. By contrast, the enlightened perspective once found must be the further object of enlightening procedures. This book accounts for this by speaking of *the enlightenment of enlightenment*.[26]

In light of this, the subject's life formation cannot be adequately conceptualized as *merely* belonging to the social order of perfection in our digital today, neither as *mere* practice nor as *mere* individualistic mode. On the contrary, the methodology of aesthetic critique shows that the subject's life formation can only be conceptualized adequately by considering the *interrelation* between these three levels of order, practice and subjectivity. And how they interrelate in our digital today can in particular become apparent, as we hope to show in the present book, by considering the archives that sketch, to put it with Kracauer, the "phenomena on the surface." As a result, we do not fall into the pitfall to only understanding society as it justifies *itself*, for example by analyzing the social norms, self-understandings, narratives or ideas that the given society provides *explicitly as self-narratives*. In our digital times, this would imply conceiving society along the lines of its self-justifications as being perfect by digital media. Kracauer's archive, by contrast, allows us to see, in and through the daily routines that subjects perform *by the by*, what Krakauer calls the "fundamental substance of the state of things." And this is the affect-centered power relations. We chose the Netflix show *Black Mirror*, the films *Her* and *Minority Report*, more recent contemporary French literature and the feminist classic *Orlando* to make these affect-centered power relations visible (for the latter two, see Part III).[27] As a side note: by citing these examples we do not mean that today's subjects should be a contemporary version of Orlando, but only that the key modes of the phenomena on the surface can be seen *in similar ways* in Woolf's *Orlando* and in our digital today. Basically, the story of Orlando centers on a figure, who over the course of several centuries changes his sex from male to female; in so doing Orlando relates herself/himself to the existing social and political orders and this is the theme that overlaps with the contexts of digital enhancement.

From this perspective, it becomes obvious that the method of aesthetic critique, understood here as functioning as a magnifying glass, must be understood as a matter of hyperconcretization. Like a magnifying glass does, aesthetic critique functions in hyperconcrete terms as it zooms in on the small things in life, that is the phenomena of the surface. This method is also part of Foucault's own style of work. In hyperconcrete ways, also Foucault zooms in on the effects that small corners of life have on subjects, like the architectural order of the panopticon that we already described in the introduction, or the seemingly random year 1639 in France that he analyzes in his lecture "Penal Theories and Institutions" to describe the mechanisms of repressions which Richelieu used against the *Nu-pieds* revolt. It goes without saying that the mode of functioning of the magnifying glass is only one effective way to make the phenomena on the surface transparent. Other effective ways are also thinkable. Instead of zooming in on the small things in life, one could imagine looking at them from a distance, such as for example looking at the open horizon of the sea of life (HL, 1, 1, 251pp), to put it with an influential Nietzschean picture. Then one does not use the style of hyperconcretization part of the metaphor of the magnifying glass, i.e., the method of zooming in on things, but instead the style of hyperabstraction.

A paradigmatic case that enables us to see how the specific phenomena of the surface provide insight into the "fundamental substance of the state of things"[28] presents Foucault's crucial interpretation of Velázquez's painting "Las Meninas." Especially Foucault's text "Die Hoffräulein" (DE 1, 32, 603pp), and also one of his key works, *The Order of Things* (1966) deal with this theme. Velázquez's painting describes the following scene: On the surface, the painting depicts the absolute monarch King Philip IV of Spain, his wife, his daughter, the infanta Margot Theresa, and his entourage, i.e., the ensemble that also gives the picture its title "Las Meninas." "Las

Meninas" means the ladies that serve the royal family. Only at first glance the picture looks like a traditional Baroque painting of a seventeenth-century royal family. But as a consequence of the very specific composition highlighted by Foucault's interpretation, Velàsquez's painting is shown to be a *non*-standard painting.

From the perspective shown here, it is a *non*-standard piece of art as it depicts the *specific* phenomena on the surface that render visible the "fundamental substance of the state of things," to again use Kracauer's term. Here the "fundamental substance of the state of things" entails the potential in and through which the subject is able to change in a self-reflexive way his attitude towards the central perspective, developing *an enlightened* attitude towards it. This process is sparked by the phenomena of the surface, by the *condensed and intensified* entanglement between the three levels of order, practices and subjectivity, in terms of affect-centered power relations as they are depicted in the painting. In the case under discussion here, this entanglement must be theoretically conceptualized between the spectator and the picture. The level of order is marked by the affect-centered power of the central perspective on the spectator. Its effect on the spectator is that she *is open to having such a will that she* sees and values the central perspective as representing what is worth being seen. This leads to what we can call quasi-automatized subjectivity that is willing to conform. In the case of Velásquez's painting, the person is willing to conform to the central perspective and to consider the central perspective as the normal way of looking at a picture. The level of enlightened subjectivity then demonstrates that the central perspective can lose its representative value from one moment to the next, allowing the subject to realize in a self-reflexive way that this perspective does *not necessarily* show what is worth to be seen. Looking at the picture, the subject becomes the specific subject who sets herself in such a relation to the central perspective that she can understand and perceive that this does not have to be *the* relevant perspective for viewing pictures properly. As the catalyst of this process can be considered the surprising fact that, in contrast to traditional Baroque paintings where one would normally expect to find the royal couple in the central perspective, Velázquez explicitly places not the royal couple, but the painter and the ladies-in-waiting at the center; he achieves this effect by using a picture in the mirror that indeed shows the royal couple but rather in the background and in the center of the picture now appears the painter and the ladies-in-waiting.

Translated to the case of digital enhancement, Foucault's interpretation of Velázquez's picture helps us to also understand that the today's "fundamental substance of the state of things" must be conceived as the subject's self-reflexive attitude towards our central perspective, that is the logic of perfection. Hence, it enables us to understand that the contemporary subject, too, no longer has to understand the central perspective, or the logic of perfection, as the supposedly true and necessary representation that these subjects until now just perceived as having the right to tell them how to live, feel and act.

THREE TYPES OF AFFECT-CENTERED POWER RELATIONS

Affective Domination, Affective Normalization and Affective Subjectivation

The previous step clarified that the methodology of aesthetic critique has the power to uncover the "fundamental substance of the state of things" (Kracauer) in the affect-centered power relations that were hitherto all too hidden. The task of the present section is now to specify the "fundamental substance of the state of things" in the types of affect-centered power relations. To this end, the section at hand discusses the three types of affective domination, affective normalization and affective subjectivation.

Aesthetic critique enables us to conceptualize the affect-centered types of power on the macro-level of order as forms of affective domination, those on the meso-level of practice as forms of affective normalization and those on the micro-level as forms of affective subjectivation. The interrelation between these three levels is made apparent by the operation of aesthetic critique, which we describe with the metaphor of the magnifying glass. The magnifying glass zooms in on the phenomena on the surface that we can now specify as the specific phenomena of affective domination, affective normalization and affective subjectivation. They show on a small scale how "the global" perspective, to put it with Foucault, of the order of perfection conflicts with "the local perspective" of the subject's alleged normal life.

Affective domination on the macro-level of the order is a matter of the hardware of digital enhancement, so to speak. Aesthetic critique enables us to better understand that the order of perfection produces patterns of affective domination, because it demonstrates that the subject is *immediately* confronted with stable and fixed structures that form the subject's will in such a way that he considers digital enhancement *the* only valuable perspective on life. "Domination" is thus understood as narrowing the subject's possibilities in life. This phenomenon is "cognitive" because the subject evaluates digital enhancement as the framework whose coordinates make life valuable. The very same phenomenon is "affective" in that the subject *is prompted* to do so. For he is open to having his will formed so that he considers digital enhancement this way. As a consequence, the subject is always on the track of perfection. In this sense, the hardware of perfection constitutes the subject as a *perfecting machine*. Analogous to the language of computer algorithms with its binary oppositions between 0 and 1, the subject is open to be programmed in his will to divide life in conformity with the binary scheme of perfect and non-perfect. "Affective" here describes the circumstance that the subject does not see that his mindset is *already* programmed by such a perfecting machine because he reacts quasi-automatically to the order of perfection. So, this subject is configured in such a way that he just has the will to follow the one-dimensional track of perfection.

Now we turn to the meso-level of practices that links the level of the order to the level of the subject. On this meso-level we are dealing with the specific phenomena of affect-centered power relations that we call, following Foucault, "affective normalization." This presents, firstly, a picture of habitualization; the subject is immediately conditioned to conceive and experience the hardware of perfection *as her normal way of life* in and through daily routines. In perceiving the order of perfection as normal, the subject is trained to participate more or less voluntarily in her programmed life. She immediately does what reflects this alleged normal understanding, so that different forms of life are concealed. Nonetheless, updates, to use today's digital language, are possible. As the supposed normal perception is generated by social practices, it is also possible to perceive life formation *differently*, or to update one's view and perceive one's previous view as outdated. But nevertheless, the affectively normalized form of life generated by the established order of perfection also has the potential to be opened by *self-reflexive* practices. In and through such practices, the subject can become aware of what she hitherto did *blindly* day by day.

On the micro-level, that is the level of the subjects, we are dealing with the affect-based power relation that is characterized by the term of affective subjectivation. Foucault is one of the key theorists of subjectivation, which describes the process through which a subject becomes who she is within the given contexts, in our case the contexts of perfection. "Affective subjectivation" describes the specific process through which a subject *is prompted* to become who she is within the given contexts. In contrast to the level of the order conceived above as the hardware, we understand the micro-level of the subject as the *software*, that is the set of immediate relations to oneself that can be made so flexible that it can also be opened for potential self-reflexive changes. Becoming who one is, therefore, does *not* occur in a freestanding manner, but in ways that are immediately determined by affect-centered relations of power. For the subject is immediately part of the order—namely, the part that at first maintains the order of perfection—and part of the practices—namely, the specific part of the practices in and through which the subject is prompted to produce and reproduce the order of perfection, like through the key practices of habitualization; so, the subject's process of becoming who she is must also be conceptualized in terms of affect-centered power relations.

In particular, the French term "*assujetissement*" reflects the entanglement between "subject and power" (SP), to put it with Foucault. Aesthetic critique thus enables the subject to perceive that she is, *at least at first*, a product of the order of perfection. But the task of the project of aesthetic critique is to conceptualize that the way the subject immediately is does not necessarily have to be subjugated to this view on her life. On the contrary, aesthetic critique enables us to see how the subject can *also* develop a *double* perspective in and through which she can *also* make a justified difference to the pre-given perspective; and this means a difference to the dominating perspectives that form her will to the extent that she wants to be a perfecting machine without being aware of this fact herself; through this double perspective, we can understand that the subject is *not* restricted to the view on her life formation generated by the logic of perfection, but can also create new views on life that are not so much governed by the logic of perfection.

The analysis of affect-centered phenomena of power also has implications on the formal level of stylistic form.[29] The form of aesthetic critique corresponds to an *essayistic form* and thus to the form whose name already includes the notion of "attempt—" the French noun "l'essai" literally means "the attempt." More specifically, due to the field of digital enhancement, the present book understands "essays" following Deleuze and Guattari as "minor literature."[30] The form of the essay is small in the sense that it does not delineate the mainstream perspective, in our case the dominant perspective of the logic of perfection, but the small, or, dissident, perspective, so to speak, that is not seen as belonging to the mainstream. The book at hand thus sees with Foucault the small form of the essay as an attempt to create the previously obscured different perspectives by which we can conceptualize how the subject can "open [her] eyes" (Foucault) and see how another, less affectively dominated life formation could be possible.

The critical-theoretical task of aesthetic critique is thus to detect phenomena of affect-centered power relations in such a conceived experimental manner on the three levels of order, practice and subjectivity; this has the further implication that the critical theorist cannot be misunderstood as a supposedly neutral observer. By contrast, the field of digital enhancement demonstrates that the critical theorist *is interested* in highlighting the affect-sensitive, power-based problems of our digital present from the outset. This interest is *not* externally ascribed to the critical theorist, but emerges in response to his *enlightened* interest. "Enlightened interest" means the specific interest in *making the difference visible* between justified perspectives on life formation in digital times and *un*justified perspectives, that is, all those perspectives that reproduce patterns of affective domination.

GENEALOGICAL CRITIQUE, CRITICALLY CONSTRUCTING AN ENLIGHTENED AGENCY AND THE MOLE'S WORK

Thus far we have shown that the methodology of aesthetic critique is able to make the previously hidden affect-centered, power-based relations between order, practice and subjectivity transparent to the subject. As a result, the subject is empowered to make a justified distinction, to invoke once again the ancient Greek understanding of "κρίνειν" (*krínein*) between her quasi-automatized way of perceiving her life conducted by the logic of perfection and new, self-reflexive views on her life that are *not* affectively dominated by her relation to the social order of perfection. In the following, we will show more specifically how this critical process works by specifying the critical methodology in terms of genealogical critique, which in this book is systematically placed in relation to the notion of enlightenment.

One might have assumed that ideology critique would have provided the required methodology for highlighting the previously invisible affect-centered, power-based relations between order, practice and subjectivity. This would be productive, too. Ideology critique is concerned with uncovering previously veiled ideas. However, this method is not the tool we need as a first instance to theoretically conceptualize the affect-centered, power-based field of digital enhancement, where, as the previous chapter showed, not so much is veiled by digital enhancement. Digital enhancement declares quite openly that it wants to perfect life; it even evokes the logic of perfection in its name if we recall that the verb "to enhance something" means "to perfect something." Thus, in the field of digital enhancement we are rather confronted with the problem that the "*history of ourselves*," to again borrow Foucault's paradigmatic formulation that is also of great importance for Tully, makes subjects open to blindly having their will formed in such a way that they seek to perfect their life. This is so *even though* they very well know about the freedom-endangering effects of the social order of perfection, such as risks to data privacy. The critical-theoretical methodology that can capture this problem horizon concerned with history is not so much ideology critique, but genealogical critique. The name genealogical critique itself implies the word of history. For Nietzsche and Foucault, genealogical critique is another name for historical critique.

The critical-theoretical method of genealogical critique needs to be traced back to Nietzsche and was developed further by Foucault who situated himself in the Nietzschean tradition. "Genealogical critique" is the methodology that theoretically conceptualizes how subjects can become aware of the historicity of their becoming in the given social and political normative order, and thus, in the case of digital enhancement, of becoming what one is in the social and political normative order of digital perfection. In detecting the histories of this process, the subject learns to understand and experience its contingencies, to wit its *non*-necessity; to recognize contingencies in one's life in turn opens up new ways of becoming what one is. Thus it makes the given perspective on life fluid and destabilizes it. But, as already mentioned in the introduction, genealogical critique is *not* to be confused with mere relativistic endeavors that would destabilize everything.

To better see why genealogical critique does *not* destabilize everything, we need to capture it in light of the notion of enlightenment. For this notion clarifies that genealogical critique comes not with a *de*constructing task, *but* with a constructing one: It analyzes *the critical construction of an enlightened agency*. The notion of enlightenment conceived in this sense is not to be misunderstood as the historic epoch of the seventeenth and eighteenth century, but as a critical stance. If the paradigmatic Kantian understanding of Enlightenment theoretically conceptualizes how the subject is able to shed light on the political, social and religious "guardians" (*Vormünder*),[31] then Nietzsche and Foucault also stand in this tradition. In this book, this line is traced to today's digital world. The kind of enlightenment that is relevant, then, implies the theoretical conceptualization of self-reflexive agency that is able to highlight affect-sensitive power relations. More specifically, constructing enlightened agency means that the subject can develop a self-reflexive attitude and become aware to what extent she was hitherto open to having her will formed by the logic of perfection. Becoming aware of this implies that the subject's view can be broadened for new, dissident, less affectively dominated perspectives on life. Genealogical critique as a mode of describing the critical construction of an enlightened agency thus enables us to make visible plural ways of life formation that were previously far too hidden by affect-sensitive power relations. These plural ways of life are justified and not just another way of life because they uncover hitherto all too hidden patterns of domination and thus pluralize the hitherto narrowed perspectives on life. In other words, in virtue of this critical-theoretical methodology it becomes apparent that the subject can recognize that the specific surface phenomena, such as her habitualized way of living, feeling and acting in conformity with the logic of perfection, is *not* natural, essential or necessary, *even though* until now it seemed to the subject to be natural, essential or necessary due to its narrowed vision in digital times. If one looks through this lens, the critical theoretical project thus has the task of *de*naturalizing and questioning supposedly necessary perspectives on life formation.

For Nietzsche and Foucault, the critical-theoretical method of genealogical critique encompasses the process of changing views, or as Nietzsche would put it, of "chang[ing] the perspectives" (EH, 6, 1, 266, trans. by the author).

Here, "changing" means shifting from an affectively dominated perspective to a new, dissident, non-affectively dominated perspective. By this understanding it is clear that "changing" or the similar word "altering" is *not* to be confused with relativistic perspectives. It rather entails a transformation of the subject that needs to be qualified as "self-reflexive." Such a change is thus *a change for the better*, for the hitherto quasi-automatized subject changes into a self-reflexive, enlightened subject. However, this is not an easy task. "Changing the perspectives" might not be so difficult while hiking, as Nietzsche showed. Standing on top of a mountain in Engadin, for example in Sils Maria, "6000 feet above Bayreuth" (EH, 6, 4, 270, trans. by the author), then one can quite easily "change one's perspective." One can look around and is therefore no longer restricted to the narrow perspective one has by contrast in the valley. So, the subject on top of the mountain has the freedom to look in any direction; it can look down to the valley, zoom in on things closer with a real telescope, but it can also place them at a distance by putting the telescope aside. Whatever the subject does in the concrete case, on the top of the mountain he has the freedom to switch between multiple perspectives. He has, so to speak, a bird's-eye view on the small little world of Bayreuth,[32] as Nietzsche stated, which was so important for the early Nietzsche before he broke with Wagner.

Obviously, in the domain of digital enhancement the subject is not able to go on top of a mountain in order to overcome its hitherto blind attachment to the pre-given order of perfection. But the picture symbolizes the contribution of aesthetic critique: now the difference does not measure "6000 feet above Bayreuth" (Nietzsche) but, so to speak, "6000 feet" (Nietzsche) between oneself and one's attachment to the perspective of the order of perfection. To initiate this dynamic, the subject must dig down to the mechanisms of her life formation subtly rooted in her affective life. Hence, instead of climbing a mountain, she goes in the opposite direction, she digs down to the mechanism that makes her form her life blindly in accordance with the logic of perfection.

Such a dynamic resembles the work of a mole as we express it. The subject, like a mole, strives to excavate the invisible mechanisms beneath the earth, or beneath her observable behavior. The image of the mole plays a prominent role in the history of philosophy. Hegel in particular used it in his lecture on the philosophy of history to describe the deep work of the spirit: "Spirit [. . .] is inwardly working—ever forward (as when Hamlet says of the ghost of his father 'Well said, old mole! Canst work I' the ground so fast?"[33] Marx also invoked the image of the mole in a similar way to illustrate the non-visible, revolutionary process. The invisible dynamics of a revolution work like the mole: "All the same the revolution is thoroughgoing. [. . .] Europe will jump up from the seat to exclaim: 'Well hast thou grabbed, old mole!'"[34]

From this perspective, we can see that the subject in digital enhancement must even go so far down that she digs beneath the surface of her easy perceivable life formation that she can observe while giving her life a form, for example in using this or that app. In the domain of digital enhancement, however, we must analyze the underlying dynamics that *already* prompt the subject to become such a "user." And that is the work of the mole. But, obviously, the contemporary digital subject does not really have to become a mole; rather, it is we as the critical theorist who needs to operate with this metaphor to understand what is involved in changing perspectives. This metaphor of the mole illustrates the dynamics of the motifs of enlightenment in the tradition of genealogical critique, which show how the subject can highlight the possible difference between the given, affectively dominated perception of the world in which she immediately lives, thinks and feels by attaching to the logic of perfection and possible new, not affectively dominated perceptions. The metaphor of the mole symbolizes the process of making this difference transparent, where the work of the mole resembles the turning point when the subject, like the whistleblower Snowden, who became aware of what he had done while working for the NSA, and in his case this meant endangering data privacy. Hence, this subject learns to reach a transparent vision like the subject on the mountain top. The subject can thereby transcend his contexts and can change from the affectively dominated, quasi-automatized contexts to enlightened contexts. But still, in order to do so, this subject cannot simply overcome his blind gaze by climbing above the clouds, although as the phenomenon of the "Maloja Snake" depicted in Olivier Assayas' film *Clouds of Sils Maria* shows—[35] when in autumn the clouds creep through mountain valleys in the canton *Graubünden* like a snake—climbing a mountain is not a guarantee of an unhindered view. The subject with whom we are dealing in the domain of digital enhancement, however, is called upon to learn a self-reflexive attitude in order to become the enlightened subject that can become aware of the affect-based power relations of which he is *initially un*aware. But if we remain with the picture of the mole, it becomes clear that the subject is continuously confronted with blind spots of affective-sensitive, power-based patterns, like the mole who surfaces from below the earth but is still blind. This means that the subject is continuously called upon to shed light on affective-sensitive power relations and search for the corners in life that are still affectively dominated.

SECOND CONCLUSION

To conclude, the methodology of aesthetic critique attempts to provide the required enlightened gaze to see how the subject in digital times is open to having her will formed in such a way that she gives her life the form that adapts to digital perfection. It is interpreted in the ancient Greek understanding of αἴσθησῖς (*aísthēsis*), which means sensory perception. Thus, the subject is already open to conforming to the digital perfection by seeing her way of living in the given social order. Sensory perception becomes critical (in the sense of κρίνειν, *krínein*) in cases where the subject is able to make a distinction between the given, affectively dominated perspective on life formation *and* new, dissident, not affectively dominated perspective. This view draws on Tully who places himself in the tradition of Foucault who himself stands in the Nietzschean tradition. It leads to a weak understanding of aesthetics that does not conceive of it as a distinct discipline, such as the dominating views on aesthetics as dealing with beauty, taste or art. By contrast, here aesthetics refers to a methodology. Yet, aesthetic critique is peculiar because it does *not* deal with *any* sensory perception, but with the *specific* methodology, which we compare with the methodology of zooming in of a magnifying glass. This method allows us to zoom in on the specific phenomena on the surface, to put it with Kracauer, and thus with the tradition in which Foucault can also be situated. An analysis of the specific phenomena on the surface provides insight into the "fundamental substance of the state of things," to cite Kracauer further. We expand this view to our digital world today and in so doing aim to develop the enlightened gaze that can make visible how subjects immediately attach to and, in the event that they become self-reflexive and enlightened, can also *de*tach themselves from the pre-given social and political order of perfection. We analyze this entanglement of the subject's immediate attachment and her possible, self-reflexive, or enlightened detachment in this chapter in terms of the affect-centered, power-based relation of order, practice and subject with its corresponding phenomena of affective domination, affective normalization and affective subjectivation. Furthermore, we sketch that the critical methodology that is best capable of highlighting this entanglement is genealogical critique. The subject in the field of digital enhancement, of course, might not necessarily be aware of such a methodological approach, but the proposed approach seeks to demonstrate that the observer requires these perspectives to understand the full scope of the subject's affect-centered, power-based life formation.

NOTES

1. Here, we are only interested in aesthetics from a philosophical perspective and *not* from the perspective of the history of art.

2. For this discussion see Martin Seel, *Ethisch-ästhetische Studien*, 3rd ed. (Frankfurt a.M.: Suhrkamp, 2019), 36pp.

3. Christoph Menke, *Kraft: Ein Grundbegriff ästhetischer Anthropologie* (Berlin: Suhrkamp, 2017).

4. Honneth, *Kritik der Macht*; Saar, *Genealogie als Kritik.*

5. In existentialism, Sartre talked about the gaze that shames another subject, see Bianchi, "Vom Scham-ersparen." In postcolonial philosophy, Frantz Fanon conceived the dominating white gaze in his far-reaching text *Black Skin, White Masks*, trans. Richard Philcox (New York: Grove Press, 2008 [1954]). It is also a central part of feminist film theory, which conceptualizes the understanding of the dominating male gaze. See Laura Mulvey, "Visual Pleasure and Narrative Cinema," *Screen* 16, no. 3 (1975): 6–18.

6. *Different to* the dominant tradition of sensory perception in Western philosophy beginning with Plato, Aristotle and passing in particular through Kant, our digital present does not call for an analysis that is primarily concerned with the epistemic status of the subject's sensory perception. Concerning this discussion: Wolfgang Welsch, *Ästhetisches Denken*, 8th ed. (Stuttgart: Reclam, 1990); see also Gernot Böhme, *Aisthetik: Vorlesungen über Ästhetik als allgemeine Wahrnehmungslehre* (München: Fink, 2001).

7. In general, on the dimension of the seeable in Foucault, basically Gilles Deleuze, *Foucault*, trans. H. Kocyba, 8th ed. (Frankfurt a.M.: Suhrkamp, 2015 [1986]), 69pp; cf. also recently John Rajchman, "Foucault's Art of Seeing," in *October* 44 (1988): 88–117. And Martin Jay, *Downcast Eyes: The Denigration of Vision in Twentieth-Century French Thought* (Berkeley/Los Angeles/London: University of California Press, 1994).

8. See also: GT, 1, preface 2, 14; HL, 90; MA 1, 2, preface 1, 13

9. For the references to optics, see especially aphorism 54 of *The Antichrist*. Here, Nietzsche speaks of the importance of "seeing in a free manner" (AC, 6, 54, 236, "Frei-Blicken-Können," trans. by the author). The instructive commentary on the *The Antichrist* is provided by Andreas Urs Sommer, *Kommentar zu "Der Anti-Christ," "Ecce Homo," "Dionysos-Dithyramben" und "Nietzsche contra Wagner"* (Berlin/Boston: de Gruyter, 2013).

10. Accordingly, the understanding of the enlightened gaze can lead to an understanding of a critical, enlightened voice.

11. In a critical perspective: Sarah Bianchi, "Aus seinem Leben ein Kunstwerk machen: Zur ästhetischen Selbstgestaltung nach Nietzsche und Foucault," in *Kongress-Akten: Vol. 4: Das ist Ästhetik!*, ed. Juliane Rebentisch, available at: http://www.dgae.de/wp-content/uploads/2017/06/Aus-seinem-Leben-ein-Kunstwerk-machen_Bianchi_DGÄ.pdf.

12. Today's discussion of the aesthetization of the lifeworld extends the criticism of the Dandy: Axel Honneth, "Soziologie: Eine Kolumne. Ästhetisierung der Lebenswelt," *Merkur* 46, no. 519 (1992): 522–27; Juliane Rebentisch, *Die Kunst der Freiheit: Zur Dialektik demokratischer Existenz* (Berlin: Suhrkamp, 2012).

13. Tully, *Politische Philosophie als kritische Praxis*. Consider especially: Forst, "Eine politische Theorie der Freiheit," 7–11.

14. Cf. Hadot, *La manière de vivre*; Nehamas, *The art of living*; Johanna Oksala, *Foucault on Freedom* (Cambridge: Cambridge University Press, 2005).

15. Rainer Forst, "Noumenal Power," The Journal of Political Philosophy 23, no. 2 (2015): 111–27. The notion of affects is not yet explored in Forst's reading of "noumenal power."

16. It is this point that is overlooked by Weber and mainly by Lukes. They consider power in mere negative terms. Weber, *Wirtschaft und Gesellschaft*, § 16. Steven Lukes, *Power: A Radical View* (Basingstoke and London: Palgrave Macmillan, 1974).

17. Siegfried Kracauer, *The Mass Ornament: Weimar Essays*, ed. Thomas Y. Levin, trans. Thomas Y. Levin (Cambridge, Mass./London: Harvard University Press, 1995 [1963, 1927]), 75.

18. Kracauer, *Mass Ornament*, 75.

19. In particular, Theodor W. Adorno's seminal text *Minima Moralia* must also be situated in this line. Theodor W. Adorno, *Minima Moralia: Reflexionen aus dem beschädigten Leben* (Frankfurt a.M.: Suhrkamp, 2003 [1951]). From the perspective of aesthetic critique developed here, Adorno, too, sketches such specific surface phenomena that exhibit in condensed ways the entanglement of order, practice and subjects. On how easily subjects identify in their practices with the social order, see Bianchi, "Unter dem und wider den 'Bann des Einheitsprinzipts.'"

20. *Brennpunkte der Erfahrung*, RGS, 15

21. The need to consider the relation between order, practice and subjects with its respective forms of power was first highlighted in Martin Saar, "Ordnung–Praxis–Subjekt. Oder: Was ist Sozialphilosophie?" *WestEnd: Neue Zeitschrift für Sozialforschung* 16, no. 2 (2019): 161–74. Whereas Saar sketches the relation in the field of social philosophy, *here* this relation is sketched for the first time for the methodology of aesthetic critique.

22. Accordingly, it would be reductive to conceive life formation as contributing to questions of life*style*. By contrast, more profoundly questions of life formation contribute to the basic processes of the subject's self-constitution.

23. For a more detailed discussion of the term "social ontology," see the chapters on Nietzsche and Foucault

24. Trans. by the author, Seyla Benhabib, *Selbst im Kontext: Kommunikative Ethik im Spannungsfeld von Feminismus, Kommunitarismus und Postmoderne*, trans. Isabella König (Frankfurt a.M.: Suhrkamp, 1995 [1992]).

25. Kant, Forst. In general, "critical constructivism" entails the understanding of constructing a better, more just society in criticizing the structural patterns of injustice that have hitherto restricted its realization. Forst, *Die noumenale Republik.*

26. It is a matter of controversy among Foucault scholars that Foucault conceptualizes patterns of power in this way. In the debate, Foucault is either understood as *only* depicting a voluntaristic subject in his later works, in other words a subject *without* contexts; alternatively, he is accused of reducing the subject to a mere automaton because he only takes the subject's contextualization into account (Honneth, *Kritik der Macht*; Menke, "Zweierlei Übung"), in other words, a subject *bound to* contexts. But then, to develop the latter objection further, he cannot provide the means also required to criticize the view of the subject as an automaton. His vocabulary, according to these objections, only further writes the subject's history as a mere positivity.

27. Especially the French cinematic tradition of the New Wave belongs to these examples, i.e., Jean-Luc Godard's films. Mireille Darc, *The Weekend*, DVD, dir. Jean-Luc Godard, Leipzig: Arthaus, 2013 [1967]. On Godard cf. Vinzenz Hediger and Rembert Hüser, ed., *Jean-Luc Godard: Film denken nach der Geschichte des Kinos* (Paderbon: Fink, 2023).

28. Foucault's interpretation is disputed. See John R. Searle, "'Las Meninas' and the Paradoxes of Pictoral Representation," *Critical Inquiry* 6, no. 3 (1980): 477–88; Habermas, *Der philosophische Diskurs der Moderne*, 303pp. We in turn use Foucault's interpretation in order to make the mutual relation between the phenomena on the surface and the "fundamental substance of the state of things" transparent. On this purpose it is required to refer to such small phenomena, to which also Velàsquez's painting belongs, that only tell *by the by* how to understand society, namely in and through those small ways which society itself does *not* describe.

29. See the section in Chapter 5 "Genealogical Critique and the Critical Construction of a De-Reified Agency."

30. Gilles Deleuze and Felix Guattari, *Kafka: Toward a Minor Literature*, trans. Dana Polan (Minneapolis/London: University of Minnesota Press, 1986). In referring to Franz Kafka the French philosophers Deleuze and Guattari characterize the form of "small literature" as being part of a minority, for example of the Jewish minority in Warsaw or in Prague. From a systematic point of view we transfer their observations into our digital today.

31. Kant, "Was ist Aufklärung?" 29.

32. From today's perspective, this point stands in opposition to Thomas Nagel, *The View from Nowhere* (Oxford et al: Oxford University Press, 1986).

33. Georg W. F. Hegel, *Lectures on the History of Philosophy*, Vol. 2, trans. E. S. Haldane and F. H. Simson (London: without publisher, 2016 [1832]).

34. Karl Marx, *The Eighteenth Brumaire of Louis Bonaparte* (Without location: e-art now, 2020 [1852]), 49.

35. Juliette Binoche. *Clouds of Sils Maria*. DVD, dir. O. Assayas, London: Artificial Eye, 2014.

PART II

The Critical-Theoretical Microanalysis of the Present Digital Time or on the Enlightenment Test

Part one outlined the hitherto too neglected problem horizon of digital enhancement with respect to the motif of the affective- and power-sensitive Janus face of subjectivity in digital enhancement. Now part two aims to develop the critical-theoretical microanalysis that echoes this problem horizon, drawing on Nietzsche and Foucault.

If it is common in the debate on digital change to refer to various kinds of tests to verify digital reality, for instance the famous *Turing Test*,[1] we propose here in light of Nietzsche and Foucault the conception of what we might call the *enlightenment test*. Turing developed the Turing Test to prove that not only human beings, but also machines can think. The problem horizon laid out here, however, shows the necessity of proving a different setting. Here, it is the question as to what extent subjects *themselves* turn into quasi-machines, to wit into such machines that are in and through their will programmed to divide their life between the binary schemes of perfect and non-perfect, like a real computer does in following the binary schemes of 0 and 1. So, while the previous chapter ironically began by stating that the enlightenment app does not need to be invented, we show that what needs to be invented is the enlightenment test.

This book *as a whole* seeks to present a kind of enlightenment test, comprising two aspects. Firstly, its task is to examine what kind of affect-centered relations of power are at stake in the concrete context of digital enhancement. Secondly, if this first part concludes that we are dealing with affective domination, then the specific context of digital enhancement must be seen as a problem, because digital enhancement has failed to keep its promise to generate freedom. Accordingly, the use of digital enhancement relevant in this concrete context is not justifiable in terms of affect-centered power relations. The upcoming chapters delineate the enlightenment test as follows: The description of the freedom-endangering side of digital enhancement shows the specific affect-centered power relations that become a problem in the field of digital enhancement. The description of the freedom-empowering side of digital enhancement then analyzes the conditions that enable the field of digital enhancement *not* to become a problem of unfreedom. Here, it is referred to the method of genealogical critique.

The enlightenment test following Nietzsche and Foucault stands in the Kantian tradition. If Hegelian approaches would show how the subject can liberate himself from the social normative order of perfection,[2] and if consequentialists would show that the social normative order of digital perfection is good as long as the outcomes are good,[3] then the Nietzschean and Foucaultian theoretical register developed here needs to be seen in the context of a Kantian approach in which especially Foucault also places himself (GSO). Basically, a Kantian approach considers two steps: Firstly, it considers the *motif* of an action as the ground that needs to be evaluated as good or bad. Secondly, if the motif is considered good, then the subject has the duty to do it and *vice versa*.[4] The Nietzschean and the Foucaultian register developed here unhesitatingly follows the first step. Here, it will be shown that the action is good if the subject's motif entails *an enlightened attitude* (see the enlightenment test). Nietzsche and Foucault, unlike Kant, do not explicate that the enlightened attitude needs to be a general law that has value for everyone. But the enlightened attitude that the subject develops is, according to Nietzsche and Foucault, nonetheless *generalizable*. Hence the subject's enlightened attitude is *not* to be confused with a *merely* particularistic one that would *only* matter for the concrete subject, but as such an attitude that *could* be generalized. Nietzsche and Foucault show this in paradigmatic concrete cases (*Einzelfälle*) that *nevertheless* have generalizable value.[5] The second step in turn would imply too much of a stretch to be *explicitly* spelled out in Nietzschean and Foucaultian terms. However, Heinrich Heine and the recent discussion that follows in his footsteps (Ernst Tugendhat) are wrong when they accuse the Kantian understanding of duty as a kind of heteronomy.[6] Heine objected to the Kantian view not to avoid establishing the Prussian officers in the subjects and thus making the subject only act in heteronomous ways;[7] Heine neglects, however, to see that according to Kant subjects can see *for themselves, and thus in a self-reflexive way* the need *to enlighten for themselves and for others* the political, social, and religious "guardians" (Kant) that hitherto dominated the subject's life, as they are *subjects among subjects*.[8] Nietzsche and

Foucault are not so foreign to this view. In their case, this point is conceived as an *enlightened interest*.

Against this background, the chapter has the task of describing the affect- and power-sensitive Janus face of subjectivity in digital enhancement with German nineteenth-century philosophy, that is Nietzsche, and with twentieth-century French philosophy, that is Foucault, who even casts himself as a "Nietzschean" (RM, 471) in the interview entitled "The Return of Morality." Here, to note again, Nietzsche and Foucault offer the theoretical perspective required to conceptualize the subject's attitude towards the logic of perfection. Thus, we do not refer to them in order to understand what optimal conditions of forms of life would look like. In this sense, with Nietzsche we introduce the present book's leitmotif and spell out the affect- and power-sensitive Janus face of subjectivity in digital enhancement in its basic grammar. The freedom-endangering side of this Janus face encompasses the motif of *epistemic automatization*. In turn, the freedom-empowering side entails the motif of *epistemic enlightenment* in the tradition of genealogical critique. The book at hand refers to the freedom-empowering side as the above mentioned enlightenment test. With Nietzsche, we gain the methodological basic approach needed to question the alleged truths that the subject took thus far for excessively granted in digital times. With Foucault, we gain the tools required to broaden the spectrum. The Foucaultian register will allow us to more broadly describe the alleged social facts that have hitherto formed the subject's blind formation of will. With Foucault, the Janus face of subjectivity in digital enhancement will be delineated first in terms of government and then in terms of the techniques of the self.

This view on Nietzsche's and Foucault's work is new among scholars. So far, the scholarly literature has in particular emphasized Foucault's critique of digital enhancement as one that especially warns us of the freedom-impairing aspects.[9] Quite often these views make reference to the panopticon. Nietzsche, in turn, is generally speaking so far situated on the other side of the spectrum as scholars use him to stress the advantages of digital enhancement.[10] In this sense, Nietzsche's texts are interpreted as part of the "overman" (Z, 4, preface 3, 14, trans. by the author) age, to put it with Nietzsche, in which robots and machine-like human beings overcome the current age of the *homo sapiens*. But these two quite common views of Foucault and Nietzsche imply *one*-dimensional readings because they overlook how the advantages and disadvantages already quite manifest in our daily life need to be taken into account *at once*. In what follows, we thus advance a new reading that develops a more nuanced perspective. This new perspective seeks to take into consideration *both* the freedom-enhancing *and* freedom-impairing aspects of digital enhancement that the present book conceptualizes in terms of affect-centered power relations.[11]

NOTES

1. Stuart M. Shieber, ed., *The Turing Test: Verbal Behavior as the Hallmark of Intelligence* (Cambridge: MIT Press, 2004).

2. Basically, on Hegelian approaches Axel Honneth, *Kampf um Anerkennung: Zur moralischen Grammatik sozialer Konflikte* (Frankfurt a.M.: Suhrkamp, 1994).

3. The classic understanding of consequentialism entails utalitarian perspectives. Cf. Jeremy Bentham, *An Introduction to the Principles of Morals and Legislation* (Garden City: Doubleday, 1961 [1789]).

4. Cf. Forst, *Die noumenale Republik*.

5. Here, we do *not* interpret Nietzsche and Foucault in nominalist ways. For from a nominalist perspective, the generalizable potential would be out of sight and one would *only* concentrate on the concrete cases (*Einzelfälle*). Jürgen Goldstein, *Nominalismus und Moderne: Zur Konstitution neuzeitlicher Subjektivität bei Hans Blumenberg und Wilhelm von Ockham* (Freiburg: Alber, 1998).

6. Ernst Tugendhat, *Vorlesungen über Ethik* (Frankfurt a.M.: Suhrkamp, 1993), 97pp.

7. Heinrich Heine, *Zur Geschichte der Religion und Philosophie in Deutschland* (Stuttgart: Reclam, 1997 [1834]).

8. See in general: Rainer Forst, *Das Recht auf Rechtfertigung: Elemente einer konstruktivistischen Theorie der Gerechtigkeit* (Frankfurt a.M.: Suhrkamp,[3] 2014), 89pp.

9. Petra Gehring, "Eine Topo-Technologie der Gefährlichkeit: Digitale Einsperrtechniken und sozialer Raum," in *Techniken der Subjektivierung*, ed. Andreas Gelhard, Thomas Alkemeyer and Norbert Ricken (München: Fink, 2013), 299pp; Susanne Krasmann et al., eds., *Sichtbarkeitsregime: Überwachung, Sicherheit und Privatheit im 21. Jahrhundert* (Wiesbaden: Springer, 2011).

10. In particular see Stefan Lorenz Sorgner, *Übermensch: Plädoyer für einen Nietzscheanischen Transhumanismus* (Basel: Schwabe, 2019).

11. Colin Koopman offers a multifaceted description of the digital world. Nonetheless, he neglects the role of affects and desire. Colin Koopman, *How We Became Our Data: A Genealogy of the Informational Person* (Chicago/London: The University of Chicago Press, 2019).

Chapter 3

On Today's "Serene Asceticism" after Nietzsche

This chapter introduces with Nietzsche the basic grammar that allows us to develop the theoretical register required to conceive the affect-centered, power-based Janus face of subjectivity in digital enhancement. If this book will later show with Foucault how the affect-centered, power-based Janus face of subjectivity in digital enhancement is generated in terms of governability and techniques of the self, for the moment we shift with Nietzsche to the basic semantics of *epistemic automatization* and *epistemic enlightenment* that place Nietzsche's formulation of "serene asceticism" (GM, 5, 8, 352; GM, 78) at the center. This view has so far been overlooked by Nietzsche scholars. The notion of "serene asceticism" (Nietzsche) stems from the third essay of Nietzsche's late work, *On the Genealogy of Morals*, that some consider as his most important text.[1]

The present book aims to demonstrate how the notion of "asceticism" is productive for reflecting the ways today's subjects form their life in digital enhancement in and through exercises. The qualification "serene" will then be considered helpful for describing how subjects do so in a self-reflexive and in a *not* quasi-automatized manner. The word "serene" thus does *not* encompass an understanding of "happy" or "gay" in its ordinary sense. Accordingly, the chapter does not say that people are happy while forming their life. Analogue to Nietzsche's understanding of "gay" in the title *Gay Science*, one of Nietzsche's key books, this chapter now underlines that "gay" refers to the subject's enlightened, self-reflexive attitude while forming their lives. This attitude is demonstrated here in terms of *epistemic enlightenment*, which the present book further delineates in terms of genealogical critique. It explicates the subject's "untimely" (UB, title, trans. by the author, "*unzeitgemäß*") perspective on his gay or serene life formation. The counterpart to this view of "gay," as the chapter will go on to argue, entails the understanding of a quasi-automatized subjectivity that the chapter describes in terms of *epistemic automatization.* It reflects the subject's "timely" (ZB, 1, 645, trans. by the author, "*zeitgemäß*") perspective on his merely routinized life formation.

While, of course, Nietzsche *himself* in no way linked the notion of "serene asceticism" to today's digital change, it is Nietzsche, too, who paved the way for systematically translating his thoughts to our digital today. For he understood himself as an "untimely" philosopher who was *not only* part of his time, but also one who claimed to make space for future generations. In this sense, Nietzsche contributes to the "philosopher[s] of the future" (J, 5, 44, 60, trans. by the author), to say it in Nietzsche's own words. One of these futures is our digital today. In systematically making Nietzsche's philosophy productive for contemporary critical thought on digital enhancement, we thus follow the path Nietzsche took first.

Against this backdrop, the present chapter is divided in three steps. The first seeks to provide with Nietzsche the understanding of digital enhancement as a contemporary, affect-centered and power-based form of asceticism in the sense of training. In the second, we sketch the freedom-endangering side of the affect- and power-sensitive Janus face of subjectivity in digital enhancement as the procedure of *epistemic automatization.* In the final step, we sketch the freedom-empowering side of the very same Janus face of subjectivity in digital enhancement. Here, we focus on the motif of *epistemic enlightenment.*

DIGITAL ENHANCEMENT AS THE PROCEDURE OF EXERCISING ONE'S OWN "AFFECT FORM" OF LIFE

First, we have to clear the ground for the hitherto overlooked understanding of digital enhancement that we gain from a systematic reading of Nietzsche's works. Unlike common understandings that conceive enhancement as the introduction of a biotechnological intervention in one's body, the problem horizon of the paradoxical condition of digital enhancement shows, by contrast, the necessity to think of digital enhancement in terms of affect- and power-sensitive forms of life marked by the logic of perfection, as the chapter "What is digital enhancement?" aimed to demonstrate. In what follows, we now provide with Nietzsche the theoretical register that seeks to make this view plausible.

With Nietzsche, it shall be shown that digital enhancement needs to be understood as *the procedure of exercising one's "affect form"* (NF 1888, 14, 121, trans. by the author) *of life*. This formulation enables us to describe how subjects are prompted to have the will to perfect their life. Simultaneously, Nietzsche also shows that it depends on the subject's attitude if and, if so, to what extent the subject is prompted to have such a will. Subjects that have an enlightened subjectivity, the present book argues, are able to resist such a will. By this, it is clear that we do not interpret Nietzsche's view on life formation in terms of perfection. By contrast, we refer to Nietzsche in order to theoretically conceptualize how subjects are able to place themselves in relation to the logic of perfection and thus to the heart of digital enhancement. Whereas in vitalist, optimist strands of newer French philosophy life is understood in merely affirmative ways (Deleuze), and in pessimist, contemporary Italian philosophy in merely negative ways (Giorgio Agamben),[2] the Nietzschean view on the "affect form" of life developed here does not contribute to either of these two lines of traditions. Unlike them, from Nietzsche we rather learn that the respective life interpretations depend on the subject's attitude towards it.

And how does the subject do so? To understand this point, we need to consult the theoretical archives that Nietzsche provides. Basically, the Nietzschean view that we need to better understand the formation of the subject's attitude to our digital today stems from his widely discussed understanding of the art of living.[3] Whereas among Nietzsche scholars the notion of affects is rather neglected in Nietzsche's notion of the art of living,[4] and whereas scholars of affect theory pay him rather little attention,[5] the present book thinks of the Nietzschean approach to the art of living in terms of affects. Accordingly, we speak of the "affect form" of life. This view stems from reading together Nietzsche's understanding of the art of living and his affect-centered view on the "will to power," one of his key features. To see how both terms relate to each other, we need to begin to consider Nietzsche's view of the art of living. We then refer to his understanding of the "will to power."

The art of living is of key relevance in Nietzsche's work from early on. It is not composed of a content, of a fixed, natural, substantial or essential pool of values, ideas or norms, but entails all the practices in and through which subjects give their life a form.[6] To describe these procedures, Nietzsche uses the key formulations "make ourselves!" (NF 1880, 7 [213], trans. by the author), "create a form from all elements" (NF 1880, 7 [213], trans. by the author), or "giv[ing] style to one's character" (GS, 290, 163). With reference to key historical figures, such as Goethe or Socrates, Nietzsche describes the practices of which life formation consists (GD, 6, 49, 151), for example the art of writing. What is relevant for the argumentation here is the hitherto largely neglected view of the art of living as being composed of "exercises" (M, 3, 22, 34, trans. by the author). The point of departure for the interpretation here is the assumption that all subjects have the potential to become what Nietzsche calls "poets of life." So be it Socrates, Goethe or everyday subjects, the procedures in and through which they all give their life form are for Nietzsche "(D)oing! Doing! Doing" (D, 22, loc. 667 of 3742, Kindle; M, 3, 22, 34).

The motifs of "training" (M, 3, 22, 34; trans. by the author) and "exercise[s]" (NF 1880, 9 [361]) are relevant in Nietzsche's entire work. The Nietzschean meaning of training needs to be analyzed here in the tradition of the ancient Greek understanding of training as asceticism, a point elaborated by Foucault (cf. the chapters on Foucault). In doing so, Foucault sharply distinguishes between the ancient Greek asceticism and the Christian variant, a line of argument *implicitly* rooted in Nietzsche. The asceticism we are dealing within digital enhancement refers to the understanding of "asceticism" in the ancient Greek sense. More specifically, it goes back to the ancient Greek verb "ἀσκέω" (*askéō*), which means "doing exercises."

In digital enhancement, such a kind of asceticism is part of daily routines. These daily routines, however, are not any kind of practices: understood in terms of the art of living developed here, they are the specific *self*-practices in and through which subjects become who they are in a given social normative order. Goethe, Homer or other authors for example became who they were by learning "the compulsion of meter, the tyranny of rhyme and rhythm" (BGE, loc. 1976 of 5195, Kindle; J, 5, 188, 108), and the "arbitrary laws" (BGE, loc. 1976 of 5195, Kindle; J, 5 188, 108). Simultaneously, these famous figures *also* became who they were via the *self*-practices of "dancing in chains" (MA 2, 2, 140, 612, trans. by the author), meaning the self-practices in and through which they became an author, *but not* in the way the tradition would have one expect. This point is central for conceiving digital enhancement with Nietzsche: For we see that training *as such* does not say anything about the way digital enhancement needs to be qualified. What counts, according to Nietzsche, is the subject's *relation from himself to himself* in digital enhancement, that is, his subjectivity. Hence, whether such training becomes a problem depends on the respective subjectivity that is part of the process of training.

This specific understanding of training leads to another view on the difference between *praxis* and *poiesis* that goes back to Aristotle. Whereas Aristotle considers *praxis* as the understanding of action that has its end in itself and *poiesis* as the notion of the techniques that are *by contrast* defined by a means–end relationship,[7] Nietzsche's understanding of the art of living shows that one needs to problematize the alleged difference between *praxis* and *poiesis* in terms of the respective subjectivities element to it. For Nietzsche does not so much stress a difference between *praxis* and *poiesis*: "The philosopher's product is his life [. . .]. That is his piece of art" (NF 1873, 7 [712], trans. by the author). With this formulation, Nietzsche describes Goethe's way of becoming an author. This way entails tendencies of *poiesis* because the subject that shapes his life in the social normative order is *at once both* subject *and* object. Goethe, to remain with Nietzsche's example, is the *subject* that gives his life the form of an author. But at the same time he is *the object* of his doing, too, because it is he who is doing something with himself, with his reason, with his senses et cetera, in order to achieve the end of becoming an author. Hence, his doing is structured by his aim of becoming an author. Generally speaking, the field of asceticism operates in similar ways. Here, the subject, too, deals with himself as a subject and an object. *Yet*, we learn from Nietzsche—and this point is crucial—that Goethe was *not* the *mere* object in this process, for he was able to develop a specific attitude towards his life formation. This point further explains why life formation is *primarily* praxis. Goethe is *not only* someone who *mechanically* shaped his life. What Nietzsche praised in Goethe in the passage above is the point that his life formation *opens up spaces for procedures* that have *a life of their own* (*Eigendynamik*). The present book considers such a life of their own (*Eigendynamik*) in terms of the subject's potential to relate *differently* to the existing dominant ends in society. In this sense, Nietzsche directs our view to the question as to what kind of subjectivity is at play in the respective procedures of life formation.

In order to understand what the work of affects in such a nuanced view on life formation is, we need to consider Nietzsche's understanding of the "will to power." Nietzsche introduces the basic understanding of the "will to power" in *Thus Spoke Zarathustra*, one of Nietzsche's later works. There, Nietzsche defines the "will to power" as "the unexhausted begetting will of life" (Z, 88). But the early Nietzsche already has a sense of what he only later named the "will to power." Here, however, he does not yet have the vocabulary in terms of power; here he puts it in the language of life. Accordingly, he tries to understand the "dark, thriving force of life" (*dunkle, treibende Kraft des Lebens*, HL, 2, 3, 269, trans. by author). The early Nietzsche lacks the language of power because this early Nietzsche was still attached to a reductive understanding of power he derived from his colleague Jacob Burckhardt in Basel; Burckhardt emphasizes power as always bad.[8] Only the late Nietzsche of the *Zarathustra* by contrast is able to understand power in plural ways. Instead of considering it merely bad because it only restricts life, Nietzsche is then able to see that power also *constitutes* life.

In the general understanding of the will to power, affects play a major role. Nietzsche explicitly calls the "will to power" an "affect form" (NF 1888, 14, 121, trans. by the author). Against biological or natural understandings of affects, Nietzsche thus advances an understanding of affects as a specific kind of power relation. This view clarifies that we are dealing with an understanding of affects that is marked as *socially constructed in and through relations of power*. Such an understanding of power is *constitutive* because it demonstrates the subject's will to live in a certain way that is constructed by the effects of power. Whereas the late Nietzsche provides the theoretical register of the "will to power" to theoretically conceptualize this view in terms of a "stimulus of wanting" (*Reize des Wollens*, NF 1884, 25 [436], trans. by the author), the middle Nietzsche is already moving in this direction when he speaks of the "feeling of power" (*Gefühl von Macht*, M, 3, 112, 101, trans. by the author; in the English translation this semantics is lacking).

How subtly the work of the power-based understanding of affects takes place becomes better understandable by alluding to the mother in the computer game "Perfect Mom" in the film *Her*, to which we already referred to in the introduction and in the first chapter. The computer game "Perfect Mom" is based on the idea of becoming an allegedly perfect mom by gaining as many perfect mom points as possible, for example by preparing a healthy meal for one's kids, by taking one's kids to school, or by looking beautiful. With reference to her we see that she, of course, knows of what she does when preparing cookies for her children and all the classmates, a healthy meal, and so on, but she does never question her actions. She is not aware of what she does on a self-reflexive level. She simply has the will to do what she does because her will is programmed to act in this manner by her daily routines. Affects, we argue, are responsible for programming the will in this specific way. Nonetheless, as the formulation of the Janus face of subjectivity in digital enhancement shows, there is still space to give an affect form to one's life *differently*.

The affect-centered view of the will to power after Nietzsche developed here differs from the predominant view among scholars. The latter stresses exclusively the freedom-enabling side implied in Nietzsche's view of the "will to power."[9] In this sense, they underline the freedom-empowering dynamics that enable the subject to entirely overcome former constraints and patterns of domination. The subject is then conceived as a triumphant "creator" (Z, 90) who can "break values" (Z, 90) and thus one who can re-evaluate the old values fully by himself. This interpretation, however, is not sufficient to address the question as to why subjects are open to having their will formed.

Although the understanding of training plays a key role in today's digital enhancement and in the overall debate on enhancement (Peter Sloterdijk), such an affect-centered, power-based understanding of training developed here differs from today's ordinary views on enhancement. Today, the subject has the feeling of living under the pressure of not sufficiently perfecting her life, the argument goes, as if the subject did not *want* enough perfection in her life. She should train better, or, as it is expressed in the computer game "Perfect Mom" if one loses perfect mom points: "Work harder" (*Her*). But on the basis of the ancient Greek understanding of training, we see that today a different approach is required to tackle the current horizon of digital enhancement. The ancient Greeks already knew that it is not sufficient to *only* focus on the existing subject's will. From the Greeks we rather learn to analyze how the subjects were and still are trained to have a specific will through effects of affect-centered power. In our time, for example we deal then with the subject's will to enhance life. Referencing contemporary issues, we illustrate what this means. Not only the digital world but also analogue life is confronted with the problem of a virus. Today, during the Covid-19 crisis, people are immediately trained to have the will not to become infected by the Covid virus. Sometimes a subject tests positive *despite* having a strong will not to become infected. In other words, what happens to the subject is not only up to the subject's will. Instead we need to realize that the subject is directly embedded in existing contexts, for instance in contexts that are determined by the power of the Covid virus.

Furthermore, we need to differentiate between our approach to training and its everyday understanding. In the field of sports we are familiar with an athlete needing to undertake training in order to improve her performance for the next competition.[10] Enhancement here is seen in its genealogy of doping. In digital enhancement, however, we deal with a different understanding as in the domain of sports. Here, we deal with *everyday subjects* who *immediately* turn, so to speak, into *ordinary athletes* that have the will to perfect life. Their competition occurs in their daily life. They desire to best perform in what they do without questioning their doing; for example, the mother in the computer game "Perfect Mom" desires to look as beautiful as possible, to cook as tasty as possible, to be liked by everyone as much as possible, and so on. Accordingly, if we speak here of training, we focus on the daily routines immediately undertaken by the ordinary subject. The example of this mother is, of course, a fictive example that might be more exaggerated than is the case in real life.

THE TRAINING EFFECT OF THE PRIESTLY BELIEF: ON EPISTEMIC AUTOMATIZATION

We now examine the freedom-endangering side of the affect- and power-sensitive Janus face of subjectivity in digital enhancement. If generally speaking it consists of *a quasi-automatized subjectivity* that conceptualizes how people are made open to having their will formed, then we provide with Nietzsche the basic grammar of how people are made open to having their will formed *via* beliefs. In doing so, the freedom-endangering side of the Janus face of subjectivity in digital enhancement will be depicted in developing the motif of *epistemic automatization*. Such a view enables us to understand that, in Nietzsche's words, "life itself has become a problem" (GS, preface 3, 5). Analogously, today's life turns into the specific problem that people live in a bubble, as one would put it in contemporary terms, which constantly makes them exercise the timely belief in the alleged value of the logic of perfection that through these exercises then turns into the subject's own belief *without* her being aware of it. To translate Nietzsche further into our digital present, it transpires that this problem is so persistent because the subjects can be conceptualized as being caught in the "walls of a prison" (WL, 1, 883, trans. by the author) although they do not feel as if that is the case. Obviously, this is a drastic formulation and might be something of an exaggeration, but it may help clear the processes of blindly forming the subjects' will.

In what follows, we will describe the exercise of beliefs with Nietzsche in terms of the priestly belief. The notion of the priest stems from Nietzsche's *Genealogy of Morals* (GM, 5, 3, 363pp).[11] There, the figure of the priest has the function of teaching subjects what they should understand as the true morals. And Nietzsche, of course, is eager to demonstrate that what these priests teach is *only* the dominant view on morals, which, however, must *not* be equated with a justified morality that does *not* smuggle in patterns of domination as dominant morals do. Priests, according to Nietzsche, do not only have the guise of human beings, as the everyday life understanding of real priests does. By contrast, he shows how the existing, dominant beliefs in society can *already* have the function of such priests. Unlike truths, a belief is not based on evidence; *nonetheless* the subject holds what he thinks to be true. For example, if one sees a ship on the water somewhere in the distance, the subject can think that there are people on the ship *although* the subject cannot prove his thoughts by really seeing people on the ship. Nietzsche, however, goes a step further and states in his text "On Truth and Lies in an Extra-Moral Sense" that there are not essential, substantial or eternal truths that have a timeless value. For him, by contrast, truths *as such* are *always* beliefs that people have *socially constructed* and taken for granted in a specific time and in a specific context. For Nietzsche, priests are key figures for establishing such socially constructed truths. However, people tend to forget that truths have a social origin.[12] Nietzsche also describes such an understanding of belief as a "superstition" (BGE, preface, loc. 672 of 5195, Kindle; J, 5, preface, 11).[13]

The present book refers to the motif of the priestly belief in order to explain the affect-centered power relations part of digital enhancement. Accordingly, we speak of *affect-centered "priestly power."* The formulation helps explain how subjects are made docile via the *training effect of beliefs*, evaluating in a blind and thus in a non-self-reflexive way the existing, dominant logic of perfection as the alleged valuable, true way of life. For it is the daily routines—as Nietzsche puts it—the habitualized "(D)oing! Doing! Doing!" (D,

22, loc. 667 of 3742, Kindle; M, 3, 22, 34), that have such a "timely" effect on the subject that he is prompted to conceive the dominant belief in society as his own belief. "Timely" refers to Nietzsche's essay "Untimely Meditations." "Untimely" means that a subject acts in a fashion that does *not* correspond to the perspective common in his time. By contrast, "timely" implies that the subject thinks and acts in correspondence with the perspective common in his time. The subject in the timely perspective has, then, the will to believe in the logic of perfection as if it is his own will that wants him to perfect life. This effect is conceptualized here as the timely effect of beliefs. Such a process is further conceptualized as "automatization." This delineates how a quasi-automatized subjectivity is born in and through exercising beliefs. Quasi-automatized subjectivity entails the understanding of the subject that is prompted by his belief machine to refer *in a machine-like manner* to the logic of perfection and this means to *only re*act in a way that the logic of perfection wants her to follow.

This kind of automatization is epistemic because the subject is programmed on a cognitive level *via* her beliefs to evaluate the logic of perfection as good. This scenario needs to be specified in terms of affective domination because the subject's potential to form his will *differently* is closed by his belief machine. The affective dimension refers to Nietzsche's affect-centered view of the "will to power" in is freedom-endangering side. The relevant formulations here are "obeying" will (NF 1885, 1 [61], trans. by the author) or the "stimulus of wanting" (*Reize des Wollens*, NF 1884, 25 [436], trans. by the author). They describe ways of being open to having a specific will *without* being aware of what one wants and thus without having a self-reflexive attitude. Accordingly, such subjects only see *one* way to form life and this entails their will to perfect life in line with their own belief that the logic of perfection is the one way that shows how to lead an allegedly good life. It is this view of affective domination that makes plausible, as we learn from Nietzsche, that subjects are open to having the will to perfect life "without breaking this long chain of the will" (GM, 35). Similar to Foucault's understanding of the knowledge–power nexus, this explains why subjects attach to the dominant knowledge of their time; this scenario is conceived here as the belief–power nexus.

This perspective on the belief–power nexus leads to a kind of *affective loop*. The term "loop" takes up the understanding of the social form of prison produced by the subject's blind belief in perfecting life. This kind of prison consists of *blind repetition*. The term "affective" signals that the subject is *not* aware of building such a prison. It is the blind work of her beliefs that makes her produce it. This in itself impacts on the subject's sensory perception of her life formation. Such a subject now only has the *one-dimensional* possibility to view life through the eyes of the dominant belief in perfection that is actually the belief of society. Thus, although the subject's life formation unfolds in front of her eyes, she has the narrowed, blind and one-dimensional gaze that is the result of her belief machine. Metaphorically speaking, one could say: the loop works like a scratched vinyl record that repeats the same refrain. Whether the subject wants to or not, her mindset is just configurated by her habitualized daily routines to simply follow the rhythm of the logic of perfection with the refrain "perfect your life."[14]

The view developed here is analogous to what Judith Butler sketched as the bad conscience in her essay *Psychic Life of Power.*[15] Butler describes how the subject produces a circle in which he is caught.[16] In general, "bad conscience" is the judging instance in and through which the subject judges her own conduct by asking herself to what extent she meets the existing dominant social norms. While we explain the functioning of the circle with reference to the affect-centered understanding of the "will to power," Butler provides a psychoanalytic perspective with the motif of "passionate attachment to regulation."[17] It renders plausible why the subject turns his will upon himself while passionately attaching himself to regulation, a procedure Butler calls internalization; in doing so the subject subjugates his will to the given social norms without being aware of what he is doing, writes Butler. This process generates the *closed* form of the will, which Butler calls "circuits"[18] and which here is called a loop. In order to sketch the turning of the societal will upon one's own will Butler uses the stylistic form of the trope that comes from the ancient Greek noun τρόπος (*trópos*) and means as well "turn." In this context, the work of the bad conscience is interpreted by Butler as the force that makes the subject continuously turning the judging will of society upon himself. By this, the loop-like dynamic is born: the subject internalizes the social norms so blindly that they *immediately* appear to the subject as *his own* norms. Any possible difference between the two, i.e., the will of society *and* the subject's will, remains *in*visible to the subject. In this sense, the bad conscience subjugates the subject imperceptibly to the judging instance of social norms and values. As a consequence, the subject himself does not perceive the need to question his perspectives on life. Seen from another angle, one can say that the work of the bad conscience has *biased* effects on the subject's life. It causes the subject to *more or less voluntarily* exclude forms of life that do *not* fit with the values provided by society.[19] Thus, the subject *himself* blindly naturalizes her life formation in ways that are not necessary.

Obviously, the language of the eternally repeated affective loop must be considered to be an exaggeration. But it nevertheless stresses an important point that is similar to what Leo Strauss had in mind when he described Spinoza's art of writing that was persecuted in his time and thus can be regarded in this sense as a dissident voice that is not part of the dominant voices of society; as a side note, Spinoza was persecuted by the Church for publishing his politico-theological writings. The similar point between Strauss and our digital time lies in the problem of making people's voices silent, such as the dissident voices of Spinoza. Strauss describes this situation as follows: "A large section of the people [. . .] have not been convinced by compulsion [. . .]. It merely paves the way for conviction by silencing contradiction."[20] Digitalization produces a similar situation today. Currently, the subject is initially confronted with the alleged truth presented by the order of perfection. Today the order of perfection prompts subjects in unnoticed ways to "silenc[ed] contradiction" (Strauss). The social expectations of perfection silently wants the subject to perceive, think, and experience the logic of perfection as the true form of life and thus this subject has no longer a will to contradict. The modes of training in terms of affect-centered power relations, and more broadly speaking of belief-based habitualization, are regarded here as the practices responsible for producing such a silence. Hence, we currently deal with the specific case, in which people are trained to immediately believe that the logic of perfection, and thus the heart of enhancement practices, is the truth that everyone needs to accustom themselves to.

Translating this perspective further into our contemporary context, we learn from the Netflix episode "Nosedive" of the series *Black Mirror* how the subject's immediate unquestioned belief in the order of perfection works (see the introduction of the present book, too). Obviously, this episode, too, is an exaggerated example, but it is helpful to outline the problem horizon in a drastic way. The affective loop is generated in the moment when the subject takes her prompted belief to be self-evident and thus as natural to such an extent that she does not see any need to question what she believes. The episode demonstrates how Lacie, the main protagonist, is just trained to be dominated by the "timely" perspective of perfection; she is caught in the affective loop of believing in the presumed truth of continuously being digitally rated for her performances, self-understandings, and identity. As a consequence, she simply subordinates every action, even the most ordinary ones like simply drinking coffee, to the social imperative of being rated. In this sense, she does not drink coffee for its own sake, but to earn the highest rates for posting pictures of lifting her cup in social media programs.[21] If she bites into a cookie, she does so *not* in order to eat the cookie, but in order to perfect the arrangement of the photo she wants to upload to the rating system. Hence, she just performs even actions like having a drink, renting an apartment, or meeting friends *in conformity with* the perspectives that are commonly valued in society as perfect and that she simply perceives as ones to which she wants to subjugate her view of life.

As Nietzsche puts it in "On Truth and Lies in an Extra-Moral Sense," we see that Lacie has forgotten in a systematic way that her unquestioned belief in the digital rating system was once invented and therefore socially constructed. For her, the truth is revealed as self-evident *although* it is obviously not self-evident. While being stuck in such a priestly belief, subjects like Lacie just turn into accomplices of the logic of perfection that produce the affective loop in which they are then continuously trapped. These subjects have forgotten not only other potential forms of life, but also the socially constructed nature of truths at all. These subjects simply cannot remember, to state with Nietzsche, that truth does *not* ground on an essential, natural, or substantial entity although the subject might be stimulated to own such a belief, but is *through and through*, and now we put it in Nietzsche's words, "a sum of human relations" (WL, 1, 1, 880, trans. by the author). To return to Lacie, we can now state: she has forgotten that other perspectives on life could be also true and she is thus blind to the fact that the digital rating system in which she is prompted to believe does not exist as such, but was once invented by people and is now confirmed in daily routines.

TO BELIEVE OR NOT TO BELIEVE, THAT IS THE QUESTION: ON EPISTEMIC ENLIGHTENMENT

Analogue to Shakespeare's play *Hamlet*, which raises the famous question "to be or not to be," the subject in digital times is called upon *to believe or not to believe* in the training effect of the belief in the logic of perfection. Up to this point, the subject, however, cannot even see that this question arises *at all*. So far the subject is caught in the belief-power nexus and *blindly* attaches to her accustomed priestly belief. The following delineates the motif of enlightened subjectivity (*aufklärerische Subjektivität*) in the tradition of genealogical critique; this shows how the subject can be empowered by the "*un*timely" training effect of beliefs to shed light on the belief–power nexus, which was hitherto not transparent to her. This perspective theoretically conceptualizes the freedom-empowering side of the affect- and power-sensitive Janus face of subjectivity in digital enhancement. It shows that the turning point does not lie in the subject's fully intentional, rational or conscious will to highlight the hitherto hidden belief–power nexus. It rather shows that this process emerges through the work of subtle contradictions that occur while the subject goes about her daily routines. This point is expressed by the formulation "untimely" training effect. But *not every* kind of contradiction leads to such an "untimely" training effect, only those that pass the enlightenment test.

In general, the understanding of enlightened subjectivity contributes to what we call, borrowing from Nietzsche, the "new enlightenment" (NF 1884, 25 [296], trans. by the author). The new enlightenment stands in the tradition that started with Kant and that conceptualizes procedures of critique in and through which the subject *is activated* to shed light on political, social and religious patterns of domination that Kant conceives as dogmatic "guardians" (Kant, "Was ist Aufklärung?," 30).[22] By this, it is clear that we use the term "enlightenment" not in the sense of the historical epoch of the Enlightenment in the seventeenth and eighteenth centuries—or even earlier, depending on one's definition[23]—but as a form of critique. The kind of enlightenment developed here is "new" in that it further pursues the way originally paved by Kant. This new understanding of enlightenment is rooted in the method of genealogical critique.[24] Nietzsche is the founding father of genealogical critique, and Foucault situates himself in his tradition. The methodology of genealogical critique is centered on the revelation that the subject's mode of becoming a subject is rooted in historical contexts and rules that initially seemed necessary to the subject *although* they were actually *not* necessary. This kind of enlightenment must be specified as epistemic. For it conceptualizes on a cognitive level that the subject gains a new, that is, self-reflexive understanding of the histories in and through which he became in quasi-automatized ways who he is. Via this new understanding, the subject learns to think *differently* and thus understands that the histories developed hitherto are *only one possibility* that was merely considered in a certain time as the way in and through which one becomes who one is, and that there could *also* be different ways. The present book calls this procedure the enlightenment test.[25]

More specifically, enlightened subjectivity viewed through the lens of genealogical critique encompasses the subject's ability to shed light on the belief–power nexus in which the subject was hitherto blindly trapped. This means that if A formerly had the subtle effect on B that B blindly thought, lived and felt differently as if without the effect of A, then A no longer has such an effect on B. A is the logic of perfection, B the subject in digital enhancement that in and through training effects of belief is accustomed to considering the logic of perfection as the true and valuable form of life. B, in becoming an enlightened subject, however, now needs to be conceptualized as someone who is no longer trapped in blind ways in the training effect of A. For B is now able to understand the entire spectrum of possible effects that A could have on B. Hence, this subject B is not only reacting in a machine-like, blind manner to the effect of A, but sees different possibilities.

This means, in terms of social ontology, that the subject becomes aware of the social-ontological character of affect-centered power relations. The social-ontological register can render plausible that the subject does not attach one-dimensionally to his blind belief in the logic of perfection, but can also develop plural views on how to live, feel, and act *otherwise* in the digital times marked by the logic of perfection. The subject thus sees *different possibilities*. This subject, in Nietzschean terms, is thus able to develop a "serene asceticism," a mode of training that is not bound to the blind belief in the logic of perfection.

In general, the theoretical register of social ontology that received high attention in recent debates makes plausible that everything that is relates to each other,[26] in the case of the present book in and through affect-centered power relations. Nietzsche was, of course, unaware of these contemporary debates on social ontology, but he described the meaning of social ontology *avant la lettre*. Thus far, current debates related to this matter have sidelined Nietzsche.[27] With Nietzsche, it is planned to demonstrate here that affect-centered power relations as such are not the problem only if the subject is not aware of the specific kind of affect-centered power relations he is part of. So, if not only the World Wide Web is a web, but also the subject's life formation is the specific web whose meshes consist of the affect form of the will to power, then the subject is *not* endangered by this specific web when he learns to develop an enlightened subjectivity. Nietzsche introduces different semantics to describe the affect-centered "will to power" that does not endanger the subject's life formation, but by contrast constitutes life formation. He speaks of the "feeling of resistance" (NF, 1884, 27 [24], trans. by the author), of the "affect of distance" (NF, 1883, 7 [106], trans. by the author), or among others of the affect that "commands" (NF, 1884, 25 [436], trans. by the author). The meaning of "feeling of resistance" is clear. "Affect of distance" describes the affect in and through which the subject can, with enlightened, self-reflexive procedures, create a distance to his blind belief in the logic of perfection. The affect that "commands" refers to Nietzsche's understanding of affects that do not obey the belief in the logic of perfection, but can by contrast command—and this means resist—the call of the belief in the logic of perfection. The common denominator of these formulations is the fact that they all describe the subject's desire to "overcome" the currently trained beliefs in the logic of perfection.

And how does this process start? It does *not* start through the subject's *mere* decision-making, but through the "untimely" effect of life with its daily routines on the subject. It conceptualizes that the subject, in and through unforeseen obstacles in life, becomes aware of his hitherto merely blind "faith" (D, preface, loc. 460 of 3742, Kindle; M, 3, preface, 12) in the traditionally and "timely" perspective, on which philosophers, but also ordinary people "have for a couple of millenia been accustomed to build as if upon the firmest of all foundations" (D, preface, loc. 460 of 3742, Kindle; M, 3, preface, 12). This process is, so to speak, the *byproduct* that emerges while one shapes one's life the way one does, which produces contradictions, however. Hence, the critical tool of genealogical critique provides the required theoretical means to show how the subject's hitherto trained blind "priestly belief" in the values that give life the form of alleged perfection is unsettled in ways that are rather surprising.[28] By this, the subject learns that the logic of perfection can by no means represent *the* origin of life formation from which one derives one's perspective on life. By contrast, he learns that there are origin*s* in the plural form and thus plural views on life formation. If the subject's view on life formation was formerly "closed" (HL 1, 1, 251, trans. by the author), then genealogical critique theoretically conceptualizes that the subject's view is now broadened in such a way that he can see the "horizon" (HL, 1, 1, 251, trans. by the author). This subject can thus see that "[t]here is another world to discover—and more than one! On to the ships [. . .]!" (GS, 163).[29] And this new world is not determined by the subject's blind belief in the logic of perfection. In other words, genealogical critique makes plausible that the subject is *no longer immanently* caught in his blind belief in the logic of perfection, but can *transcend* the context of such beliefs. If the subject is able to understand and experience his attitude towards the dominant beliefs in society in this self-reflexive fashion, then, according to Nietzsche, he can make a difference and see "an opposition between a world in which until now we were at home with our venerations [. . .] and another world that we ourselves are" (GS, 203). In this sense, genealogical critique highlights the gap between the "timely" views, that are dominated by the logic of perfection, and possible different perspectives, i.e., the "untimely" ones.

Simultaneously this approach clarifies the need to defend genealogical critique against two objections often raised by left Hegelians: the objection of seemingly unsettling everything, thus of being *only* relativistic, therefore senseless; secondly, genealogical critique must be defended against the objection of being merely particular.[30] In opposition to the first Hegelian reproach that genealogical critique theoretically conceptualizes how every norm is unsettled, one must insist that genealogical critique does *not* lead to a so-called relativistic nowhere that would liquefy *any* foundations and thus would question any subject's view on life.[31] Nietzsche's understanding of genealogical critique *only* seeks to question what was hitherto perceived, understood, and experienced by the subject *blindly* and thus in ways that here is called, drawing on Nietzsche, *affective domination*. By understanding the historicity of the perspectives that hitherto affectively dominated his own view on life formation, the subject simultaneously understands the contingency of how things emerge, and in so doing he is able to learn to create new, different, not affectively dominating views on life formation. Instead of leading to a relativistic nowhere, the argument here attempts to demonstrate that by criticizing patterns of affective domination, genealogical critique describes *the critical construction of an enlightened agency*. In this sense, genealogical critique contributes to the tradition of *critical constructivism* that started with Kant.

Secondly, genealogical critique has been criticized as excessively focused on *particular* perspectives. Against this critique it must be stated that to be effective in detecting patterns of affective domination, one can*not* do other than start with the particular perspective in order to investigate whether the historically formed norms, values, and self-understandings, in and through which the subject so far just lives, feels and acts, have in fact been considered by the subject *herself* as having an unquestioned and thus dominating value for his life formation. This question can *only* be raised by the subject *herself*. Otherwise she would speak for others and thus would exercise power over them, in other words, she would exercise domination and replicate *anew* patterns of power that she actually seeks to detect. Nonetheless, genealogical critique is *not merely* particularistic, because in detecting patterns of affect-centered domination it provides a perspective that has a *generalizable* character. It could thus also be justified by other subjects.

A prominent example in our time would be the turning point in Snowden's life. By looking at this turning point, one can illustrate how the theoretical methodology of genealogical critique works. Snowden became an employee of the NSA. We can now see that an *un*timely training effect empowered Snowden to turn from a strong proponent of the NSA, an institution in the US praising digital technologies, to one of its harshest opponents. Obviously, subjects like Snowden, might not be aware of the theoretical register, but the critical theorist needs the theoretical lens to describe the potential changeability of the subject's life. In his autobiography *Permanent Record*, Snowden describes the subtle modes of his change. Accordingly, he did not make "any decisions."[32] The turning point, therefore, cannot be situated in mere rational, conscious or intentional decision-making, but, in our terms, by affects, or in Snowden's words: such decisions "[a]re made subconsciously and only express themselves consciously once fully formed."[33] If one thinks of Snowden's own life, one can observe that his change goes hand in hand with existential risks, even of Snowden being forced to abandon his former life, his family, and his house, and compelling him to live a life in exile persecuted by the NSA, his former employer. So, his case provides a glimpse at the unforeseeable existential risks that might come with such a process. In other words, such a change can imply a position of "discomfort" (ED), to put it with Foucault.

More specifically, Snowden's change began in his affective life when he haphazardly held a talk in Tokyo about surveillance technology. While speaking at a conference, he firmly believed in the binary scheme of good cop/bad cop, so to speak. He believed himself to be a good one whereas seeing the Chinese as the people who spy on their own people. At this point, he was not able to see that American technological systems like PRISM work similarly; he was simply too uncritical to perceive the US spying system as structurally similar to its Chinese counterpart. Nonetheless, the theme of his talk already worked in Snowden's affective life. In his autobiography he notes that in the nights after the talk he could not sleep very well. But at the time he did not ask himself why this was so. Only three years later did he become aware of the entanglements in which he was trapped. With Nietzsche we see that he was able to perceive the "walls of the prison of his belief" (Nietzsche). Snowden's awareness emerged instantaneously without warning. "[I]t struck me"[34] is the phrase Snowden uses to describe how he was prompted to see clearly what his work for the NSA actually did and thus what digital technology in the US—and not simply elsewhere—was capable of. Thus he became aware that not only China is spying on their people, but also US government institutions using technology he helped to develop, such as Cloud technology that promises to perfect the subject's digital life by storing data in a "cloud" but in reality helps *also* to survey people.

If the social-ontological register is helpful for accounting for the fact that affect-centered power relations encompass a plural spectrum from being affectively dominated to not being affectively dominated, we nevertheless need to clarify that the social-ontological register developed here *only* considers the "software," and thus the subject's attachment to the given order. The "hardware," i.e., the order *itself*, remains untouched. To illustrate the consequences of this view: during the Covid-19 crisis, the subject is called upon to enhance his health by socially distancing. From the point of view of the subject's software, i.e., his mindset, attitudes, his way of living, feeling, and thinking, the subject can set himself in relation to the given perspective of social distancing. The subject can thus create an entirely new set of perspectives on how to relate himself to the given perspective. He can for example start to work in a home office, conduct home schooling, or gather digitally with friends. But these new perspectives do not necessarily affect at the social-ontological level the hardware, that is the pre-given orders *as such*. In the case of the meatpacking industry, e.g., the German Tönnies factory,[35] we can further problematize the social-ontological approach at the level of the subject as developed here. Tönnies became a target of German discussions because of the Covid-19 outbreak in one of the factories near Gütersloh in summer 2020 that became one of the largest in Europe with over 2,000 infected workers. It became obvious that the working conditions, and thus the order, were incompatible with social distancing. Moreover the factory's air conditioning system caused the virus to spread rapidly. In this case, we see that the subject's mindset, or software, can be programmed to consider social distancing as vital, but that the existing hardware—air conditioning—cannot accommodate to the situation resulting from systems failure. Moreover, workers employed in the meatpacking industry are often faced with precarious living conditions. Many are migrant workers from Poland,

Bulgaria, and Romania.[36] Their income does not suffice to rent apartments; in most cases they are forced to live in crowded collective housing. The order of these living conditions structurally undermines the subject's software, which is geared toward health improvements. The social-ontological approach can nevertheless pave the way for a conceptualization of the necessity of hardware changes. Here, collective strategies are required. Subjects can for instance organize unions, establish committees, and so on. Still, the starting point needs to be conceived in the subject's software, his attitude towards the given order, and thus his awareness that the status quo must not be the last word but can be changed.

To further understand how the subject's software so to speak, i.e., his attitude, can be transformed into an enlightened one, we consider Nietzsche's discussion of specific examples, e.g., Homer and Goethe, or Christians. In his discussion of these individuals, Nietzsche demonstrates how the subject can begin to understand that his own life formation is not necessarily subjugated to the "timely" perspective. These subjects can be empowered, as Nietzsche puts it, to "dance in chains" (Nietzsche) and thus to dance within the existing logic of life formation with all its timely, affectively dominating training effects. Obviously, today's subjects need not to turn themselves into Goethe or Christians. From a systematic perspective, however, an analysis of the development of their respective attitudes is helpful to understand the basic dynamics that create a subject's turning point. These cases are thus helpful to conceptualize that ordinary subjects like factory workers, could understand and experience that there might be different views of work during the Covid-19 pandemic.

With reference to Nietzsche's discussion of Christianity, we seek to demonstrate how the subject develops the potential to create in a self-reflexive way an "untimely" perspective on his life that is not subordinated to the preformed "timely" perspective. To describe Nietzsche's understanding of the term "untimely" in his *Untimely Meditations* of the mid-1870s, we need to connect the middle Nietzsche with his later works, and more specifically with the figure of the "slave revolt in morality" (GM, 5, 7, 268, trans. by the author), which Nietzsche introduces in his late text *On the Genealogy of Morals*. This enables us to clarify how subjects in digital times are able to change their perspectives and thus to "revolt" like the whistleblowers Snowden, Kaiser, Manning or Assange did when they questioned the given perspective. "Change one's perspectives" is linked here to the subject's immediate awareness that the "world that is relevant to us" (BGE, 34, loc. 1242 of 5195, Kindle; J, 5, 34, 54) implies that there is not one dominant perspective but multiple ones that empower the subject to begin to see differently. In this discussion, we are not concerned with the epistemic status of the change in perspectives as Nietzsche scholars often do.[37] In light of the subject's positioning himself in digital enhancement, we refer to the notion of the "change in perspective" in order to highlight how the subject can broaden his hitherto closed and thus dominated view of life and instead see the potential openness of his life formation. The subject that has developed such an awareness is, in a self-reflexive way, able to undertake a "revaluation of values" (BGE, 203, loc. 2242 of 5195, Kindle; J, 5, 203, 126); thus he is able to re-evaluate his thus far blindly habitualized attachment to the "timely" perspective and can "give impetus to opposed valuations" (BGE, 203, loc. 2242 of 5195, Kindle; J, 5, 203, 126). In this way, the subject can perceive that the "timely" perspective loses the value thus far taken for granted. Thus he is able to understand and experience the "timely" perspective not as the last word, but as a "provisional perspective" (BGE, 2, loc. 703 of 5195, Kindle; J, 5, 2, 16). Nietzsche describes with reference to early Christianity the emergence of the "untimely," i.e., a heretic, perspective. As a result, Christians can be seen as an example of subjects who are able to develop a different faith within the "timely" perspective of the pagan religions of the Roman Empire.

Nietzsche's discussion of early Christianity is useful for our own analysis in that it can help us describe the development of the self-reflexive possibility of an "untimely" perspective evident in the cases of whistleblowers like Kaiser and Snowden. Structurally, they share with early Christians the problem that they are confronted with a dominant, "timely" perspective that exerts power over their view of religious, political, and social life. Early Christians were persecuted in the Roman Empire. Today's whistleblowers, too, are confronted with the dominant perspective according to which digital media is seen as perfecting life. At first, Kaiser and Snowden supported the dominating digital settings. Kaiser worked for Cambridge Analytica (cf. part one), the organization responsible for manipulating the voters' opinion in the US elections of 2016 and the Brexit referendum. Snowden's case was referenced above. But then, both became able to distance themselves from the dominant "timely" perspective and turned into subjects with an enlightened attitude that we call with Nietzsche "free spirits" (MA 1, 2, preface 2, 15, trans. by the author),[38] they are able to distance themselves and initiate a "revaluation of values," which Nietzsche was fond of. These "free spirits," the whistleblowers, like the early Christians, no longer felt the necessity to *blindly* attach themselves to the timely perspective as "bound spirits" (HH 1, 226, loc. 2430 of 4697, Kindle; MA 1, 2, 226, 190).

From a systematic perspective, the Christian example is important because it illustrates two points: first, it implies the dynamic element of "changes in perspectives." Second, it highlights the structural problem confronting subjects with an enlightened attitude. With reference to Nietzsche we can clarify that even subjects with an enlightened attitude are *not* free from also anew essentializing, naturalizing, and substantializing their newly gained perspectives on life. Regarding the first point, we see with Nietzsche that the early Christians are able to conceive and experience that the "timely" perspective does not constitute *the* truth of what religion is all about, i.e., a natural, essential, or substantial understanding of truth, but that religious truth*s* needs to be described in the plural form. In other words, the gesture of "changing the perspectives" is rooted in a process of denaturalizing, de-essentializing, and desubstantializing dynamics. In this way, the subject recognizes that truths are socially practiced and that the daily habitualized social practices of truth *only appear* to the subject as self-evident and thus as necessary, *although* they are not. By perceiving the plural character of truths, early Christians could immediately initiate the above mentioned "re-evaluation of values" and develop their different view on religions. The second point clarifies, however, that Christians themselves immediately turned into the dominating group that *itself* oppressed others.[39] Thus, with reference to the history of Christianity, we see the necessity of *the enlightenment of enlightenment*. There is not a fixed subject with an enlightened attitude that forever has an enlightened view on life formation in and through which the affect-based power relations are transparent to him. By contrast, the enlightened subject is constantly called to further enlighten new patterns of affect-centered domination. Christians lacked this kind of enlightenment of enlightenment. Their enlightened view of life was problematic because they immediately took their perspective in excessively absolute terms and thus generated its own blind spots because they failed to recognize how they were prompted to turn from oppressed into oppressors who excluded pagan perspectives. The crusades in the Middle Age or the missionary work during colonialism illustrate the violent aspects of Christianity. Analogously, we can state that subjects in digital times who were once able to gain an enlightened view are *not* eternally freed from patterns that reproduced affect-centered dominating structures. Accordingly, the subject with an enlightened attitude is constantly called upon to shed light on its enlightened perspective in order to render visible its possible own complicity with the patterns of affect-centered domination.

With reference to Nietzsche's view of Goethe, we further outline the turning point in and through which the subject turns from one whose attitude is subjugated to the given perspective to a subject whose attitude is opposed to his own will to subjugate to the "timely" perspective. In general, Nietzsche praises Goethe emphatically both for his life and work. Interpreting this from a perspective of "serene asceticism," we see that Goethe, according to Nietzsche, is able to perceive the difference between the "timely" perspective of how to blindly become oneself conventionally, and the "untimely" perspective of how to become oneself in a self-reflexive manner and thus in a mode we describe with Nietzsche as "serene" and "untimely."[40] Accordingly, Nietzsche considers Goethe in the passage "Skirmishes of an Untimely Man" of *Twilights of the Idols* and praises him as follows: Goethe "fought against the separation of reason, sensibility, feeling, will [. . .], he disciplined himself to wholeness, he created himself" (TI, 49, 222).[41] Goethe can be seen in this sense as able to be opposed to the division and separations that are produced by the "timely" perspective.

Goethe likewise praised the subject's potential to create plural perspectives on his life. His emphasis on the subject that is able to become oneself by the motif of "die and become"[42] (trans. by the author, "*[s]tirb und werde*") is interpreted here in light of Goethe's *Urworte. Orphisch* (*Orphic Primal Words*) as a Nietzschean mode of the "serene asceticism."[43] The timely, dominating perspective appears, then, as the forms of "δαιμων"[44] (*daemōn*, in English: demon) and "ἀνάγκη"[45] (*anánkē*, in English: necessity) that immediately structure how one can feel forced to give one's life form in blind fashion. In this context, "δαιμων" (*daemōn*) is conceived as the eternal laws that just prescribe the subject's life. Via "ἀνάγκη" (*anánkē*) the subject feels forced in a similar direction. His life formation appears to her in an immediately formed way that unfolds along the lines of the direction foreseen by the stars. However, Goethe does not describe a fully determined picture of a subject's life formation. On the contrary, he leaves room for a perspective we describe in the Nietzschean sense as the "untimely" one that allows the subject a "serene" attachment to the "timely" perspective, one we call with Goethe the perspective of "ἀνάγκη" (*anánkē*) and "δαιμων" (*daemōn*). The affect of hope ("ἐλπίς,"[46] *elpís*) is according to Goethe responsible for such immediately formed, freedom-enabling dynamics.[47] Thus defined, hope shows that the "timely" perspective that appeared to the subject in dominant patterns as an insurmountable wall can be overcome and therefore be seen as changeable. When the subject thus has hope that he must not always be the same, then such a hopeful subject can immediately develop what we call with Nietzsche "serene asceticism." The subject is then able, to just see the open sea, and to thus understand that another tomorrow, another "Daybreak" (D) could be possible.

Nonetheless, the hopeful subject is far from seeing clearly what the content of this "daybreak" could be. Otherwise the subject simply reproduces the patterns of domination he tries to criticize. His new view is only a "provisional perspective" (Nietzsche). Not coincidentally, Nietzsche ends his work *Daybreak* with a question mark, which underlines the open character that the subject's hope in the possibility of an "untimely" perspective necessarily implies. The subject thus asks without giving a fixed answer: "And whither then we go? Would we cross the sea? [. . .] Or my brothers. Or?" (D, 575, loc. 3475 of 3742, Kindle; M, 3, 575, 330). Analogously, subjects in digital times can be compelled by the affect of hope to start travelling on the open sea of the digital world. In doing so while learning from Goethe and Nietzsche, their view on the broad horizon immediately becomes a problem in case they misunderstand what they then see as an eternal perspective and thus forget its merely provisional character.

THIRD CONCLUSION

This chapter has sketched with Nietzsche the basic grammar that allows us to conceptualize the so far rather neglected affect-centered, power-based Janus face of digital enhancement. To this end, we introduced with Nietzsche the hitherto forgotten motif "serene asceticism" which originates in one of Nietzsche's most important texts, *On the Genealogy of Morals*. In referencing this motif, we addressed three points: first, the Nietzschean understanding of asceticism enabled us to conceive today's field of digital enhancement broadly as a specific facet of life formation that is part of training. In this sense, we described digital enhancement as the daily routines and habitualized practices in and through which the subject desires to give form to his life by blindly attaching himself to the logic of perfection in digital times. This broad understanding of 'asceticism' is rooted in the ancient Greek verb "ἀσκέω" (*askeō*), which means "I do exercises." Foucault made this understanding of training explicit, but implicitly it is already rooted in Nietzsche's work. On this basis, we also developed the basic grammar of the affect-centered, power-based Janus face of subjectivity in digital enhancement. To this end, we have outlined along the lines of Nietzsche's hitherto neglected understanding of the "will to power" as an "affect form" how the subject *both desires blindly* to give his life form in daily routines by blindly attaching himself to the logic of perfection *and* at the same time is able to highlight his own distancing from his hitherto blind attachment to the logic of perfection. The former describes the freedom-impairing side of the Janus face of subjectivity in digital enhancement, which we specified by the term *epistemic automatization*. The latter refers to its freedom-empowering side, which we delineated with the motif of *epistemic enlightened* in the tradition of genealogical critique. In order to illustrate what is at stake in the digital arena, we referred to contemporary cases, such as the Covid 19 pandemic and the whistleblowers Snowden and Kaiser in the context of the NSA-scandal, the Cambridge Analytica scandal, and the Brexit referendum.

NOTES

1. Saar, *Genealogie als Kritik*. An instructive commentary on the *Genealogy of Morals* provides Andreas Urs Sommer, *Kommentar zu Nietzsches "Zur Genealogie der Moral"* (Berlin/Boston: de Gruyter, 2019).

2. See Gilles Deleuze, "Das Leben als Kunstwerk," in *Gilles Deleuze: Unterhandlungen 1972–1990* (Frankfurt a. M.: Suhrkamp, 1993), 136pp. Giorgio Agamben, *Homo sacer: Die souveräne Macht und das nackte Leben* (Frankfurt a. M.: Suhrkamp, 2002 [1995]), 12pp. On these reductive understandings of life, see Saar, *Immanenz der Macht*, 204pp.

3. Nehamas, *The Art of Living*.

4. In the debate over the art of living the role of affects has thus far not been at the center of the classic studies. Nonetheless, there has recently been growing interest in the field of affects. But the understanding of affects developed there mainly focuses on the concrete approach to affects in the sense of emotions, such as love, fear of living or the existential feeling of suffering. Loukidelis, ed., *Nietzsche und die Lebenskunst.* So here, we lack the vocabulary that we require to describe the anonymous mechanisms of power in and through which subjects immediately attach to and detach from the order of perfection that she immediately evaluates as good on a cognitive level.

5. Protevi, "Affect and Life in Spinoza, Nietzsche, and Bergson."

6. Here, when Nietzsche speaks of art, he does not do so in the sense of the original art (*originäre Kunst*). "Original art" refers *merely* to the artist in relation to his work, for example a poet and her poem.

7. Rahel Jaeggi, *Entfremdung: Zur Aktualität eines sozialphilosophischen Problems* (Frankfurt a.M./New York: Campus, 2005), 225.

8. See Jacob Burckhardt, *Weltgeschichtliche Betrachtungen*: *Kritische Gesamtausgabe*, Vol. 10, ed. Jacob Burckhardt-Stiftung, (München: Beck, 2000pp), 376.

9. Of course, Volker Gerhardt's brilliant study on Nietzsche's understanding of the will to power is a milestone in Nietzsche scholarship. However, he is more interested in the freedom-enabling elements in Nietzsche's notion of the will to power. Nonetheless, this study gives a highly instructive overview of how to read power in Nietzsche. Volker Gerhardt, *Vom Willen zur Macht: Anthropologie und Metaphysik der Macht am exemplarischen Fall Friedrich Nietzsche* (Berlin/New York: de Gruyter, 1996).

10. On the notion of training in times of enhancement: Peter Sloterdijk, *Du musst dein Leben ändern: Über Religion, Artistik und Anthropotechnik* (Frankfurt a.M.: Suhrkamp, 2009). Sloterdijk, however, develops a one-dimensional view that overlooks the freedom-impairing aspects; see the Foucault chapter; also consider Sarah Bianchi, "'Maschinen-Cultur' und Anerkennung: Über Nietzsches Verständnis von sozialer Individualität," *Nietzscheforschung* 25 (2018): 435–48; on the notion of "machine-culture" in Nietzsche see Renate Reschke, *Denkumbrüche mit Nietzsche: Zur anspornenden Verachtung der Zeit* (Berlin: Akademie-Verlag, 2000), 49pp.

11. This motif is grounded in Renaissance philosophy; see Spinoza's understanding of superstition. Spinoza, TTP.

12. WL, 1, 1, 878pp.

13. In today's discussions the theme of fake news, that is false information, is widely spread. Cf. Nils Kumkar, *Alternative Fakten: Zur Praxis der kommunikativen Erkenntnisverweigerung* (Berlin: Suhrkamp, 2022). Of course, this theme is very important. But Nietzsche's critique begins a step earlier. More broadly, he is interested in understanding *the conditions* in and through which knowledge *at all* and *not only* false information gains power in society.

14. *In contrast to* what is nowadays called positivism, Nietzsche demonstrates that there are no positive, given, and objective facts. On the contrary, facts are socially constructed because according to Nietzsche, they were once invented.

15. In general on Butler: Paula Irene Villa, *Judith Butler* (Frankfurt a.M./New York, Campus, 2003); Eva von Redecker, *Zur Aktualität von Judith Butler: Einleitung in ihr Werk* (Wiesbaden: VS Verlag für Sozialwissenschaften, 2011).

16. Butler, *Psychic Life of Power*. On the relationship between Nietzsche and Butler see Sarah Bianchi, "Verletzlichkeit und Leben: Nietzsche im Kontext der Anerkennungsdebatte," in *Oldenburger Jahrbuch für Philosophie*, ed. M. Bauer und N. Baratella (Oldenburg: BIS-Verlag, 2019), 111–38.

17. Butler, *Psychic Life of Power*, 66.

18. Butler, *Psychic Life of Power*, 63.

19. In this sense, Butler speaks of melancholy. Butler, *Psychic Life of Power,* 132pp. To translate the term into current vocabulary, "melancholy" arises when a subject comes about to live his life for a too long time only *in subjugation to* the given, "timely" perspective *without* considering the possibility of "untimely," and thus different perspectives.

20. Leo Strauss, *Persecution and the Art of Writing* (Chicago/London: The University of Chicago Press, 1988 [1952]), 22p.

21. This is not only valid for Lacie. With Moeller and D'Ambrosio we see that "today people are increasingly speaking, dressing, and acting as if a vide of them might [. . .] be uploaded for dozens, hundreds." Hans-Georg Moeller and Paul J. D'Ambrosio, *You and Your Profile: Identity after Authenticity* (New York: Columbia University Press, 2021), 253p.

22. In general, on the notion of enligthenment in Kant: Sarah Bianchi, ed., "Schwer punkt: Zu Kants 300. Geburtstag: Aufklärung, Freiheit und Nicht-Beherrschung," *Zeitschrift für Praktische Philosophie* 11, no. 1 (2024), forthcoming.

23. See Jonathan Israel, *Radical Enlightenment: Philosophy and the Making of Modernity 1650–1750* (Oxford: Oxford University Press, 2002).

24. Although "new" might make one think of a rupture between Kant and Nietzsche, that would be a misunderstanding. "New" only stresses a new aspect of the key understanding of enlightenment developed by Kant.

25. In current debates, there is talk of a third enlightenment, see Richard J. Bernstein, "Hilary Putnam: The Pragmatist Enlightenment," *Graduate Faculty Philosophy Journal* 38, no. 2 (2017): 373–83; recently, Michael Hampe, *Die Dritte Aufklärung* (Berlin: Nicolai Verlag, 2018). Following Hampe, the first was the Socratic reflections, the second was Kantian thought, and the third is the present day. The book at hand, however, seeks to demonstrate the continuity in enlightenment thinking. Furthermore, in current debates the notion of enlightenment is accused of being

European: Nikita Dhawan, "Die Aufklärung vor den Europäer*innen retten," in *Normative Ordnungen*, ed. Rainer Forst and Klaus Günther (Berlin: Suhrkamp, 2021), 191–208. But the book at hand aims to show, with reference to European enlightenment thinkers, that the notion of enlightenment entails a self-reflexive attitude, which is transferrable—in a context-sensitive manner—to global questions. Hence, it does not lead to the problem that is described as cultural parochialism in contemporary debate, i.e., the theoretical conceptualization of ignoring cultural differences, cf. Melissa Williams, ed., *Deparochializing Political Theory* (Cambridge: Cambridge University Press, 2020).

26. Cf. in general on the debate on social ontology Ian Hacking, *Historical Ontology* (Cambridge, MA: Harvard University Press, 2004); Robert Nichols, *The World of Freedom: Heidegger, Foucault and the Politics of Historical Ontology* (Stanford: Stanford University Press, 2014).

27. Johanna Oksala, "Foucault's Politicization of Ontology," *Continental Philosophy Review* 43, no. 4 (2010): 445–66.

28. In this sense, the mode of genealogical critique advanced here highlights *aspects* in and through which the subject is caught in his current perspectives, as Owen puts it, "Kritik und Gefangenschaft," 125.

29. In the Foucault chapter, the aesthetic category of the play will be introduced systematically as a freedom-empowering gesture.

30. Axel Honneth, "Genealogie als Kritik," in *Michel Foucault: Zwischenbilanz einer Rezeption: Frankfurter Foucault-Konferenz 2001*, ed. Axel Honneth and Martin Saar, 3rd ed. (Frankfurt a.M.: Suhrkamp, 2017), 117pp.

31. From this perspective it is clear that we do not defend postfundamentalist views such as those developed by Oliver Marchart, *Das unmögliche Objekt: Eine postfundamentalistische Theorie der Gesellschaft* (Berlin: Suhrkamp, 2013).

32. Snowden, *Permanent Record*, 214.

33. Snowden, *Permanent Record*, 214.

34. Snowden, *Permanent Record*, 214.

35. Gavin Lee, "Corona-Virus: What Went Wrong at Germany's Gütersloh Meat Factory?," *BBC News*, June 25, 2020, available at: https://www.bbc.com/news/world-europe-53177628.

36. Alois Berger, "Billiges Fleisch, unhaltbare Zustände," *Deutschlandfunk*, May 5, 2021, available at: https://www.deutschlandfunk.de/schlachthoefe-in-europa-billiges-fleisch-unhaltbare.724.de.html?dram:article_id=481030.

37. James Conant, "The Dialectic of Perspectivism II," *Nordic Journal of Philosophy* 7, no. 1 (2006); Nehamas, *The Art of Living*, 146pp.

38. *Human, All too Human: A Book for Free Spirits* deals with "free spirits," as the title indicates. Later in BGE, Nietzsche takes up the thread of "free spirits." In *On the Genealogy of Morals* (which Nietzsche considers as comment to BGE) Nietzsche delineates different modes of how to become a free spirit. Sommer, *Kommentar zu Nietzsches "Zur Genealogie der Moral."*

39. Nietzsche discusses the case of Christianity in the context of the "slave revolt of morality." On the basic double character of the "slave revolt of morality" that both

creates freedom and closes spaces of freedom, see especially David Owen, *Nietzsche, Politics & Modernity* (London et al.: Sage Publications, 1995), 76.

40. The general relationship between Goethe and the art of living is drawn in Rüdiger Safranski, *Goethe: Das Kunstwerk des Lebens* (München: Carl Hanser, 2013) and consider, too: Alexander Nehamas, "How one becomes what one is," in *The Existentialists: Critical Essays on Kierkegaard, Nietzsche, Heidegger, and Sartre*, ed. Charles B. Guignon (Lanham et al: Rowman & Littlefield Publishers), 73–100.

41. How Nietzsche praises Goethe is sketched in Sarah Bianchi, "'Oh Voltaire! Oh Humanität! Oh Blödsinn!' Über den Zusammenhang von Anerkennung und menschlichem Selbstverständnis bei Nietzsche," *Zeitschrift für praktische Philosophie* 4, no. 1 (2017), 15pp.

42. Johann W. Goethe, "Selige Sehnsucht," in *Westöstlicher Divan. Buch des Sängers, Goethe Sämtliche Werke, Gedichte und Epen. Volume 2, Hamburger Ausgabe* (München: Beck, 2000), 19.

43. Johann W. Goethe, "Urworte: Orphisch," in *Goethe Sämtliche Werke. Gedichte und Epen. Volume 1, Hamburger Ausgabe* (München: Beck, 2000), 412pp.

44. Goethe, "Urworte," 359.

45. Goethe "Urworte," 360.

46. Goethe, "Urworte," 360.

47. Concerning the affect of hope, consider basically: Geuss, "Die Hoffnung," 232–7.

Chapter 4

Knowledge/Power, Affect/ Life Formation

The Affect- and Power-Sensitive Janus Face of Subjectivity in Digital Enhancement after Foucault

Having sketched the critical-theoretical microanalysis of the affect- and power-sensitive Janus face of subjectivity in digital enhancement with reference to Nietzsche, we will now continue this line of argument with Foucault and thus situate him in the tradition in which he also situated himself (VN). But, of course, Foucault did *not* do so in order to become—to put it in Nietzschean terms—Nietzsche's "disciple" (FW, 3, 32, 403, trans. by the author). Nietzsche would also reject such a blind and uncritical follower, to update this to today's language of digital change, and Foucault does reject such a view, too. By contrast, "Nietzschean" implies, according to Foucault, being *at the same time* a partner and an adversary (SP, 219p). It is *only* in this sense that Foucault calls himself "Nietzschean." With Nietzsche, we described the affect- and power-sensitive Janus face of subjectivity in digital enhancement in terms of belief; with Foucault, we broaden the focus and shift to the wider semantics of social facts. Generally speaking, social facts are data that are considered as self-evident by society. They are therefore not questioned. With reference to this point, this book delineates how the logic of perfection *first* turns into a social fact, which is so self-evident to the subject that she has the blind will to conduct her life along these lines, an entanglement that the present book describes as the affect-centered knowledge-power nexus.

Against this backdrop, the overall chapter is divided into two main parts. *Firstly*, we will begin by outlining the affect- and power-sensitive Janus face of subjectivity in digital enhancement following Foucault in terms of the notion of governability, and thus of the key semantics that Foucault himself regards as a "guideline" (DE 3/255, 900, trans. by the author) of his work. *Secondly*, we turn to the semantics of reification, a crucial theme in contemporary discussions in social philosophy. Foucault has not often been linked to this theme, *although* it is of central importance in his work.[1] This focus makes it clear that, like Foucault, we are not interested in delineating the optimal conditions of life formation in digital times; rather, we refer to Foucault in order to understand how the subject positions herself in relation to the logic of digital perfection. Also at this point it should be noted that, of course, the subject in digital enhancement is not necessarily aware of the conceptual register that we provide; but in order to understand the affect- and power-sensitive Janus face of subjectivity in digital enhancement, we require this view.

. . . IN TERMS OF GOVERNABILITY

In what follows we will begin to outline the affect- and power-sensitive Janus face of subjectivity in digital enhancement following Foucault in terms of governability. Against this background, the first section on Foucault involves five steps. The first develops Foucault's power-based approach to the notion of affects. The second describes a new view on digital enhancement as "affect management." The third and fourth steps stress for the first time the affect- and power-sensitive Janus face of subjectivity in digital enhancement in terms of governability. Accordingly, the third step addresses the freedom-endangering side of this Janus face in sketching the motif of *blindly governed subjectivity*. As will be shown, blindly governed subjectivity entails an understanding of subjectivity as, to put it in Foucault's words, the specific "seeing machine" (DP, 207) that we call, following Foucault, the *panoptic machine*. The fourth step goes on to analyze the freedom-empowering side involving the motif of *self*-governed subjectivity. "Self-governed subjectivity" encompasses the theme of enlightenment in the tradition of genealogical critique. In the fifth step the present book will summarize the main points.

IT IS NOT OF WHAT "BOTANISTS OR BIOLOGISTS SPEAK": FOUCAULT'S POWER-SENSITIVE APPROACH TO THE NOTION OF AFFECTS

To begin with, we clarify what we mean when we talk about the notion of affects in Foucault, and thus refer to a conception that has featured prominently in philosophy, social science and social psychology over the past three decades; so far, however, this debate has not paid very much attention to Foucault's work.[2] And even Foucault scholars have not paid very much attention to affects in Foucault.[3] One reason might be the fact that Foucault explicitly does *not* develop a general theory of affects, but a *microanalysis of affects* that is concerned with specific case studies that the present book describes in the subject's concrete life formation of our digital times. By this, he provides the required, but hitherto neglected, perspective that enables us to conceive affects as the name for the specific relations of power that are not so transparent to the subject. In the present book we translate Foucault's perspective on affects to the concrete case study of digital enhancement—which he, of course, did not analyze. By this perspective, it is clear from the outset that affects, according to Foucault, are *not* a natural process, it is not a matter of Freudian "drives"[4] or of what "botanists or biologists speak" (CF, 194) in the ironic formulations of Foucault's interview partner in "Confessions of the Flesh." On the contrary, it is the product that is socially constructed by power relations.

In general, Foucault's approach to affects (or desire; we use the terms interchangeably) goes back to his paradigmatic description of the dispositive. This description is representative of Foucault's basic approach to affect-centered power relations in general. Essentially, a "dispositive" is a net-like relation of power. In the interview "The Confessions of the Flesh" (1977), Foucault describes the elements of the "dispositive" as the "heterogeneous ensemble consisting of discourses, institutions, architectural forms [. . .]" (CF, 194). He goes on to state that the "apparatus itself (the dispositive) is the system of relations that can be established between these elements" (CF, 194). And this "system of relations" now functions in such a way that it leads people to value their life *in quasi-automatic reactions* to the given contexts. It is this quasi-automatic aspect of power relations that we specify as affects, or, in the language of the dispositive, as the *dispositive of desire*.

We now translate this view into our digital today, in order to describe the affect-centered power relations that are present here. We therefore speak of the digital "dispositive of desire." This signals that the subject, in our digital today, is prompted to pre-value the logic of perfection as a self-evident social fact, which is *not* questioned. For it seems necessary to the subject to be governed in his life formation by this logic of perfection, by virtue of its mere existence. Although one might think that affects and cognitive aspects are totally different, the dispositive of desire and its digital variant also have central cognitive aspects. On a cognitive level, affects make the subject *immediately* evaluate his quasi-automatic response to the logic of perfection as rational and thus as good, *without* being aware of this quasi-automatic cognitive evaluation. As a consequence, the subject does *not* realize that society, which is composed of discourses, institutions etc. and thus of the components of the *dispositive*, is blindly imposing on him the self-understandings and justifications that are produced by the logic of perfection.

If it is true that affects are following Foucault the name for netlike relations of power that initially make the subject act, live, and feel in quasi-automatic ways, then they are characterized by the following three features: they are relational, "intentional and nonsubjective" (HS 1, 94) and productive. Firstly, their "strict relational character" (HS 1, 94) signals that the logic of perfection is ubiquitous. In taking the ontological question seriously, which we started to introduce with Nietzsche in terms of social ontology, it is clear that we are not only dealing with relations between persons, but with relations between everything that exists, i.e., subjects, things, institutions, animals, etc.[5] Affect-centered relations of power are thus those forces that immediately establish a ubiquitous net among everything that is; through this net, the different forces (*Kräfte*) react to each other. The dimensions of affects mark the specific aspect in and through which these different forces react to each other *quasi-automatically*. Thus, the relational aspect describes a net whose meshes are quasi-automatically knit by cause and effect. In digital times, cause and effect are produced by the logic of perfection. Hence, in digital times we not only have the virtual World Wide Web, but also the net of affect-centered power relations.

The second feature entails, to put it with Foucault, the aspect of what is "intentional and nonsubjective" (Foucault). By this Foucault does not understand intentions as they are traditionally understood in philosophy, as what subjects *seek* to do. This traditional view would be conceivable along the lines of the standard history of power, of which the Weberian action theoretical model of power is a paradigmatic conception. In general, Weber focuses on the subject's will and is concerned with the subject's intention-based actions. The standard Weberian action-theoretical model conceives of relations of power as "every chance within a social relationship, of enforcing one's own will even against resistance."[6] Hence, his blind spot lies in understanding the power-based conditions of will formation, which this book addresses in terms of affects. This means that Weber cannot provide an understanding of the social-theoretical conditions in and through which the subject's will is *already* blindly formed and thus socially constructed by affect-centered relations of power. In contrast to Weber's view, we learn from this that the subject's will is *not* a fixed unity, but the product of affect-centered power relations. Furthermore, this feature shows that *not only* agents that have a will stand in affect-centered relations of power with other agents that have a will, but already with *anonymous*, "nonsubjective," structures. In the field of digital enhancement, these "nonsubjective" structures are another name for the effects of the logic of perfection on the subject. Thus it would be shortsighted to analyze solely agent-based power relations in our digital times, for example the power of the concrete CEOs working at Facebook, Netflix, Amazon or Google who intend to make the world an alleged perfect place; on the contrary, from Foucault we learn to understand the affect-centered relations of power in richer ways by conceptualizing the anonymous strategies of the logic of perfection that immediately circulate in our digital times as having an "intentional and nonsubjective" (Foucault) character. In this sense, the intentional character of "the rationality of power is characterized by tactics [. . .] becoming connected to one another" (HS 1, 94). The different ways in which they relate to each other form a process that works *as if* it points in a specific direction—in other words, this process anonymously *provokes* directions that conduct the subject's way of being, living and feeling in digital times.

The third feature of affects, namely their productive character, makes the constitutive character of power relations plausible (HS 1, 95). This makes it clear that with Foucault we are *not* contributing to the ordinary understanding of productive in the utilitarian sense of "useful." In contrast to this traditional understanding, "productive" indicates the constitutive aspect of affect-centered power relations that *constitute* new relations. Hence, the relevant understanding of affect-centered power is not a negative one, which could only conceptualize restrictions. The peculiarity of the productive understanding of power, by contrast, lies in the fact that settings of power generate *further* settings of power. Here, affects function as the catalyst that continuously gives rise to new relations of power. It is in this sense that Foucault provides a positive understanding of affect-centered power relations.

The Field of Digital Enhancement as Affect Management: On Governable Life

While Foucault uses the conception of governability to differentiate it from state-based forms of government and to outline *not only* how populations are conducted, *but also* how souls, the mad or prisoners are governed in his time, that is in the 1970s, we carry Foucault's line of thought forward to our digital present and ask how the today's subject's life formation is immediately conducted by the logic of perfection. Foucault himself suggested such systematic interpretations in formulating: "We live in the era of a governmentality discovered in the eighteenth century" (STP, 109). And today we propose to say in analogue ways: We *still* live in the era of a governmentality. Our present era is marked by what we call "affect management," to borrow Foucault's term. Basically, "affect management" is characterized, as this book analyzes, by the twin concepts of *conduct and conducting oneself.* It encompasses a perspective that has received little attention so far: the way subjects *are made open to being conducted by the logic of perfection*, which is so subtle that they also *want to conduct themselves* according to the logic of perfection.

"Affect management" is based on Foucault's notion of "governmental management" (STP, 107). Foucault addressed this term in his lectures on the notion of government held at the end of the 1970s at the Collège de France (STP). "Governmental" traces back to the notion of "governmentality," that is, the "neologism"[7] developed by Roland Barthès and which Foucault also employed. The term "governmentality" derives from the substantivized adjective "governmental" and is *not* composed of the French verb "gouverner" and the noun "mentalité," as one might initially suspect.[8] Foucault uses this term to describe all those processes in and through which subjects *can be* structurally conducted "from the government of children, of souls, of communities, of families, of the sick" (SP, 221). This specific understanding is represented in the motif "governability." Hence, "government" *does not* refer to "political structures or to the management of states" (SP, 221).

In order to understand the mechanisms of governability in his own time, the 1970s, Foucault read texts from the sixteenth to the eighteenth century on the concept of government. With this he sought to show how, for example, the introduction of statistical measurement of the birth rate is a tactic that can govern the population "omnes et singulatim" (Foucault), all and every single one. With this perspective two points become obvious: firstly, the notion of government, according to Foucault, is not to be misunderstood in terms of centralized state activities, but rather needs to be conceived in *de*centralized ways, namely as the strategies that can reach the smallest corners of the subject's everyday life. Secondly, the notion of government does not refer to persons, i.e., to politicians, chancellors or ministers, as one might think. Rather, with Foucault the term needs to be understood as *anonymous* strategies. In the case of digital enhancement, these anonymous strategies resemble the logic of perfection.

The term "management," like the notion of government, is *not* used by Foucault in the usual sense. Whereas the notion of management formerly belonged only to the realm of commercial enterprise, Foucault shows that every social practice becomes a matter of management, including social relations such as family bonds or friendships, and more broadly speaking life formation altogether. In other words: management is conceived as all those social and cultural practices in and through which everyday subjects can be conducted in their life formation. Accordingly, *affect* management is then defined as the specific social and cultural practices in and through which everyday subjects *are rendered open* to being governed (STP, 125pp). In the digital present, the effects of the logic of perfection on the subject are such that he becomes open to being governed by the very same logic of perfection.

It is in this context that the understanding of digital enhancement as "affect management" is situated. Thus far the concept of affect management has *not* been used in the debate over enhancement, although the more general notion of management is widespread in this debate.[9] But as the debate over enhancement thus far has not paid much attention to the notion of affects, this omission is not surprising. In what follows, we now address this gap, without of course claiming to have the last word on the subject.[10] We now develop a central perspective on this term in order to describe how the effects of the today's logic of perfection on the subject needs to be conceptualized in analogous ways, and thus as a strategy that governs today's subjects' life formation both as a whole and for every individual, like a shepherd, to use Foucault's image, is able to guide his flock as a whole *as well as* every single sheep. Obviously, our digital present *is not* characterized by real shepherds and real flocks, but in our digital present the logic of perfection functions in analogous ways. The *topos* of affect management thus helps us to conceptualize that in digital enhancement ordinary subjects are immediately delighted to perform as well as possible in their life, and thus want to be blindly governed by the logic of perfection; furthermore, they have the will, too, to immediately govern themselves along these lines: they just want to be so "fit" that they can manage everything in their life, like the mother in the computer game "Perfect Mom" of the film *Her*; she immediately wants to manage everything at the same moment, for example cooking, looking beautiful and raising children. Theodore, the protagonist in the film *Her*, also shows how life formation simply becomes a matter of affect management. Because he feels lonely, the thought just comes in his head to manage his feelings by purchasing a technological device that promises to solve his social problem.

In order to understand the effects that the logic of perfection has on the subject, we now specify the practices of "affect management" with Foucault as the self-practices that contribute to the subject's self-constitution and his life formation altogether. These self-practices refer to a notion known in the history of philosophy as "subjectivation." It basically means to become a subject in and through a given context. This interpretation is rooted in a combined reading of the Foucault of the notion of governability and the later Foucault of the art of living, a line of thought that has been somewhat neglected by Foucault scholars so far. In thus reading the late Foucault of *The History of Sexuality* together with the earlier Foucault of *The History of Government* we take up a line of thought that Foucault *himself* already suggested. Due to Foucault's sudden death in 1984, he was not able to expand on this line himself; however, the fourth volume of *The History of Sexuality*, entitled *Confessions of the Flesh*, which was only recently published (for greater detail on its history, see the second chapter on Foucault), makes initial hints in this direction explicit.[11]

The term "subjectivation" stems from Foucault's teacher, Althusser, and was then developed further especially by Foucault and has been taken up in current feminist debates by Butler in particular.[12] With Foucault we see that we deal with a *praxeological* understanding of subjectivation: The subject becomes who she is in and through the specific practices in the given contexts, which Foucault specifies as *self*-practices. The later Foucault gives many examples of such self-practices, in the context of health, politics, friendship, family life and so on (HS 1–4). Given the practice-based understanding of self-constitution, it would be wrong, according to Foucault, to conceive the subject as if it were an eternal essence or a substance.

"Affect management" now refers to the specific case of subjectivation that the book at hand calls affective subjectivation (see also the chapter on aesthetic critique). Accordingly, we are dealing with the subtle practices of self-constitution by which we describe how subjects *are prompted to* become who they are within the given social order. In today's context of digital enhancement, this notion of affective subjectivation provides us with the necessary theoretical tools to explain why subjects are *so blindly* attached to the logic of perfection. Affective subjectivation theoretically conceptualizes that this is the case because subjects are structurally constituted in their immediate being by the order of perfection and also structurally constitute themselves along these lines in their immediately practiced life formation. Hence, they do not see their attachment to the logic of perfection. More specifically, the practice-based understanding of "affect management" provides the following understanding of digital enhancement which has been lacking up to now: The logic of perfection, which we will call A, has such subtle effects on the subject B, that B *immediately* becomes such a subject in the given contexts that she is open to being blindly governed in her life formation by the logic of perfection. Furthermore, on a cognitive level, the subject B immediately understands and experiences her blind attachment to the logic of perfection, A, *as working in her best interests*.

ON THE TWO SIDES OF DIGITAL ENHANCEMENT: DESIRING TO BE BLINDLY GOVERNED *AND* DESIRING SELF-GOVERNMENT

In the following, the hitherto neglected affect- and power-sensitive Janus face of subjectivity in digital enhancement involving a freedom-endangering and a freedom-empowering side will be conceptualized drawing on Foucault. The freedom-endangering side of this Janus face will be explained in terms of *blindly governed subjectivity*. Here we will refer to the notion of the panoptic machine developed by Foucault in *Discipline and Punish*; the present book develops the understanding of the panoptic machine as the "seeing machine" (Foucault) that makes the subject wish to be blindly governed by the logic of perfection, having seen her life in a machine-like manner through the lens of perfection. The notion of the "machine" is central to Foucault. He does not use it in the ordinary way to describe real machines. *Instead*, he refers to machines such as the "machine of the signifiers" (WC, 23, trans. by the author), in order to show how the usual, normal and general perspective on life formation allows the subject to *be programmed in such a way* that he *initially* considers this usual, normal and general perspective as *the only valid perspective*. The freedom-empowering side of the Janus face of subjectivity in digital enhancement will be described with reference to the motif of *self-governed subjectivity*. The motif of self-government goes back to the later Foucault (GSO). Self-governed subjectivity, as this book hopes to show, involves the motif of enlightenment in the tradition of genealogical critique. It theoretically conceptualizes the subject's desire *not to be* governed by the logic of perfection.[13]

With Foucault, then, we describe *two sides* of the subject's desire: *both* the subject's desire to be blindly governed by the logic of perfection *and* his desire to be governed *differently*, that is, *not* to be governed by the logic of perfection. Therefore, it is clear that we are *not* starting from an either-or model of social agency, and thus from a subject that is conceived as *either* unfree *or* as free, but from a both/and model of social agency. Accordingly, the subject is conceptualized as *both* unfree *and* free. The attributes "unfree" *and* "free" should not be understood as implying that we want to investigate *the content* of freedom in digital times. We are *not* interested in praising digital platforms or in suggesting that there are fixed, absolute values in the digital field which are always good. Reflecting on such a content-based understanding of freedom leads into an impasse we seek to avoid. As Foucault showed in *What is Enlightenment?*, those questions lead to an "empty dream" (WE, 316).

Furthermore, a content-based understanding would *not* guarantee that the subjects *themselves* are aware of what they are doing in the digital field. Autocratic regimes like China could define certain contents or values as good in digital times, but then the subject's capacity for self-reflexivity would be neglected. Hence, the theoretical conceptualization of digital enhancement requires an appropriate understanding of freedom, which is related to the motif of the subject's self-reflexive ability and, broadly speaking, to the motif of critique. The present book considers this need and develops *the understanding of self-reflexive freedom* in terms of enlightened freedom. This understanding involves *the enlightened procedures in and through which* the subject learns to become aware of the logic of perfection and to take *a self-reflexive approach* to it. Freedom, then, is conceptualized as the subject's broadened view on his life formation in digital enhancement, which is not determined by the logic of perfection; unfreedom is the narrowed view on life formation in digital enhancement, marked by the logic of perfection.

So, *in contrast to* the classic philosophy of the sovereign subject, as propounded by Thomas Hobbes, John Locke or Pierre Bayle,[14] which theoretically conceptualizes the free subject as a subject that has things *fully* in his own hands (*die Dinge in der eigenen Hand haben*), the field of digital enhancement shows that subjects do *not* have things fully in their own hands. *To different degrees*, they depend on the logic of perfection. Thus, it would be simplistic to understand freedom and unfreedom as opposites, as the classic philosophy of the sovereign subject would suggest.[15] Nonetheless, it would also be a misunderstanding to conceive the subject's social agency as always unfree and free in equal measure. Instead, as this book seeks to demonstrate, the desire for self-government enables the subject *to transcend to different degrees* the contexts that are determined by the logic of perfection and thus lead to situations of unfreedom. Hence, enlightened subjectivity is described as the road to freedom. This theoretically conceptualizes the possibility of leaving the situations of unfreedom behind, *although* the subject is in *constant* danger of getting into new situations of unfreedom. Therefore, this book highlights the need to consider the conception of the *enlightenment of enlightenment*. It clarifies that the understanding of enlightened subjectivity does *not* reach an end point, giving rise to alleged ultimate justifications (*Letztbegründungen*), but implies a continuous calling into "question" (GSO, 12) of patterns of unfreedom, to use the words of the late Foucault.

It may sound unfamiliar to speak of the subject's desire for unfreedom and for freedom. *Nonetheless*, this way of speaking belongs to the classic canon of political philosophy. It is already part of ancient philosophy (especially in Plato,[16] if one considers the dialogue of Phaidros) and then recurs in Renaissance political philosophy (in particular in Niccolò Machiavelli). But this archive of the history of philosophy is less frequently visited than its major relevance would deserve. But Foucault was familiar with this archive (STP, 87pp).[17] Machiavelli introduces the notion of desire at the beginning of chapter nine of his *Prince*, one of the classical texts in political philosophy. There, he describes the notion of desire as belonging to "two distinct parties,"[18] as he puts it: it is part of "either the people or the nobles."[19] He goes on to write that "the people do not wish to be ruled nor oppressed by the nobles."[20] On the contrary, Machiavelli argues that it is the nobles who wish to oppress the people.[21] While Machiavelli refers to the notion of desire as belonging *to two different parties*, in Foucault the theme of desire belongs *to the same subject*. From this perspective, the formulation of the notion of desires loses its appearance of strangeness. For Machiavelli shows how this view can be put in quite ordinary language: We deal with subjects that have different ambitions and strivings while they do things. What they do, so to speak, does not so much emerge from a uniform or coherent perspective;

their doing rather consists of *heterogeneous* perspectives. In speaking now of the subject's desire, we reformulate a perspective on such heterogeneous strivings and ambitions that one has while giving one's life a form. From this perspective we conceptualize that the present-day subject in times of digital enhancement does not have one uniform or coherent view of his life formation either, but *multiple* ambitions, strivings and perspectives that are *at the same time* present while he gives his life a form: *at first* the subject desires to be blindly governed by the logic of perfection. *Later*, the subject *can learn* to desire self-government and thus to desire *differently*.

. . . in the Panoptic Machine: On Blindly Governed Subjectivity

The freedom-impairing side of the affect-centered, power-based Janus face of subjectivity in digital enhancement is specified here as *blindly governed subjectivity*. Here the present book follows Foucault's theoretical concept that *not only* real prisons incarcerate people, but also the prison as "social form" (PS, 227) does so. If Foucault describes the social prisons of "factories, schools, barracks, hospitals" (DP, 228), then the contemporary field of digital enhancement stands in the same line. As this chapter shows, this is because the subject sets herself *in such a machine-like relation* to the field of digital enhancement that she immediately sees her life with the lens of the logic of perfection; as a consequence, she *blindly* governs herself according to the logic of perfection. Hence, it is the subject's blind gaze that socially constructs "the social form" of prison. Obviously, the notion of the prison must be understood as an exaggeration, not an accurate representation of today's digital field; nonetheless, it enables us to highlight the main characteristics of the problem.[22]

In general, this interpretation draws on two main resources. The first is Foucault's major work *Discipline and Punish*. This seminal text provides the requisite perspective on the "panopticon," to which we add the previously too rarely considered perspective of affects sketched by Foucault himself. Thus we sketch the mechanisms that immediately program the "event in the 'history of the human mind'" (DP, 216).[23] In order to understand the contemporary field of digital enhancement, this resource must be combined in new ways with a second resource, namely Foucault's lectures on the notion of government that he delivered in the late 1970s at the Collège de France. For only this second perspective enables us to understand what immediately happens in the subject's mind in our present-day digital enhancement. By this the present book shows that contemporary subjects are *not* disciplined by any threat of repressions or sanctions to participate in the field of digital enhancement, as one might think if one *only* considered the Foucault of *Discipline and Punish*. But the semantics of government allows us to conceptualize how contemporary subjects are so thrilled to be part of the digital field that they *immediately* see and value it as *the* rational way of being, living and feeling. It also shows that they fail to consider that this supposedly self-evident social fact of how to live, feel and act might *not* be in their best interest, *although* it promises to be so. Basically, this is what the present book refers to as the "panoptic machine."

The panopticon literally refers to a conception of total visibility as described in the introduction. The word "panopticon" is composed of two ancient Greek words: "pan" (*πάν*) which means "everything" and "optikós" (*ὀπτῐκός*) which in turn means "visible," so that together they form the notion of "seeing everything." In the debate on digital enhancement, commentators often speak of the so-called 360° visibility generated by digital media.[24] It is emphasized that Big Brother is watching the subject all the time and everywhere she goes, while reading books online, doing online shopping or relaxing and listening to music; even the smart TVs in people's homes are in the meantime so smart that they record what people in the room are talking about, as reported recently by the German newspaper *Süddeutsche Zeitung*.[25] So far, however, the perspective of affects is not much present in this debate. This desideratum must be considered as a major gap. It means that the subject's peculiar blindness in digital enhancement cannot be conceptualized. As a result, it cannot be made sufficiently plausible that the subjects in the field of digital enhancement *are blind* to the "panoptic machine" to which they themselves belong, and thus unable to realize that they view their life not self-reflexively, but in a machine-like manner, from the lens of the logic of perfection. The present book seeks to fill this gap. As emphasized in the introduction, "blind" does not refer to the rare cases when subjects say that they were so blind they failed to see things. Instead we follow Foucault's understanding: that "blind" describes the procedures through which the subject's will is formed, subsequently determining what the subject *herself* wants.

The literal meaning of the word "panopticon" is also part of other settings in our digital reality. One of the major big data companies in the Silicon Valley, as we learn from the whistleblower Wylie, includes in its name the word "Palantír," which also means "seeing everything." It is a reference to John R. R. Tolkien's books *The Fellowship of the Ring*,[26] where it is used to describe stones that can see everything. In our digital world, we do not encounter real stones that can see things, but a metaphoric stone, and more specifically the net, namely the World Wide Web, that also has eyes, so to speak. Furthermore, the metaphor of seeing everything is also reminiscent of the semantics of the "God's eye view."[27] This metaphor is often used in the debate on digitalization to describe the effectiveness of social media programs like Facebook:[28] social media can see every action that the user leaves on the respective pages (see the description of how like buttons function in the chapter on digital enhancement)—in other words, they can see the subject's digital "Doppelgänger."[29]

Government: A Peculiar Pair of "Surveillance" and "Discipline"

If we now interpret such a nuanced view on the panopticon in terms of government, we need to specify the perspective on the notion of government. For the notion of government should *not* to be understood in terms of the pair "control" and "self-governance." According to Foucault, the word "control" stems from English moral rules. It applies, for example, to religious or social groups in which "control" is part of their methods of self-governance. For the people who are controlled *choose freely* to be governed by the moral norms and rules of the respective social or religious groups.

By contrast, the semantics of government relevant in the field of digital enhancement should be understood in terms of the pair "surveillance" and "discipline." The word "surveillance" stems from the French field of incarceration. In *Discipline and Punish* Foucault delineates the different surveillance measures that formed the eighteenth-and nineteenth-century incarceration system. In contrast to "control," "disciplinary surveillance" applies when the respective subject is conceived as an outsider, like the inmate in a prison, and *not* as part of a group, as in the case of the social and religious or moral groups to which the word "control" refers. The outsider or inmate does *not* want to be part of the juridical rules and norms that condemn him to prison. But the seeing order of the panopticon—the ringlike order in virtue of which every inmate can be seen by the guard in the central tower without they being able to see the guard in turn—*makes* him disciplining himself in ways that are *not chosen* by him in a free manner, but he was *made functional* to discipline to what the "punitive society" (PS) wants him to be. Correspondingly, this subject does *not choose* to be disciplined as the controlled subject who belongs to a social, religious or moral group does; rather, it is *made docile* by the threat of sanctions or punishment if it does not submit to the punitive order.[30]

Now, the term "government" developed here reconsiders the pair "surveillance" and "discipline" in the peculiar way that mirrors the understanding of the programmatic unity between being blindly governed by the logic of perfection and conducting oneself along the lines of perfection described above. Hence, government in this sense cannot be *mis*understood as conceptualizing the subject's free choice. Nonetheless, it can*not* be translated one-to-one into the pair "surveillance" and "discipline" either. For the crucial point of the programmatic unity between being blindly governed by the logic of perfection and conducting oneself in conformity with that same logic consists in the circumstance that the subject *is explicitly not* disciplined by the threat of sanctions or repression.

Nonetheless, in our digital age this subject must expect a certain punishment. And this punishment consists in the circumstance that she is not so much liked if she does not fulfill the social expectations to perfect her life, one of the most important currencies in the age of followers on Twitter, Instagram or Facebook. These punishment mechanisms work *so subtly* because they are rooted in the subject's affective life. As the understanding of blindly governed subjectivity shows, it is affects that make the subject *open* to *quasi-automatically* and thus blindly react to these social expectations in ways that are *not* transparent to the subject. Thus, the disciplinary mechanisms do not operate in virtue of what the subject has only in her head—filled, for example, with all the well-known facts about spying technology—, but by what quasi-automatically displays in the subject's mind, and thus in virtue of the mechanisms that *immediately* program the subject's mind so that it blurs her gaze in unnoticed ways. This subject just wants to be blindly governed by the view of perfection. In this sense, the subject is open to being surveilled in her life formation by the lens of perfection, which she then simply accepts as her own view, by which she then conducts her life. This understanding is also "cognitive," because *at this stage* the subject evaluates the logic of perfection as a self-evident, rational social fact, which operates in her best interest.

Like a Machine: Blindly Dividing Life into the Binary Scheme of Perfect and Non-Perfect

We now have the theoretical tools required to understand the freedom-endangering side of the affect- and power-sensitive Janus face of subjectivity in digital enhancement: the motif of *blindly governed subjectivity*. This describes the procedures of the panoptic machine, which makes the subject willing to be conducted—with *eyes wide open*—by the logic of perfection, *and* also open to conducting himself according to the logic of perfection. Here, we explicitly speak *not* of "governing oneself," but of *conducting* oneself. For *governing oneself* alludes to the notion of self-government. But here the present book is describing, with reference to Foucault, the freedom-empowering side of subjectivity in digital enhancement.

In this sense, contemporary subjects' will to be blindly governed by the logic of perfection and to conduct themselves in this way resembles a machine. Like a machine, they are so programmed by the logic of perfection that they want to divide their lives into binary schemes—not, of course, the binary schemes of 0s and 1s, as a real machine does, but into the binary scheme provided by the panoptic machine that divides life into perfect and nonperfect. And subjects, like Lacie from the Netflix series "Nosedive," show that such subjects, of course, desire to fit into the scheme of the alleged perfect life. So what Lacie will see if she "opens [her] eyes"—to again translate Foucault's formulation in his lecture *Penal Theories and Institutions* into our contemporary digital reality—is what the eyes of the given view of digital enhancement wants her to see; and, as should be clear from the foregoing passages, this is the logic of perfection immediately valued by the subject as the rational and reasonable way to live, feel and act. In this sense, the subject, to put it with Foucault, is "vis-à-vis government [. . .] both aware of what it wants and unaware of what is being done to it" (STP, 105). It is the "panoptic machine" that blurs the subject's gaze, making him see the logic of perfection as a rational social fact that simply guides people's lives. In this sense, the subject is "aware of what [he] wants." But he is "unaware of what is being done to [him]," because he does not yet have the eyes to perceive the freedom-endangering "price" of the panoptic machine.

The twin concepts of "being blindly governed" and "conducting oneself blindly" describe the subject's situation in the digital era as, in Foucault's words, the prison as "social form" (Foucault). Understood in this sense, the contemporary prison is *not* a substance or essence, it is *not* an "anthropological constant" (PS, 252), but is socially constructed by affect-centered power relations. Its social construction is the result of the subject's *immediate* attachment to the logic of perfection, and thus of the circumstance that we describe in terms of the knowledge-power nexus. The knowledge-power nexus involves an understanding of how power and knowledge stabilize and produce each other. Basically, the knowledge-power nexus is central to Foucault's entire work. Foucault first took an interest in this nexus in his inaugural lecture in December 1970, *The Order of Discourse* (OD). He later showed how the systems of knowledge and power structurally stabilize each other in the fields of madness, criminality, discipline or pleasure. The field of digital enhancement shows by itself as being part of this tradition. Here we show, with Foucault, how the systems of knowledge and power stabilize each other in the domain of the supposedly perfect life formation.

In general, the knowledge-power nexus describes the relations of power in and through which the subject considers and experiences the episteme, the dominant knowledge of a specific time, as a self-evident social fact, which therefore has an unquestionable value. "Power" makes plausible the subject's attachment to the episteme. This book analyzes a specific knowledge-power nexus in the age of digital enhancement. "Episteme," in our time, encompasses the dominant knowledge marked by the logic of perfection. "Power" is the description of the subject's attachment to this episteme. "Affects" now specify the power relations in terms of blindness. Thus, they describe how the subject attaches himself blindly to the episteme of the logic of perfection. This understanding of affects stems from the basic understanding delineated above, which goes back to Foucault's conception of the dispositive. It is, to quote Foucault, "the said and the unsaid" (CF, 194) that *prompt* the subject to be formed in his will and to then conduct himself according to this will formation. In the age of digital enhancement, we deal with the specific "said and the unsaid" in the context of perfection. As a consequence, the subject *is open* to having the blind will to attach himself to the logic of perfection. This entanglement is "cognitive" in the sense that the subject has the blind will to qualify the episteme of the logic of perfection as a self-evident social fact, which values a supposedly perfect life.

About the "Social Prison" of the Assujettissement

In order to better understand how the subject's immediate attachment to the logic of perfection creates such a firm attachment that can be compared with Foucault to the prison as "social form" (Foucault), we refer to Foucault's understanding of "*assujettissement*." It offers the description for rendering plausible that subjects in digital enhancement are trapped in such closed views that we describe them as being incarcerated.

Foucault introduces the understanding of "*assujettissement*" in *Discipline and Punish*; this understanding has a paradigmatic status, which is relevant not only to *Discipline and Punish* but to all Foucault's works. Butler in particular draws on this interpretation. The French term "*assujettissement*" implies the entanglement of subjugation and becoming a subject: it is composed of the noun "*sujet*," which means "subject," and the verb "*assujetir*" which corresponds to the English verb "to subjugate." In this sense, *assujettissement* describes the power relations in and through which the subject becomes who she is *while* being subjugated to the given contexts. In this sense, Foucault gives the following quasi-definition of what he understands by *assujettissement*.

Due to the difficulty of precisely translating the term "assujetissement" into English, let us consider the French definition: "*Un assujettissement réel naît mécaniquement d'une relation fictive*" (*Surveiller et Punir*, loc. 4361 of 6954; DP, 202). The important word here is "mechanically." It helps to explain the fact that subjects relate to the logic of perfection *in a programmed manner*. This means that they are *not* able to reflect on what they are doing, but simply *re*act to what the logic of perfection wants them to do, and this means having a desire to enhance their lives. This is why Foucault can then continue to describe the term *assujettissement* as follows: "*De sorte qu'il* *n'est pas nécessaire d'avoir recours à des moyens de force pour contraindre le condamné à la bonne conduit [. . .]*"(*Surveiller et Punir*, loc. 4361 of 6954; DP, 202).

In order to explain why no sanctions or means of repression are required to ensure alleged "good" behavior, the present book considers with reference to Foucault, the notion of affects. In this sense, we speak of "affective *assujettissement*," a widely underestimated conception up to now. "Affective *assujettissement*" refers to the specific processes in and through which the subject *immediately* has the *feeling* of becoming who she is by *fully* attaching to the given contexts, such as, in today's digital field, the subject's full attachment to the logic of perfection. In other words, "*assujettissement*" describes the specific processes of *felt* domination. It theoretically conceptualizes that the subject *feels* the alleged necessity to immediately subjugate herself to the given contexts of perfection and, in doing so, she closes off the possibility of creating different views on her life. In this sense, the term "affective *assujettissement*" contributes a central perspective to the understanding of affective domination that we introduced in the chapter on aesthetic critique: This subject *immediately* attaches herself *so closely* to the episteme of perfection that she does *not see* any view that differs from the given view of the order of perfection. This subject *fully* identifies with the usual view of life in digital times in ways that are immediately determined by the logic of perfection. As a consequence, other views on life are *totally* invisible. This closed view operates similarly to what Carl Jung had in mind when he said more or less "If you don't know who you are, the world will tell you who you are." This is also valid in our digital era. Now we can say subjects only have ears to hear the sound of perfection generated by the noise of digital enhancement (more specifically on acoustical metaphors, see the second chapter on Foucault).

The case of the panoptic machine even confronts us with a more intensified case than the one described by Jung, which deals with subjects who at least raise a question. These subjects are eager to know who they are. In the field of digital enhancement, however, we are confronted with the radical case that contemporary subjects do *not* ask the world who they should become. The digital world *just* tells them and they *simply* listen without being really aware of what they are doing. In other words, these subjects are so subjugated to what social media wants them to be that they are *simply open* to what social media wants "to make people see[ing]" (The Great Hack). This is especially the case in the Cambridge Analytica Scandal. To recap, the Cambridge Analytica scandal showed us that the company Cambridge Analytica used social media to make people see their lives in the ways that the company was paid to promote—and in the 2016 US presidential campaign, that meant in particular in the way paid for by the Trump campaign and in the Brexit referendum in the way the Leave campaign paid for (on this point, see also the first part).

Affective Internalization: On the Guard of the Watchtower in the Subject

In order to specify how the procedures of affective *assujettissement* function, we consider them through forms of affective internalization; this term enables us to gain a better understanding of the subject that blindly creates the "social prison." Basically, "internalization" describes the specific processes in and through which subjects perceive external norms, in our case the social expectation to perfect life, as their own norms, without being aware of the difference between the external norms and their attachment to them (cf. the chapter on Nietzsche). "Affective" indicates that subjects undertake this process *immediately and thus blindly*, that is, without being pushed by external constraints. With Foucault, we can also say that these subjects turn the will of society, i.e., in our case the social expectations to perfect life, upon themselves *without* noticing what they are doing. As a consequence, the subject's view and the will of society are seen by the subject as identical, *although* of course they are *not*. But the subject does not have the eyes to see the possible difference. In this way, the subject blindly naturalizes, essentializes and substantializes her perspective on life, by taking her life formation as an alleged social fact, which of course it actually is not.

If, in addition, we think in terms of the architectural plan of the panopticon, we can perceive how the modes of affective internalization work in the specific panoptic machine that we follow Foucault in calling the "seeing machine" (Foucault). As mentioned in the introduction, the panopticon is in the first place characterized as the real prison whose architectural plan was invented by Bentham. At the center of the architectural plan is a watchtower and around this tower the cells in which the inmates are incarcerated are arranged in a ring-like order. The key element of this architectural structure is that the inmate "is seen, but he does not see" (DP, 200). This seeing order has the effect that the subject always behaves *as if* it is being observed—in other words, the threat of sanctions or norms disciplines the subjects to fulfill the norms generated by the watchtower. In this sense, as Harcourt outlines, the panopticon functions as the "reverse of the spectacle."[31] Whereas in the ancient Greek amphitheaters, the many watched the few, in the case of the panopticon the few, that is the guards, watch the many, i.e., the inmates. In contrast to the Greek amphitheater when the actors on the stage knew that they were seen, the prisoners in the panopticon do not know whether they are being observed or not.

In translating this architectural plan into the semantics of affective internalization, it is clear that today we are not dealing with guards who compel people to perfect their lives, it is rather the subjects *themselves* who have their guard in their own minds, through which they watch and survey their entire life formation. This guard can be seen as the internalized logic of perfection, which the subject simply takes as her own view without being able to perceive that it is actually the dominant episteme in society. As a consequence, the subject immediately internalizes the social expectations of perfection in such a way that she quasi-automatically surveys her own life more or less voluntarily with the eye of perfection that she, however, mistakenly conceived and experienced as her own eyes rather than the eyes that society actually subtly imposes upon her. So she feels quite self-evident in doing what she just does, but in so doing she fails to see that she blindly essentializes, substantializes and naturalizes her view on life.

Transcending the Panoptic Machine: On Self-Governed Subjectivity

The following section delineates the motif of *self-governed subjectivity*, in order to theoretically conceptualize how the subject *can transcend the panoptic machine* and thus become self-reflexively aware of the ways in which she was previously blindly governed by the logic of perfection, and also blindly conducted herself along the lines of the logic of perfection.[32] "Self-governed subjectivity" as developed here encompasses procedures of enlightenment in the tradition of genealogical critique, in and through which the subject "calls into question" (GSO, 13) her "today" (GSO, 11), to use the relevant Foucauldian terms. A subject with this subjectivity can learn to *un*govern her blindly governed life formation. This procedure contributes to what this book calls the enlightenment test. So, there is no pill that can help the subject to shed light on her blind attachment to the logic of perfection, but instead it is the subject who is called to test her attachment to the logic of perfection and to prove the power relations at stake.[33]

The interpretation takes its basic orientation from Foucault's late texts on enlightenment, in particular from "What is Critique?" and "What is Enlightenment?." While this later Foucault is concerned with ethics, or with ethos in the sense of attitude, the understanding of enlightenment developed here contributes to this view by speaking of the *enlightened ethos*. However, we need to consider Foucault's understanding of enlightenment in the light of his entire oeuvre in which the notions of knowledge, power and ethics, or as just mentioned ethos, play a crucial role from the beginning. By this it is clear, too, that the understanding of enlightenment in the tradition of genealogical critique developed here does not understand the notion of enlightenment as the historic epoch of the Enlightenment, which, depending on one's periodization, either extends from the seventeenth to the eighteenth century or began already in the sixteenth century.[34] For, as Foucault emphasizes, it would be impoverished to conceive of the notion of enlightenment as only "situated on a calendar" (WE, 309). Rather, this needs to be understood as a self-reflexive procedure.

If Kant paved the way for the crucial, self-reflexive understanding of enlightenment that consists of the subject's activation, allowing him to detect previously hidden dogmatic patterns or relations of domination, e.g., by social, political or religious "guardians," then Foucault also positions himself in this tradition. In particular, the first lectures of his lecture series *Government of Self and Others* give rise to this view (GSO, 1pp). Foucault goes further down this road and develops the theoretical methodology of genealogical critique that was initially devised by Nietzsche. As we already saw in the chapter on Nietzsche, genealogical critique is the critical method that theoretically conceptualizes how the subject can be activated, enabling her to cast light on the previously concealed historicity of her life formation. This allows her to understand and experience the possibility of different ways of being, feeling and living. This book now demonstrates a new reading of genealogical critique, which it situates in the tradition of *critical constructivism* that started with Kant. Genealogical critique, understood in this sense, theoretically conceptualizes that *criticizing* patterns of domination can *construct* an enlightened agency.[35]

In general, enlightened agency involves an understanding of self-reflexive subjectivity, in and through which subjects can learn to relate themselves to themselves in the context of a specific time that is *not* blindly determined by the history that shapes these contexts. In an age of digital enhancement, this encompasses the understanding that the subject's self-relation in digital times is *not* determined by the historically formed contexts of the logic of perfection. On the contrary, the subject can transcend these contexts. This view is *not relativistic*, as genealogical critique is often accused of being, but involves an *implicit normativity*. As shown in the introduction and in the chapter on Nietzsche, the implicit normativity shows that *not* every norm is liquefied; *only* problematic norms are made fluid. Norms are problematic if they have so far hindered the creation of different, dissident views, and excluded other subjects from this possibility of generating dissident views. In this sense, genealogical critique theoretically conceptualizes subtle, previously concealed situations of domination *that justify what is not justifiable*. What is not justifiable is the voices that are muted by the affect-centered, power-based conditions of digital enhancement (for example, the panoptic machine), which only make space for voices that are in line with the dominant voice of the logic of perfection.

And how can the subject uncover these forms of domination, in particular of affective domination? Here, the method of genealogical critique shows how the subject can *un*learn how to *blindly* establish the link between being blindly governed *and* conducting oneself blindly. For it is this blindly undertaken link that makes the subject want to live by the logic of perfection. The key Foucaultian term for this is "desubjectivation" (WC, 15, trans. by the author). In general, "desubjectivation" describes the self-reflexive procedures by which the subject *un*does the process of becoming who she is in the given contexts. This conception is "affective" insofar as it includes the procedures in and through which the subject unties her blind attachment to the logic of perfection. It is also cognitive, as the subject unknots the evaluations that come with the blind attachment. Hence, she no longer merely evaluates the logic of perfection as the optimal way of life, but gains an *enlightened and thus a clear view* on her previous qualifications. This transparency opens up new possibilities for previously hidden qualifications. Translating this view in the context of digital enhancement, we speak of *enlightened "desubjectivation."* This involves the subject's realization that her life formation does not have to be blindly governed by the binary, machine-like schemes that divide life into perfect and non-perfect. This enlightened view opens up new and different perspectives on the subject's life formation, so that the subject's view on life *is pluralized*. In the same way as described above, enlightened "desubjectivation" also has affective and cognitive aspects. According to this view, the subject learns to understand and perceive that she does not only become who she is while desiring to be governed by the logic of perfection; she is also who she is while desiring to resist her very own attachment to the logic of perfection—that is, while desiring self-government. So, the subject does not immediately listen to what the "world tells them to be," to return to Jung's formulation; this subject can also resist this seductive voice and create new views on her life that are not determined by the given accounts of how one should live, feel and act in times of digital enhancement. However, the subject with an enlightened subjectivity does *not always* have a clear view of his own attachment to the logic of perfection, that is, for example, of the panoptic machine; rather, the subject is continuously confronted with new patterns of affective domination. It is therefore necessary to consider a *sustainable* understanding of enlightenment, based on the motif of *enlightenment of enlightenment*. For it is only by conceiving the self-reflexive procedure of enlightenment as an ongoing, never-ending process, which lacks any ultimate justification (*Letztbegründung*), that the subject can keep shedding light on new patterns of affective domination.

So, it is clear that the enlightened remedy for the subject's situation in digital times, that of being in the panoptic machine, *cannot* be expected to come from the supposed "outsider," as suggested by Herbert Marcuse, but from the subject who *is in the* panoptic machine that immediately blurs her gaze. In Foucault's words, this means: "Where there is power, there is resistance" (HS 1, 94).[36] If this view is to be more than a mere positive fact (a criticism often made of Foucault), then it must be able to show that the understanding of genealogical critique, intertwined with the conception of enlightenment developed here, is a hitherto underappreciated theoretical resource for making plausible how the enlightened subject can learn to self-reflexively criticize her blindly governed subjectivity.

Genealogical critique can do this insofar as it theoretically conceptualizes how enlightened subjects can learn to become aware that the "history of ourselves" is based on *historically* formed conditions, whose evaluation needs to be considered as depending on its respective times. In understanding this point, the subject can realize that there could be different "histor[ies] of ourselves," and can understand and experience the *contingencies of life formation*, that is, the non-necessity of becoming who one is in the historically shaped contexts. Contingency is therefore *not* to be confused with coincidence. Accordingly, the subject can learn to see that there is *not only one* "mode of relating to contemporary reality" (WE, 309), as the panoptic machine formerly prompted the subject to think and perceive, but *plural* modes of doing so. As a consequence, the subject can become aware of his previously restricted gaze and can now broaden his view to see the different possibilities of becoming who one is in the historically given contexts. In the age of digital enhancement, this implies the understanding that the ways in which the subject has thought, lived, and felt so far do not necessarily have to be marked by the logic of perfection.

Enlightened agency is "voluntaristic" because the subject is *not only dependent* on the given contexts marked by the logic of perfection, but *can learn to become in*dependent of the logic of perfection. This point does *not* deny the socio-theoretical account of the affect-centered power relations at stake in the context of digital enhancement; nonetheless, it shows that the subject is *not* merely an accidental by-product and thus a mere effect of the logic of perfection. At the same time, genealogical critique makes it clear that the subject cannot fully rationally, intentionally or consciously decide to gain a new view on her life formation, independent of the logic of perfection. As the case of Snowden showed, it is a subtle, not entirely rational, intentional or conscious process, which starts by the by, in and through the ongoing contradictions in life—which subjects stumble over, again and again. In Snowden's case, the turning point was his talk in Tokyo on Chinese spying technology. It was only after this that he slowly started to realize that he, in working for the NSA, was also contributing to spying technologies. After this, he was no longer able to consider and experience ordinary digital technologies as an allegedly necessary means to perfect people's lives. Up to then, he had been aware of the problems of data privacy that come with these technologies, for example the smart TV that is able to record what people in the room are saying.

The interpretation of the freedom-empowering side of affects developed here goes back to Foucault's understanding of the dispositive of desire. Whereas in the previous passages we outlined the dispositive of desire in its freedom-impairing aspects, we now take up the thread again and analyze its freedom-empowering aspects with the help of the notion of enlightenment in the tradition of genealogical critique. To recall, by a dispositive Foucault understands "the system of relations that can be established between these elements" (Foucault). The specific dispositive of desire in its freedom-empowering side shows that the subject is *not only* prompted to immediately attach to the episteme of the logic of perfection; the dispositive can also empower the subject to *de*tach from its previous attachment to the episteme of the logic of perfection. The interview "The Confessions of the Flesh" in which Foucault introduces the term "dispositive" enables us to understand how this "double process" (CF, 195) functions. There, Foucault writes: "[T]here is a process of functional overdetermination" (CF, 195). This refers to the freedom-impairing side. In that interview, Foucault describes the freedom-empowering side as the potential "interplay of shifts of position and modifications of function which can also vary very widely" (CF, 195). This formulation shows that the dispositive once established does *not* function forever in the same way; "unforeseen" (CF, 195) shifts and changes are *also* possible: "Thus, a particular discourse can figure at one time as the program of an institution, and at another it can function as a means of justifying" (CF, 194p). The heterogeneity involves the potential for possible changes. For the leitmotif of the logic of perfection, and thus the logic that at first *immediately* structures the subject's digital present, is so heterogeneous at the different points in the subject's life that it creates unforeseeable effects, too, as Snowden's case shows. These effects can give rise to new views on life that have a life on their own (*Eigenlogik*) and thus cannot be simply subsumed under the logic of perfection. For, to remain in Foucault's picture, the mesh of the net of the dispositive might be indeed tightly knotted, but it is not necessarily knotted for all time in the same way. The mesh is defined by the thread of history, and thus by social practices that could also be otherwise. And the freedom-empowering aspects start at the point where the subject realizes that she is confronted to a life formation that is shaped by affect-centered, power-based social practices and hence is not a self-evident social fact. As a result of this realization, the subject can desire self-government, which means not being governed by the logic of perfection, but governed *differently*; in this way, the subject can start to desubstantialize, deessentialize and denaturalize her own view on how to live, feel and act in digital times.

In other words, the notion of enlightenment in the tradition of genealogical critique enables us to conceptualize that the affect- and power-sensitive Janus face of subjectivity in digital enhancement entails *both* settings of the subject's desire for *un*freedom *and* settings of the subject's desire for freedom. In general, considering unfreedom and freedom together is usually conceived on a theoretical level in terms of the social-ontological register, which has been widespread in the social-philosophical debates for the last thirty years.[37] This enables us to understand that in the field of digital enhancement unfreedom, or power, and freedom are *not* opposed, as the classic Weberian conception of power might suggest.[38] By contrast, it enables us to theoretically conceptualize that the dichotomous understanding of power and freedom is reductive because the latter reduces settings of power to settings of domination. But this makes the affect- and power-sensitive Janus face of subjectivity in digital enhancement invisible. It is the social-ontological perspective introduced in the chapter on Nietzsche that reveals *both* the freedom-impairing *and* the freedom-empowering side of the affect- and power-sensitive Janus face. Nonetheless, Weber does indeed have a point, so to speak. Obviously, the systems of knowledge and power, and thus the order that produces patterns of domination, *still* exist. But the perspective of social ontology enables us to understand *the potentiality* of the subject's attachment to these systems of knowledge and power. Hence, this perspective enables us to understand how until now the subject *only felt* dominated by the very same systems of knowledge and power. The social-ontological register, however, enables us to make plausible that *in reality* they do *not* necessarily have dominating effects on the subject. We hoped to show this point by referring to the formation of the subject's desire for an enlightened attitude. From this perspective, we thus analyzed how the subject can self-reflexively generate a new perspective on her life that enables her to see that how she relates herself to the episteme of her time—that is, the logic of perfection in our case—does not eternally have to be the way it is: Her attachment is not a "mysterious substance" (SP, 216), but changeable.[39]

To translate the scenario into the language of the knowledge/power nexus: the notion of enlightenment in the tradition of genealogical critique makes the specific social-ontological register in terms of affects transparent which we require to conceptualize how the subject can see that her quasi-automatic attachment to the logic of perfection, which she previously regarded as necessary, is indeed *not* necessary, and could also be otherwise. Thus we see that the subject recognizes the change*ability* of her hitherto immediately formed attachment to the order of perfection, an attachment that we conceive with Foucault as the knowledge/power nexus. This subject is able to perceive her attitude towards her digital present as "a field of possibles, of openings, of indecisions, reversals and possible dislocations" (WC, 66). If the freedom-impairing side showed that the subject is prompted to conceive and experience her life formation in the not necessarily closed form that looks like a "social prison," then the theoretical register of social ontology shows that the subject can also perceive her digital reality in its "multiplicity of force relations" (HS 1, 93). So this subject with an enlightened subjectivity can see that there can be a "more or less open field of possibilities" (SP, 221) and *not only* the reflex-like reactions by which she previously narrowed her perspective on the forms of life. In this sense, the subject no longer understands and experiences the field of digital enhancement as a self-evident social fact, as a substance, essence or nature that one simply does not question. On the contrary, the subject learns to broaden her gaze and see the plurality of views on her digital present.

FOURTH CONCLUSION

In this first chapter on Foucault, we provided the hitherto neglected theoretical tools for understanding the affect- and power-sensitive Janus face of subjectivity in digital enhancement in terms of government. Following Foucault, we showed that the affect- and power-sensitive Janus face of subjectivity in digital enhancement entails two sides, consisting of the subject's desire for being blindly governed *and* her desire for self-government. The freedom-impairing side has therefore been analyzed as blindly governed subjectivity, which makes the subject willing to be mechanically conducted by the logic of perfection, and conduct herself according to this logic, an attachment that we called, using Foucault's terms, the freedom-impairing aspect of the knowledge/power nexus. According to this analysis, the subject blindly perceives the episteme, or the dominant knowledge, of perfection as the rational way of living, feeling and thinking in digital times. This side of the Janus face is "cognitive" because the subject mechanically qualifies his blind attachment to the logic of perfection as a self-evident social fact, which shows him how to rationally shape his life. The freedom-empowering side in turn was sketched with Foucault as self-governed subjectivity, that is, the freedom-empowering aspect of the knowledge-power nexus. The present book delineated this view by combining a reading of the critical method of genealogical critique and the motif of enlightenment. In this way, we described that the subject is not only subjugated to her view in and through which she perceives, understands and experiences the order of perfection rationally as working in her best interest, but can also become aware of her blind attachment, and thus she can govern herself in an enlightened, self-reflexive manner. Through this approach, we hoped to show that the subject can question her desire to be governed by the logic of perfection, and thus *un*learn to be blindly governed by the logic of perfection.

NOTES

1. On this topic, see Bianchi, "Mikropolitiken der Aufklärung."

2. Mainly, the debate on affects refers to Spinoza or to Deleuze. Cf. Melissa Gregg and Gregory J. Seigworth, *The Affect Theory Reader* (Durham: Duke University Press, 2010); Massumi, *Politics of Affect*; Saar, *Immanenz der Macht*.

3. In this direction: Mühlhoff, *Immersive Macht*.

4. Sigmund Freud, *Triebe und Triebschicksale: Psychologie des Unbewussten* (Frankfurt a.M.: Fischer, 2000 [1915]).

5. Viewed through this lens, it is clear that we are not dealing with an anthropocentric understanding of power that sees relations of power as existing exclusively among human beings. Although human beings do not have a pre-eminent position in the world, they are the only beings who are able to reflect on their conditions of life, at least as far as we know.

6. Max Weber, *Economy and Society*, trans. Keith Tribe (Cambridge, Mass./London: Harvard University Press, 2019), 134.

7. Roland Barthes, *Mythen des Alltags* (Frankfurt a.M.: Suhrkamp, 1964 [1957]), 114.

8. Cf. Thomas Lemke, Susanne Krasmann and Ulrich Bröckling, "Gouvernementalität, Neoliberalismus und Selbsttechnolgien," in *Gouvernementalität der Gegenwart: Studien zur Ökonomisierung des Sozialen*, ed. Thomas Lemke, Susanne Krasmann and Ulrich Bröckling, 8th ed. (Frankfurt a.M.: Suhrkamp, 2019), 8.

9. Schmid, *Mit sich selbst befreundet sein*, 152pp. The concept also points to the contemporary debates over neoliberalism Wagner, *Selbstoptimierung*, 92pp; Ulrich Bröckling, *Das unternehmerische Selbst: Soziologie einer Subjektivierungsform*, 7th ed. (Frankfurt a.M.: Suhrkamp, 2007).

10. In the debate on affects the conception of affect management is common as well. See the section "Managing Affects" in Gregg and Seigworth, eds., *The Affect Theory Reader*, 229pp. *Nonetheless*, the power-based approach developed here that entails the cognitive level, too, is *not* described in the reader.

11. The link needs to be conceived in Foucault's understanding of "pastoral power" (STP, 115) and thus in the phenomena of power with which Foucault also describes the Christian genealogy of the regimes of government in the sixteenth, seventeenth, and eighteenth centuries. Foucault understands the archaic phenomena of "pastoral power" as the specific relations of power in and through which subjects, like the shepherd, conduct other subjects that resemble the shepherd's flock; and this metaphorical flock guides itself in correspondence to the metaphorical shepherd's guidance. This formulation takes up the aspect of the programmatic unity of the regimes of government, which entails both being governed, in our case by the logic of perfection, and also conducting oneself in conformity with the logic of perfection.

12. Butler, *Psychic Life of Power*.

13. The name of this desire stems from Foucault's decisive talk "What is Critique?" (WC, 45).

14. Pierre Bayle, *Toleranz: Ein philosophischer Kommentar*, ed. Eva Buddeberg and Rainer Forst, trans. Eva Buddeberg (Berlin: Suhrkamp, 2016); Thomas Hobbes, *Leviathan oder Stoff, Form und Gewalt eines kirchlichen und bürgerlichen Staates*, trans. Walter Euchner (Frankfurt a.M.: Suhrkamp, 1984 [1651]); John Locke, *Zweiter Kommentar über die Regierung*, trans. Hans Jörn Hoffmann (Frankfurt a.M.: Suhrkamp, 2008).

15. It would be a misunderstanding to conceive John Rawl's differentiation between "concept" and "conception" as an either-or model. In general, "concept" is the basic meaning of a term. It has normative implications. On the contrary, "conception" refers to different interpretations of the different aspects of the "concept." It does not include normative implications. In the case of the concept of solidarity and its overlap with the concept of justice, we can see the entanglement between concept *and* conception: Is solidarity a conception of justice or is justice a conception of solidarity? Rainer Forst, *Toleranz im Konflikt: Geschichte, Gehalt und Gegenwart eines umstrittenen Begriffs* (Frankfurt a.M.: Suhrkamp, 2003), 30pp.

16. Plato, *Phaidros*, trans. Thomas Paulsen and Rudolf Rehn (Hamburg: Meiner, 2019 [around 370 BCE]).

17. Sean Erwin, "Political Technique, the Conflict of Umori, and Foucault's Reading of Machiavelli in 'Sécurité, Territoire, Population,'" *Foucault Studies* 9 (2015): 172–90.

18. Niccolò Machiavelli, *The Prince*, trans. William K. Marriott (Berlin: epubli, 2018), loc. 675, Kindle.

19. Machiavelli, *The Prince*, loc. 675, Kindle.

20. Machiavelli, *The Prince*, loc. 675, Kindle.

21. For a productive reading of Machiavelli in democratic terms see John McCormick, *Machiavellian Democracy* (Cambridge: Cambridge University Press, 2011).

22. Generally speaking, the social understanding of the prison is a prominent *topos* in contemporary debate: Butler, *Psychic Life of Power*; Warren Montag, "The Soul is the Prison of the Body: Althusser and Foucault 1970–1975," *Yale French Studies*, no. 88 (1995): 53–77.

23. Hence, we are *not* interested in understanding the "microphysics of power" (DP, 38, German edition, trans. by the author). Gehring for example stresses the physical aspects. Gehring, "Eine Topo-Technologie der Gefährlichkeit," 299. Obviously, as Foucault shows, they exist, too. But the interesting point in the field of digital enhancement starts with the notion of affects and thus it starts as we interpret it with what happens in the subject's preprogrammed mind, to again refer to Foucault's formulations.

24. Ulrich Bröckling, "Das demokratisierte Panopticon: Subjektivierung und Kontrolle im 360 Grad Feedback," in *Michel Foucault: Zwischenbilanz einer Rezeption. Frankfurter Foucault-Konferenz 2001*, ed. Axel Honneth and Martin Saar (Frankfurt a.M.: Suhrkamp, 2003), 77–93.

25. See Hakan Tanriverdi, "Aufregung um Spracherkennung. Samsung hört mit—aber nur manchmal," *Süddeutsche Zeitung*, February 9, 2015, https://www

.sueddeutsche.de/digital/aufregung-um-spracherkennung-samsung-hoert-mit-aber-nur-manchmal-1.2341288.

26. Wylie, Mindf*ck, 83p.

27. Wylie, Mindf*ck, 83.

28. This view is disputed, see: Petra Gehring, "The Inverted Eye: Panopticon and Panopticism. Revisited," in *Foucault Studies* no. 23 (2017): 42–62; Reckwitz, "Die Transformation der Sichtbarkeitsordnung," 197pp.

29. Harcourt, *Exposed*, 141.

30. Harcourt, *Exposed*, 86pp.

31. Harcourt, *Exposed*, 83.

32. Although the conceptions of "immanence" and "transcendence" are usually conceived as being opposites, our digital reality shows that they need to be understood together. This means that the subject is, *at first*, an immanent part of the logic of perfection, but can subsequently learn to transcend the logic of perfection in certain contexts of her life formation.

33. In contrast to Oksala's interpretation, we can see with the methodology of genealogical critique that the subject is able to criticize the current conditions of his episteme. Cf. Oksala, *Foucault on Freedom*, 76.

34. Israel, *Radikalaufklärung*.

35. There is dispute as to whether genealogical critique is a method that *theoretically* conceptualizes changes in subjects, or one that also leads to changes *itself*, see Axel Honneth, "Einleitung: Genealogie als Kritik," in *Michel Foucault: Zwischenbilanz einer Rezeption: Frankfurter Foucault-Konferenz 2001*, ed. Axel Honneth and Martin Saar (Frankfurt a.M.: Suhrkamp, 2003), 119p. This book *only* refers to the first understanding, and thus conceives genealogical critique as a *theoretical tool* that conceptualizes the subject's learning process, in and through which she can become aware of her previous, blind attachment to the logic of perfection in digital times.

36. On the wide-ranging debate around these passages Schubert, *Freiheit als Kritik*, 85pp.

37. In the next chapter on Foucault, we will sketch in greater detail Foucault's specific approach to the notion of social ontology and thus to the debate that is widespread over the last thirty years.

38. To recap, the Weberian understanding considers power relations in terms of domination. Weber, *Wirtschaft und Gesellschaft*, § 16.

39. The social-ontological register thus refers to the *non*-standard reading of power that begins in Western ancient philosophy with Aristotle and is taken up in modern philosophy by Spinoza. By the term "dynamis," plural phenomena of power that are *not* opposed to phenomena of freedom can be conceived with Aristotle and with Spinoza in terms of "potentia."

Chapter 5

On the De-Reifying Test

The Janus Face of Subjectivity in Digital Enhancement in Terms of Affective Imaginary Reification

In the previous sections we began to render Nietzsche's and Foucault's "philosophy of the present" productive for contemporary critical thoughts on digital enhancement by delineating the basic grammar of the affect- and power-sensitive Janus face of subjectivity in digital enhancement. Furthermore, from the previous chapters it should be clear, too, that here we do *not* refer to Foucault's works in order to better grasp the optimal conditions of life formation. We rather refer to these texts in order to theoretically conceptualize the ways in and through which *the subject sets herself in relation to the logic of perfection.* In the upcoming part on Foucault we develop this line of inquiry further. To this end, we consider the problem of how things become even more complicated in digital enhancement through the case of affective imaginary reification. This case shows the intensified problem of subjects becoming blind in and through *their affect-centered imaginations*. Now subjects are *already* open to having their will formed through the way they *imagine* how to think, live and feel in digital enhancement. To problematize this nuanced situation in digital enhancement, this book outlines the affect- and power-sensitive Janus face of subjectivity in digital enhancement in terms of affective, imaginary reification.

The freedom-endangering side will be depicted by referring to the three key modes of habitualization, normalization and passivization. Accordingly, the present book problematizes the motifs of the habitualizing machine, the normalizing machine and the silencing machine. Additionally, the freedom-empowering side of the Janus face of subjectivity in digital enhancement will be conceptualized by formulating the motif of the de-reified ethos in the tradition of genealogical critique and the notion of enlightenment. This freedom-empowering side contributes to what here is called the *dereifying test*. The de-reifying test is a specific kind of enlightenment test. It also consists of two aspects: Firstly, its task is to verify the kind of affect-based power relations at stake in the concrete contexts of digital enhancement. Here, it needs to check the specific relations of power in terms of reification. Secondly, if the first step concludes that this is a case of affective, imaginary reification, then this shows that digital enhancement has become a problem. According to this diagnosis, the use of digital enhancement in these specific contexts is not justifiable in terms of affective, imaginary reification. The chapter's architecture presents the de-reifying test as follows: The freedom-endangering side seeks to show, in terms of affective, imaginary reification, how digital enhancement can become a problem. The freedom-empowering side then hopes to demonstrate, in terms of de-reification, that digital enhancement does not *have* to become a problem.

Against this backdrop, the chapter is divided into three steps. The first sketches digital enhancement as an affect-centered training laboratory and thus picks up the thread we have left earlier. The second step delineates the freedom-endangering side of the Janus face of subjectivity in digital enhancement in terms of affective imaginary reification. And the third clarifies the freedom-empowering side in terms of affective *de*reification.

This interpretation stems from continuing the reading of Foucault we introduced in the previous chapter. Accordingly, the study goes on to combine Foucault's conceptions of the regime of government mainly part of his *History of Government* on the one hand and the art of living in particular expressed in his *History of Sexuality* on the other hand. To rehearse, due to Foucault's early death in the mid 1980s, Foucault *himself* was not able to pursue this line of inquiry; but the fourth volume of Foucault's *History of Sexuality*—recently published to the surprise of the Foucaultian community—[1] paves the way to such a reading.[2] Moreover, in order to sketch digital enhancement as an affect-centered training laboratory, we needed in particular to consider Foucault's *History of Sexuality*. Preliminary discussions of the topic are addressed in his lectures *Truth and Subjectivity*, *Hermeneutics of the Subject* or *On the Government of the Living*. They can be considered as "*pistes de recherches*" leading to *The History of Sexuality*, which grew out of them.[3] In stressing the connection between the subject's life formation and the systems of knowledge and power, it was necessary to refer to Foucault's earlier works, too, that focus on the entanglement between knowledge, power, and subject.

The perspective on the specific Janus face of subjectivity in digital enhancement developed here is based on a *systematic* interpretation of Foucault's descriptive analyses of the art of living. Foucault describes the motif of the art of living in the context of antiquity and early Christianity.[4] In this context, he analyzes the subject's attitude to a set of norms in the field of sexual morals. Naturally, digital enhancement in today's society is differently characterized and has its own norms, or better: social expectations that can acquire normative power for certain subjects. *Nonetheless*, both areas are *not mutually exclusive*. They are *systematically* connected by the question if, and if so, *to what extent the subject is able to set herself in relation to the existing pool of norms and social expectations*.

Such a systematic reading grounds in Foucault's own approach. Foucault does *not* approach the epochs of antiquity and early Christianity as an end in themselves, but with a *systematic* interest. He is interested in systematically understanding the "forms and modalities" (HS 2, 6) in and through which the subject's "experience" (HS 2, 4) of sexuality develops in different historic contexts. In contrast to the common hypothesis that Christians developed new and stricter rules than the Greeks and Romans, Foucault argues that Christians *continued* to use ancient, Stoic rules via its practices of conscience. What changed over time, as Foucault shows, was *not* the normative arsenal, but the subject's attitude to the existing rules. The further we approach Christianity, the closer we come to the subject that is more open to conform to these existing rules. From an analogue perspective, we refer to Foucault's analysis of antiquity and early Christianity in order to answer the question as to how the subject's conceived and experienced attitude towards digital enhancement is conditioned by the existing social expectations concerning a perfect life. We analyze this question, as Foucault did in the ancient and early Christian context, by referring to affect-centered power relations in terms of affective imaginary reification. According to Foucault, this question is of "unquestionable importance" (HS 2, 10), a statement which applies *not only* to Foucault's era of the 1970s and 1980s, *but also* to our own time because the history of the subject's attitude has *not* ended following Foucault's own formulations (HS 2, 11).

From this systematic perspective, the present book's approach is a contribution to what Foucault calls the "history of subjectivity" (ST, 88), that is to the specific history of the subject's "relation to self" (HS 2, 6); Foucault describes the subject's "relation to self" interchangeably as ethos or as attitude. If Foucault himself showed how subjectivity emerges "through the divisions between the mad and the nonmad, the sick and nonsick, delinquents and nondelinquents" (ST, 88), then digital enhancement problematizes the specific "relations with oneself" (ST, 88) through today's division between the supposedly perfect and the supposedly *non*-perfect. A problematization of this division enables us to understand the affect-centered, power-based conditions, in Foucaultian terms,"in which human beings 'problematize' what they are, what they do, and the world in which they live" (HS 2, 10), and translated to today's context we must add: what they are and what they do in reference to the logic of perfection. This problematization contributes to Foucault's "history of thought" (HS 2, 10) and, therefore, to the key terrain of Foucault's research trajectories which he advanced at his chair for *The History of Systems of Thought* at the Collège de France in Paris.

At this point we need to comment on Foucault's own method of reading because he is often criticized for barely commenting and criticizing his object of analysis, as in the debate over neoliberalism.[5] He also very rarely offers commentaries in his *History of Sexuality*, and thus in the key text we consider as decisive in our interpretation of digital enhancement as a contemporary form of the art of living. The texts sometimes even seem refractory, since Foucault offers such a close reading of the ancient and early Christian sources. This is especially true for the fourth volume *Confessions of the Flesh*. There are almost no references to recent literature. As a result, one could argue as in the case of the debate of neoliberalism, that Foucault does not explicate his critique in *The History of Sexuality*. But similarly to what strands of the debate on neoliberalism have already pointed out, Foucault's sparsely commented descriptive analysis of the *Confessions of the Flesh* and the overall *History of Sexuality* must also be seen as a form of critique.[6] Accordingly, we can see that when Foucault reads ancient and early Christian texts, he does *not* do so in order to become their advocate.[7] On the contrary, he undertakes his reading to lay bare the affect-centered, power-based conditions of life formation that have until today produced settings of unfreedom and of freedom. And he does so *not* in order to stabilize these affect-centered, power-based conditions of life formation, but in order to confront the reader with this problem horizon, and in doing so to open up new spaces for different and deviant perspectives on how to give life a form.

DIGITAL ENHANCEMENT AS AN AFFECT-CENTERED TRAINING LABORATORY

The present chapter has the task of explaining today's digital enhancement as the specific object of analysis that can be understood as *an affect-centered training laboratory*. This interpretation stems from translating Foucault's reading of the art of living in terms of affect-centered power relations into our digital present.

In interpreting digital enhancement as a contemporary form of the art of living, we take up the thread left earlier in the chapter on Nietzsche. As already seen, in general the motif of the art of living entails the understanding of giving life a form. This motif has a long tradition: it emerged in antiquity, crossed Weimar Classicism and nineteenth-century German philosophy and newer French philosophy, and is part of today's debate on digital change.[8] Foucault is one of the key figures in the general debate on the art of living. He stands in the Nietzschean tradition introduced earlier. When the book at hand situates digital enhancement in the tradition of the art of living, we see the "longue durée" of digital enhancement, to speak again with the French historian Bloch. Accordingly, the socio-cultural phenomenon of digital enhancement cannot only be conceived as a contemporary fad. On the contrary, it is anchored already in the social practices in and through which subjects exercise their life formation for centuries.[9]

The corpus of texts on which this interpretation is based refers especially to the late Foucault. It is in particular the late Foucault of *The History of Sexuality* that stresses the field of ethics, and thus the territory, of the art of living. Likewise, Foucault also calls the art of living "arts of existence" (HS 2, 10), the "aesthetics of existence" (HS 2, 12) or the "techniques of the self" (ST, 87). In *The History of Sexuality* Foucault examines the birth of the art of living in antiquity and early Christianity. At the time, the art of living was called "ἐπιμέλεια ἑαυτοῦ" (*epimeleia heaoutou*) in ancient Greek context, and "cura sui" in the Roman and early Christian territory; thus the motif encapsulates the theme "care of the self," too. In order to understand the art of living, Foucault reads especially the ancient authors, such as Plato, Aristotle or Seneca, as well as the early church fathers, in particular Clemens von Alexandria, Cassian, and Augustine.[10]

The History of Sexuality entails four volumes that seek to provide an answer to the question as to how the "desiring man" (HS 2, 5) has shaped his sexual life for centuries. Instead of speaking of the "desiring man" (HS 2, 5),[11] today we had better say the desiring subject.[12] In this context, Foucault especially discusses the cases of confessions, vows or marriage. The first volume, *The Will to Knowledge*, sketches the social practices in and through which the modern subject formed his sexual life by attaching himself to the medical and legal framework invented in the eighteenth and nineteenth centuries. The second volume, *The Use of Pleasure*, shows how sexual life was differently practiced in Greek antiquity around the sixth century BCE. In this context, Foucault stresses in particular the motif of self-mastery. The third volume, *The Care of the Self*, goes on to describe how sexual life in the first two centuries CE was formed, in particular by the examination of conscience. Volume four, entitled *Confessions of the Flesh*, shifts to the early Christian period of the first to fifth century.[13]

In order to understand that life formation, or the art of living, is *not* to be confused with questions of *mere* life style, but *instead* is *structurally* part of the systems of knowledge and power, it is necessary to *also* refer to the earlier Foucault. This approach is seldom undertaken. But it is only by connecting the art of living to the understanding of the systems of knowledge and power that it becomes understandable that the subject forms his life *in relation to* the systems of knowledge and power and *not* in an alleged freestanding manner. This reading goes back to Foucault himself, who states that the notion of the subject is at the center of his work (SM, 81); with the notion of the subject, Foucault shows how people can *set themselves in relation* to the orders of knowledge and power. It is the movement of "setting oneself in relation to the order" that Foucault calls "ethics" and that characterizes the art of living. In the language of the late Foucault, "knowledge" encompasses the understanding of the moral code. The moral code entails the pool of the existing rules that prescribe how to conduct an allegedly usual or normal life. For the late Foucault, "power" is, then, the name for the effect on the subject's life formation that the moral code generates.

Having clarified the corpus of texts relevant for the interpretation at hand, we will now proceed to translate Foucault's understanding of the art of living in today's context of digital enhancement. Here, we focus on the specific understanding of the art of living that is part of the field of training. Thus, we take up a view we already started to develop in the chapter on Nietzsche. This view is explicitly outlined by Foucault in referring to the ancient Greek meaning of training in the sense of "ασκέω" (*askéō*, in English: *I do training*). Thus, Foucault uses the language of "ascetics" (HS 2, 28) in a broad sense. The narrow sense would be traced back to the understanding of asceticism in the Christian tradition. In this Christian tradition, it would have the meaning of renunciation, with its "bad connotations"(DE 4/293, 202, trans. by the author). Foucault especially rejects this meaning as being blind to all the "practice[s] of the self" (ECSPF, 282) in and through which one exercises one's own life formation. Such an understanding of training is characterized in particular by the social and cultural practices in and through which one "works on oneself" (ECSPF, German ed., 254, trans. by the author). In translating this view to our digital times, it is especially the exercise of daily routines that characterizes today's training. In this sense, we can say with Foucault that digital enhancement turns into the training field in which subjects become the high performers that do their best every day in order to succeed in achieving an allegedly perfect life.

Now, digital enhancement proves to be the specific training field that we theoretically conceptualize as *an affect-centered training laboratory*. The term "laboratory" highlights how subjects try, test, and experiment with how to "give form to their behavior" (HS 2, 23). They do not do so by following an absolute plan. What Foucault reveals in reading the ancient Greek texts and what is also relevant in times of digital enhancement is the subjects' *search* for caring for one's own life formation by *not* departing from the general perspective of how subjects usually or normally "give form to their behavior," but instead by beginning with their subjective point of view. Nonetheless, this understanding does *not* lead to an *a*moralism. Amoralism would imply that the subjects were trying out anything, as if there were *no* rules. *However*, the ethical subject, Foucault observes, *only* tests and experiments with *justifiable* forms of life. Justifiable forms of life are all those that do *not* produce settings of domination in and through which a subject gains power *over* another subject. It is for this reason that the experiment is conceptualized here as *an enlightened experiment*. Enlightenment, to recall, is understood as the specific kind of critique that shows how the subject can develop such a clear gaze that the affect-centered relations of power and domination are no longer intransparent, but visible to her.[14]

But Foucault comes to the conclusion that the subject plays with marked cards, so to speak. When they want to play their subjective card, Foucault demonstrates, however, the different degrees to which subjects already have the blind will to play the general card. So, instead of having a chance to *really* test her own subjective perspective, the subject ends up starting to different degrees with the general perspective. This view is problematic, as the subject does so in ways that are *not* transparent to herself. To account for this problem, we refer to the notion of affects. Accordingly, the training laboratory shows itself to be affect-centered. The notion of affects enables us to theoretically conceptualize how subjects in today's field of digital enhancement, like the subjects in the ancient and early Christian times that Foucault analyzes, are trained in such a fashion that they have *a blind* will to conduct life. The subjects that Foucault analyzes have *the will to knowledge*, to refer to the title of Foucault's first volume of the *History of Sexuality*, that is, the blind will to form life *in conformity with* what the dominant knowledge of sexual life suggests they should do. In similar ways, today's subjects have the blind will to enhance life *in conformity with* what the dominant knowledge of digital life suggests they should do; and this view entails the understanding of blindly considering digital means as providing an allegedly perfect life. The affect-centered training laboratory is "cognitive," too, because subjects value digital enhancement as the allegedly perfect way of life.

The notion of affects developed here is thus *not* to be confused with the common reading of the desiring subject in the late Foucault. Usually, when it comes to the notion of desire in the late Foucault, it is common to mostly consider the understanding of the desiring subject in terms of sexuality.[15] But in the key interview "On the Genealogy of Ethics" Foucault clarifies: "I must confess that I am much more interested in problems about techniques of the self and things like that than sex" (GE, 253). Accordingly, we refer to the notion of the desiring subject *in a broad sense*. This broad sense involves the theoretical conceptualization of the power relations that are part of the subject's way of exercising life formation. In so doing, the present book outlines the emergence of the desiring subject in and through the procedures that Foucault calls "subjectivation" (HS 2, 29). To recap, "subjectivation" is the power-based process in and through which the subject becomes a subject in the given contexts. To theoretically conceptualize the blind aspects of this process, we speak of affective subjectivation. "Affective subjectivation" is thus described with reference to Foucault as the understanding of the subject *as being open to having the blind will to become who she is in and through ascetic procedures*. Hence, we see that the subject, as Foucault puts it, "is not given to us" (GE, 262), it is not an invariant "constant" (HS 2, 4); rather, she becomes who she is in the contexts of her time in and through daily, routinized exercises. But *initially* she is not particularly aware of these exercises.

Put differently, affect-centered power relations entail the *training effect* that A has on B to the extent that B is prompted to exercise his life formation in such a way that he blindly wants to live, feel and think in the ways he was directed by the training effect of A. A in this case is the moral code of which Foucault speaks. It encompasses all the norms around marriages, vows or confessions in the field of sexuality. In our digital present, it entails the social expectations of the logic of perfection. B is the desiring subject that is part of Foucault's *History of Sexuality*; in our context it refers to the subject in digital times. The affect-centered, power-based understanding of the desiring subject developed here thus clarifies that the subject *is prompted* to *re*act to the pool of social expectations *without* being aware of being prompted.

This view on affects is basically anchored in Foucault's understanding of the dispositive introduced in the previous chapter. When the dispositive is "the said as much as the unsaid," then the late Foucault of *The History of Sexuality* describes all the subject's *immediate* reactions to what is said and unsaid in the field of sexuality. The field of digital enhancement proves to be structured in similar ways. Here, we are dealing with the subject's immediate reactions to what is said and unsaid in the field of life formation that is marked by the logic of perfection. In the ancient and early Christian times, these reactions initially make the subject *immediately* value the moral code as the allegedly perfect way of living. In our times, the subject at first immediately values the logic of perfection as supposedly the best way to live. Due to this "prevalent influence" (CF, 195), the dispositive entails cognitive aspects, too. For the subjects immediately value the moral code as supposedly good, in the case of the late Foucault, or the logic of perfection as good, in the case of digital enhancement. The basic lesson we learn from this affect-centered, power-based understanding of the desiring subject shows that the subject's desire is *socially constructed* by affect-centered power relations and *not* a natural entity as one would assume following the psychoanalytic tradition of Freud's notion of "drives."[16] While today's common debate considers to what extent the subject's desire is part of nature or culture, that is, to what extent it is given or socially constructed, the preceding discussion clarifies that we need to understand the notion of desire as culture and *not* as nature.

Viewed from this perspective, "affects" thus enable us to understand the paradox that like in the ancient and Christian times, today's subjects also do the practices of subjectiviation *in a more or less free way despite* the well-known freedom-endangering effects that come with it—in our digital present the knowledge about the issues of data privacy and in the ancient and early Christian times the issues of self-mastery or the conduct of the conscience: the subject gives her life a form, and—translating Foucault into our digital today—"without any relation with the juridical per se, with an authoritarian system, with a disciplinary structure" (GE, 260). Thus, the subject is not constrained to give her life form in and via "a system of coercion" (ECSPF, 291). Yet she does so *despite* its well-known freedom-impairing effects. And the notion of affects enables us to account for this paradoxical situation by theoretically conceptualizing that the subjects' will is socially produced by power relations in such a way that they *blindly want* to enhance life, to update Foucault's title *The Will to Knowledge* for today's digital world.

To further specify the object of analysis of digital enhancement as an affect-centered training laboratory, let us consider Foucault's tripartite understanding of morals in his late works. First of all, morals, following the Foucault of *The Use of Pleasure*, are "an ensemble of values and rules of behavior" (HS 2, 36, transl. by the author). Transferred to the case of digital enhancement, morals are the ensemble of the social expectation that stem from the logic of perfection. From this basic understanding of morals Foucault, distinguishes between three sub-understandings. Firstly, morals comprise all the "real behavior" (HS 2, 36, trans. by the author) that is observable. In Foucault's case, it is the subject's care for his "aphrodisia," the subject's sexual desire, or for his "flesh," to use the Christian term, that is observable and that Foucault considers as his object of analysis. In digital enhancement, we advance with Foucault the understanding of the desiring subject in terms of "askesis" as the name for the specific affect-centered power relations as described above. Thus, "morals" entail all the behavior that is trained to such an extent that it obtains the guise of *a reflex* by which subjects are prompted to blindly use digital means as much as possible. The second understanding entails the existing pool of moral rules that Foucault calls the moral code. It encompasses the prescriptions that have such an effect on the subjects that they want to conduct their life *in conformity with* these moral rules. The third understanding encompasses the subject's ethos, that is, his attitude, towards the existing moral code. This can range from submissive attitudes to critical attitudes towards it that this book calls enlightened attitudes.

The third understanding has further four sub-categories. Here, Foucault specifies that the ethos consists of an ethical substance, of modes of subjectivation, of an ethical work and of a telos. The first describes the object of analysis, that is, the subject's care for his "aphrodisia." In the case of digital enhancement, it is the subject's care for the social expectations that comes with the logic of perfection. The second shows that this care concerns the subject's entire way of living; it is not just a superficial question of life-style. By contrast, it is a procedure in and through which the subject becomes a subject in the context of the moral code. In this sense, Foucault writes: "no constitution of the moral subject without modes of subjectivation and without ascetics or practices of the self" (HS 2, 39p, trans. by the author). In analogue ways, digital enhancement also matters for the subject's entire life formation. The third point entails the ethical work, that is, the social practices in and through which the subject seeks to achieve his moral conduct. In the times of digital enhancement, the motif of the ethical work entails all the social practices that the subject undertakes to enhance his life by using digital media. The fourth feature consists of the telos. In the case of Foucault's analysis, the telos encompasses the understanding of the goal of ethical life. In our times, this goal is the alleged perfection of life by using digital means as much as possible.

The understanding of digital enhancement as an affect-centered training laboratory now clarifies that the understanding of the telos is not naturally given, but is the product of the subject's blind will to enhance life. This blind will makes them conceive and experience digital media as the fancy tools that one *just* needs in life to perform best. In this sense the dispositive of desire renders plausible that the policemen in the film *Minority Report* call themselves "clergy" and not cops, as one would expect: from the perspective developed here, we see that they are in the dispositive of desire to do so. They blindly believe in the necessity to perfect life via the method of predictive policing. Like the monks in Foucault's fourth volume of *The History of Sexuality* who did *not* have to be carried over the threshold of the monastery, these policemen—or better, the "clergy" in uniform—did *not* have to be carried over the threshold either, where the threshold in question is obviously not that of a real monastery, but that of their digitalized police station. They are *simply* in the dispositive of desire which makes them to believe in the alleged social facts according to which the method of predictive policing would perfect life.

UNDER THE SPELL OF IMAGINATIONS: ON THE HABITUALIZING MACHINE, THE NORMALIZING MACHINE AND THE SILENCING MACHINE

In what follows, we analyze the freedom-endangering side of the affect- and power-sensitive Janus face of subjectivity in digital enhancement *under the spell of imaginations*. To this end, we spell out the type of affective imaginary reification and outline three interwoven, specific procedures of this type of reification: the habitualizing machine, the normalizing machine and the silencing machine, all three of which produce quasi-automatized subjectivity.

DOING ONE-SIDED OBJECTIFICATION: THE BASIC GRAMMAR OF THE AFFECTIVE, IMAGINARY TYPE OF REIFICATION

It might initially surprise the reader that we seek to draw the type of affective imaginary reification in interpreting Foucault's work. After all scholars rarely associate the notion of reification with Foucault because he does not address the term extensively.[17] Generally speaking, the term of reification is associated with Western Marxism,[18] first appearing in Georg Lukács' essay collection *History and Class Consciousness*, published in Weimar Republic.[19] While suffering a period of neglect after World War II, the term has recently reappeared in philosophical discussions and currently constitutes one of the central themes of social philosophy.[20] Following Foucault the specific understanding of reification underscored thus far entails the notion of blindly undertaken *one-sided objectification*. More specifically, the hitherto rather neglected type of affective imaginary reification involves the understanding of quasi-automatized subjectivity in and through which the subject *immediately imagines* himself as the one-sided object of the given contexts. The term "immediately" takes up the thread of affect-centered power relations addressed earlier. It refers to the specific affective type of subjectivation in and through which the subject *is prompted to* become who she is in the given contexts as *only* the object of these contexts. "Imaginary" is derived from the Latin word "imago," which means "picture."[21] But it would be shortsighted to reduce the understanding of the imaginary to the notion of "pictures." It involves the understanding of the subject's intrasubjective world of images. Unlike the usual view, which holds that imaginations are *not* part of concept-based thinking, Foucault shows that imaginations *are* part of it. According to Foucault, concept-based thinking belongs to the field of the systems of knowledge and power, imaginations to the intrasubjective world of images. These two domains are connected as follows: The intrasubjective world of images *is structured by* the systems of knowledge and power, insofar as it includes the images produced by shows, films et cetera that are composed of the knowledge-power-nexus, which the subject at first blindly imagines as valuable. So, the intrasubjective world of images involves the understanding of the knowledge-power nexus. It is in this sense that the intrasubjective world of images specifies the type of affective subjectivation that we already delineated in the context of the knowledge-power nexus. Thus, it theoretically conceptualizes that the subject's intrasubjective world of images *itself* prompts the subject to become a specific subject that has a specific gaze composed of specific imaginations. The mother in the computer game "Perfect Mom" in the film *Her*, for example, is confronted with images that show an attractive mother, a beautiful house, a modern kitchen, and so on. These pictures must be understood in the enriched sense of images, because the mother develops an intrasubjective world of images that makes her immediately produce an allegedly perfect image of what a perfect life looks like. As a consequence, she is prompted to immediately attach to this intrasubjective world of images. This interpretation stems from Foucault's notion of the dispositive as shown earlier. Here, the dispositive is part of the field of imagined life formation. It is this field that then has the immediate effect on the subject of prevaluing the imagined life formation as good. If the imaginary is in feminist debates mostly understood as freedom-impairing and in the Kantian tradition in freedom-enabling ways,[22] then we conceive with Foucault the affect-based imaginary as twofold: it *initially* has a freedom-endangering side, but it *also* leaves room for freedom-empowering aspects.

More specifically, from Foucault we learn that affective imaginary reification consists of practices and thus needs to be conceptualized as praxeological or practice-based. Hence, *blindly undertaken, one-sided objectification is performed* and is *not* an essence or substance. It describes the subject's quasi-automatized attitude in and through which he immediately imagines something, namely his life formation, as a one-sided object *although* he himself is *not* a one-sided object. When Foucault clarifies that the subject's attitude consists of "a mode of relating to contemporary reality" (WE, 309), the subject with an affect-centered, imagination-based reified attitude does not see that he is dealing with "a mode" and thus with "a certain way of acting" (ECSPF, 286). He imagines himself as an automaton that does not act but only *re*-acts. That he himself is doing something, namely objectifying himself in a one-sided way by ways of imaginations, is not transparent to him. Accordingly, such an imagination-based automaton-like attitude makes the subject become the imagined one-sided object of the given contexts that has the will to only take the place reserved for him by society. That the subject *could also choose* a seat is out of question in the territory of the affect-centered, imagination-based reified subjectivity. Following Ernaux, we see that these subjects lead an automaton's life because, to translate Ernaux's terms into the context of digital enhancement, they "never stopped *wanting* to click on 'save' and keep all the photos and films, viewable on the spot."[23] That they were not able to stop is the result, as we say following Foucault, of the subjects' will that is socially programmed by the affective imaginary type of reification.

Basically, with Foucault we see that the praxeological type of affective imaginary reification consists of three aspects: *firstly*, patterns of affective imaginary reification are historically produced, i.e., by the "history of ourselves" that is marked in our digital present by the logic of perfection. *Secondly*, affective imaginary reification is part of a variety of layers. They unfold within the three "axes" (GE, 203, trans. by the author) of knowledge, power, and self-constitution, leaving space for various life formations that range from patterns of subjects that are fully conceived as affect-centered, imagination-based one-sided objects, as in patterns of domination, to arrangements of affect-centered, imagination-based *de*-reification, to translate Foucault's concept of "desubjectivation" in the context of reification. "Knowledge" describes the dominant knowledge, that is, the episteme, in and through which the subject understands himself as the one-sided object of the given contexts—in the case of digital enhancement, for example, as the one-sided object of the knowledge of how to best shape life in conformity with the logic of perfection. In the context of affect-centered, imagination-based reification, however, we are dealing not with knowledge, but with the subject's intrasubjective area of imaginations. In times of digital enhancement the dominant imaginations entail the imaginations in and through which one just shapes his life in conformity with the logic of perfection. "Power" is the subject's attachment to these imaginations. Here, we deal with the specific power relations that are the name for affects. Following this view, the subject *blindly* performs imagination-based one-sided objectification. "Self-constitution" clarifies that the subject *herself* shapes her life formation as an affect-centered, imagination-based one-sided object. This entanglement of imaginations, power, and self-constitution is understood here as the imaginary–power nexus in analogy to Foucault's understanding of the knowledge-power nexus described in the previous chapters. The extent to which the subject objectifies herself in an affect-centered, imagination-based one-sided way depends then on the following question we can raise with the late Foucault: is the subject prompted to tend within her life formation in the direction of the "axes" of the imaginary-power nexus or rather to the "axis" of the self-constitution? *Thirdly*, the subject is reified *and also* reifies herself. With the late Foucault, we gain an enriched understanding of the subject's active participation in the processes of affective, imaginary reification. *Indeed* the subject is reified by extant systems of imaginations and power. *But* she becomes *also* the accomplice of these reifying systems of imaginations and power by *also* constituting herself as the one-sided object of these systems.[24]

To specify the understanding of the problematic dynamics that are at work in the affective, imaginary type of reification, we refer to the motif of *exclusionary unification*. From this perspective, we can better understand that the affective, imaginary reification is a problem because the subject *is prompted* to unify the dominant perspective on imagining life with possible deviant perspectives on imiganing life formation differently *by immediately excluding the different, deviant perspectives*. Such a process of unification emerges insofar one-sidedly and thus in exclusionary ways as the subject immediately imagines herself as someone that *just values* the given perspective as the necessary one. As a result the given perspective prevails over different perspective on life formation. In reality, the given perspective might *not* necessarily gain predominance, but due to the forces (*Kräfte*) we conceptualize with the affective imaginary type of reification, the subject *feels the necessity to do so.* As a consequence, the subject is *only* the one-sided object of the dominant perspective on life formation.

The film *Her* clarifies the problem horizon as follows: The computer game "Perfect Mom" depicted in the film conveys images of what it means to be a perfect mother and players can win so-called perfect mom points, for example by preparing a healthy breakfast for the children, by making other women jealous, or by bringing home-baked cookies to school. The problem starts when the players do *not* question what they are actually playing, and thus they are initially not aware of the one-dimensional unifying process: the subject with an affective-based, imagination-centered reified attitude is *not* able to perceive that his immediately practiced unification is from the outset dominated by the given perspective on how to usually imagine life. As a result these subjects take the socially constructed images of being a perfect mother for granted and thus believe *reflexively* that these randomly assembled images actually convey what it means to be a perfect mother.[25] These images thus appear to the subject as the nature, essence, or substance of what images of being a good mother needs to consist of. In immediately imagining these pictures as *the only* valid perspective, the subject *himself* turns into a one-sided object.

From this perspective, it is clear that the understanding of affective imaginary reification developed here is explicitly *not* situated in the standard history of the notion of the imaginary. If we consult the archives, then we encounter first and foremost the standard reading rooted in psychoanalytic theory, especially in Jacques Lacan, but also in Freud and Carl Gustav Jung.[26] Lacan concentrates on processes of imaginary identification in the formation of the ego. From this perspective, the imaginary generates a twofold process: on the one hand, it prompts processes of disintegration and thus of differentiation and, on the other, it causes the subject to build a unity of the self by binding himself to the images of a wished-for object that he is not yet.[27] Whereas Lacan defines the imaginary as an *un*conscious and *a*-rational process, with Foucault we see *in turn* that the subject knows of its rationale. For the subject knows pretty well what he is doing. Theodore of the film *Her*, for example, knows well that he is buying a digital operating system called Samantha in order to feel less lonely. But he does *not* know the consequences that this process implies for his life. He is thus *not* aware of the affect-centered, imagination-based objectifying forces (*Kräfte*) that are at work in this process, as he fails to notice that he becomes the one-sided object of Samantha by immediately attaching himself to her seductions.

THE SPELL OF AFFECTIVE IMAGINARY REIFICATION THAT PRODUCES A MACHINE-LIKE SUBJECTIVITY

This chapter delineates how *the spell of affective imaginary reification produces a machine-like subjectivity*. We sketch this type of subjectivity first as a habitualizing machine, second as a normalizing machine and third as a silencing machine. The three types of the machine-like subjectivity are interwoven as follows: When one immediately habitualizes imaginationes of life, they turn into a normal part of one's own life. In being prompted to imagine something as usual, normal, and thus as self-evident diminishes the subject's active potential to imagine in other ways than the normal or usual images of life suggest. In this sense, we speak of passivization. It passivizes the different voices in life formation and thus it puts them on mute. As a consequence, they are silenced.

As already seen in the previous chapter on the panoptic machine, the notion of the machine is central to Foucault. The motifs of the habitualizing, the normalizing and the silencing machine stand in the tradition of the panoptic machine. They refer to the specific machine that Foucault calls the "machine of the signifier" (WC, 23, trans. by the author).[28] Basically, the notion of the signifier goes back to semiotics and was developed by the French structuralist Ferdinand de Saussure in the nineteenth century. Saussure shows that signs are composed of the difference between the signifier and the signified.[29] The "signifier" theoretically conceptualizes the form of the sign, that is the word, whereas the "signified" describes the object of the form. The way "signifier" and "signified" are bound to each other is then understood as the sign. Now Foucault develops the understanding of the "machine of the signifier" in order to describe basically how the general perspective on life formation forces the subject *to be programmed in such a way that he initially merely* considers this general perspective as *the only valuable perspective* in life. Here, the "signifier" corresponds to the formative process in which the general perspective on life formation dominates. The "signified" in turn represents the object as the subject's concrete life formation that is dominated by the general perspective. The habitualizing, normalizing and the silencing machine are, then, specifications of this basic view. They show how procedures of habitualization, of normalization and of silencing dominate the subject's way of imagining his life.

Under the Spell of Habitualization: "Training! Training! Training!" (Nietzsche)

The theme of *being under the spell of habitualization* entails the motif of the habitualizing machine. The habitualizing machine encompasses the type of subjectivity in and through which the subject is blind to the fact that due to her daily routines she is the affect-centered, imagination-based one-sided object of "business as usual." The corpus of text relevant to the motif of habitualization entails the works of the late Foucault and in particular *The History of Sexuality*. In broadening the question to include the subject's life formation, we see that it is related to Foucault's entire work. It leads to the core question of how the subject is habitualized to stabilize *and also to question* the divisions between mad and non-mad, between normal or abnormal, or between delinquent or non-delinquent. Translated to the context of digital enhancement, these divisions must be enlarged by also considering the alleged division between perfect and non-perfect.

Within the context of habitualization, digital enhancement is conceived as the specific contemporary form of the art of living that is part of training routines.[30] In the previous chapter, we stressed the relationship between the art of living and training with reference to Nietzsche. With Foucault, we reconsidered this path in the earlier sections. To reiterate: the motif of routinezed training is shared not only by Nietzsche and Foucault, but also present in the contemporary debate of enhancement. Whereas in the past this motif was reserved for sports, and in particular for competitive sports, in the era of digital enhancement *everyone* is called upon to become an athlete, that is, to become an ordinary competitor who needs to perform best in everyday life and thus to be perfectly "fit." An extreme understanding of enhancement as training is presented by Sloterdijk, who speaks of "anthropotechniques"[31] rather than—following Nietzsche and Foucault—of the art of living. In contrast to the perspective developed here with Nietzsche and Foucault, Sloterdijk embraces enhancement as a modern training camp *without* being concerned about the reifying aspects that constitute it. By ignoring this problem, Sloterdijk addresses the elitist project around the motif of producing oneself in *You Must Change Your Life: On Anthropotechnics*. In so doing, he supports reification when he writes: "It is time to reveal humans as the beings who *result* from repetition."[32] This is problematic because Sloterdijk notes that human beings are the *mere* result of repetition. Indeed, we depart from a similar view, we conceive a produced life formation, too. By contrast, we stress, however, the necessity to criticize the fact that the subject appears at first in digital enhancement as a result of the training procedures *instead of* praising it as Sloterdijk does.[33]

Against this background, we now specify the procedures of affect-centered, imagination-based reification in the context of digital enhancement described as routinized training. Here, the subject's immediately habitualized attachment to the imaginations of an allegedly perfect life corresponds to the motif of the imaginary-power nexus. This view implies the following: the subject is so routinized in his life formation that he attaches blindly to the imaginations that depict what a supposedly usual life looks like and that the subject automatically values as good. Thus, she imagines to do what she does quasi-automatically, or, to put it more succinctly: *she is caught in a reflex*. We interpret this dynamic as producing the subject's status as the unnoticed one-sided object of her daily routines. Her life formation is so extensively automatized and thus so self-evident that she simply follows what she imagines to do day to day in a repetitive routine. "Business as usual" becomes the subject's unquestioned life perspective. This dynamic of affective, imaginary reification entails three key aspects: *firstly*, the subject is historically produced as an unnoticed one-sided object via accustomed images deemed valid in digital enhancement. *Secondly*, reification is exercised on multiple layers of imaginations, power, and self-constitution. This multi-faceted exercise is retaken by the formulation of the imaginary-power nexus. Thirdly, the subject is constituted as an unnoticed one-sided object *and also* constitutes herself this way. Hence, the subject is involved in the process of her own reification. Instead of being innocent, she becomes the accomplice, though in defense of the subject we need to add: the subject has no clue of her complicity.

This multi-faceted dynamic is characterized by a persistent mechanism: the more the subject exercises her life formation, the more intense she adapts to the usually given imaginations of how to give life an alleged perfect form. As a consequence, the subject intensifies and stabilizes her affect-centered, imagionation-based reified attitude in digital enhancement. Within this context, the subject's desire for conformity with the imagined usually given perspective of the logic of perfection is understood as the glue that binds the subject persistently to the trained practices of life formation. As a result, she cannot simply stop after one session and instead wants to repeat what she does not only once or twice, but three times. It is the unquestioned routine that starts with the series of doing something three times and, to return to Nietzsche, to imagine life while engaging in "[t]raining! Training! Training!."[34] Hence, such a subject *herself* weakens her enlightened potential to question what she does by imagining to do things repeatedly. Instead of "opening one's eyes" and seeing what one does, to return to Foucault's formulation, the subject closes her eyes without noticing what she does: the "chain" (HL, 1, 59) of her imagined life formation, to put it with Nietzsche, simply "runs with" (HL, 1, 59) her, to further express it with a key motif in Nietzsche's collection of aphorisms *The Wanderer and His Shadow*: the wanderer is continuously followed by his shadow. From the perspective of affective imaginary reification, we conceptualize that this "chain" does *not only simply* "runs with" her; rather, the subject *herself* becomes the unnoticed accomplice: the subject *is prompted to activate* her reified attitude and thus becomes the one-sided object without having any clue of this entanglement. She simply imagines that life consists of maintaining the habitualized mechanisms, or the "chain" in Nietzsche's sense, that directs the subject to the place in society that is reserved by society for her.

To illustrate the problem of affective imaginary reification in the context of training, let us recall the computer game, *Perfect Mom* of the film *Her*. The game conveys the imagination of what we should expect from a "perfect mom": she should be so fit that she can handle everything—job, children and a stylish personal appearance—simultaneously. If she is not able to do so, then, according to the common understanding, she should be perceived as *insufficiently fit*. Accordingly, a mother who imagines being a wonderful mother differently is *quasi-automatically de*valorized and she becomes the accomplice of such a *de*valorizing process. This mother would thus turn herself more or less voluntarily into the one-sided object of her habitualized self-imaginaries because she expects herself to be a "fit" mother who can fulfill all conventional expectations. The expectations in question are no longer simply the expectations of society, but *also* the imagined *self*-expectations she grew up with. In this sense, the mother becomes uncritical because she imagines being a good mother by adapting to these historically produced and socially constructed expectations, which acquire for her, however, the status of a self-evident fact that *has* to be this way because it has *always* been the case. It is no longer possible for her to think outside the box, outside her narrow, habitualized imaginations of what it means to be a good mother. Thus, her habitualized way of being a mother appears to her as natural, *even though* it is not natural.

With reference to *Minority Report,* we further investigate the problem of exercising life formation in the context of predictive policing, shifting to the territory of blindly predicting future crimes in ways that need *not* to be blind. But the subject at this point of the discussion cannot perceive this potential. In the scenes of the films, the policemen imagine themselves as the one-sided object of digitally-supported predictions because they are prompted to believe in imaginations that diminishing the crime rate must be grounded in predictive methods of crime detection. While policemen thus have their eyes wide open, they are blind to the fact that they are relating to predictive practices in objectiving ways, even though their attitude is obviously *not* a thing. However, the policemen's relation between themselves and their imaginations of how to build a safe society has become so habitualized that they *simply* can*not* see their imagined commitment. And their actual work leaves them hardly any space to rethink their determined imaginations that render the subject the object of their work. As the film demonstrates, while the predictions are occurring the policemen always repeat the same procedures quasi-automatically to such an extent that it cannot cross their minds to imagine creating a safe society differently. In every case, the scene unfolds in standardized ways: the "precogs," the figures who render the previsions, gasp and shake, then a red ball emerges with the name either of the perpetrator or the victim, the policemen take the ball, and digitally project the visions of the future crime on a wall, zoom, pre-view, and rewind these pictures. In repeating the same movements, the policemen make themselves into the muted object of the scenario as they act like automata, even though in reality there is no need to do so. These policemen have acquired a blind routine of repeating the same actions every time, so that the images and pictures of these scenes have become an unquestioned part of their cultural life. They just do it. Hence, the policemen are unable to criticize the trained imaginations of predicting future crimes.

Under the Spell of Normalization: "Hey! You! There!" (Althusser)

We now turn to the affective imaginary type of reification through processes of normalization. In this chapter, the process of normalization enables us to conceptualize how the subject becomes the unnoticed one-sided object in the digital order that generates persistent imaginations according to which the logic of perfection appears as necessary and thus as normal. The present book describes this entanglement as the type of subjectivity that resembles a *normalizing machine*. In and through this subjectivity, the subject only imagines a stereotypical "back to normal" as the relevant perspective on her life. From this lens we speak of the subject as being *under the spell of normalization*.

In general, norms play a key role in Foucault's work from early on. Here, it is important to stress that the late Foucault was interested in uncovering the multifaceted power of normalization, *in contrast to* the early Foucault, who was interested in the norms' disciplining effects. Foucault analyzed this latter aspect in his early lectures *Abnormal* as pertaining to the motif of medical norms and in *Punitive Society* related to legal norms (ABM; PS) which lead then to one of Foucault's key books, *Discipline and Punish* (DP). *By contrast*, in Foucault's late writings from the 1980s, that relate to the motif of the art of living, Foucault cares for the multi-faceted role of norms and thus raises the question *to what extent* the subject can create a critical attitude towards the pre-existing arsenal of norms. This scenario is in particular developed in *The History of Sexuality*.[35]

Moreover, we need to specify the object of analysis examined here, stressing the relation between digital enhancement and normalization. It might be surprising to consider normalization as a key part of digital enhancement. In general, the context of digital enhancement is characterized by a desire for singularity. Lacie, to return to the protagonist in the Netflix episode "Nosedive" of the series *Black Mirror*, explicitly seeks to imagine herself in singular ways, wanting to be recognized as someone who takes extraordinary pictures while drinking coffee or playing with her cuddly toy, all of which is intended to demonstrate her uniqueness. But if we advance a specific understanding of what is considered normal, her actions are characterized by a *nuanced* understanding of normality, one that factors in the social expectations demanded from subjects seeking to perfect their life. Such an approach can conceptualize Lacie's specific desire for normality by revealing her desire for conformity with the given perspective of social expectations. To reiterate: this desire does not result from any natural drive, as one could think of in Freudian terms. With Foucault we detect that it is socially constructed by the dispositive of desire. The dispositive of desire causes Lacie to imagine, conceive and experience the logic of perfection as a normal perspective one *only* applies in one's life. As a consequence, she imagines the given perspective *simply* as normal. To conceptualize this dynamic, we speak of the specific object of analysis in terms of normalization.

And how does the process of normalization turn into a specific problem of affective imaginary reification? Generally speaking, we need to note that it is practiced. Corresponding to the three aspects of affective, imaginary reification, the type of affective, imaginary normalization is characterized by the following three facets: the subject is historically produced and normalized as an unnoticed object; it is furthermore historically produced and normalized on various layers as an unnoticed object, i.e., of imaginations, power, and self-constitution; thirdly, the subject is *both* normalized by systems of imaginations and power *and additionally* normalizes herself. Hence, the subject plays an active part in the process of normalization.

To understand the specific dynamics, we need to return to the classic concept of normalization in the sense of Althusser's scene of interpellation,[36] which exerted a major influence on Foucault. In general, the scene of interpellation deals with a pedestrian on the street that turns around because a policeman calls "Hey! You! There!" Basically, most of the debate centers on the question as to why the pedestrian turns around.[37] Usually, the answer is that the pedestrian turns around because he is made functionable by the policemen's call. He fulfills his function to obey a state's authority. Here, we offer a hitherto underscored affect-centered, imagination-based interpretation. In doing so, we analyze how affects glue the pedestrian to the call of the policeman in ways that are not transparent to the subject. Translated to the case of digital enhancement, the pedestrians are everyday life subjects and the policeman the social expectations to lead an allegedly perfect life by using digital means as much as possible. Affects now enable us to theoretically conceptualize that the subject *is open* to turning around *in a relatively free manner* to the policeman, or the social expectations, who *merely* shouts "Hey! You! There!" In other words: she *simply* turns around. It is in this sense that the understanding of affect-based normalization enters the stage. For the subject simply does what she *immediately* understands as allegedly normal behavior. Accordingly, we can say the policeman is not only shouting "Hey! You! There!" The subject is prompted to understand by this call "Hey! You! There! You're normal! So, react normally!" In similar ways, the today's subject in digital enhancement simply turns around to the social expectations of using digital media in life as much as possible because she *just* considers these social expectations as paving the way to an alleged normal life.

"Imaginations" now enable us to specify that the policeman is not really calling on the street; now we see that he is already part of the subject's intrasubjective world of images. By this we can specify: This time, it is not a policeman calling, but "normal self-imaginaries." They produce the spell of normalization. By way of such alleged normal self-imaginaries, the subject calls *herself* to turn around, which means to pursue one's own self-imaginaries *in conformity with* the given imaginations of an alleged normal life formation. This spell causes the subject to become the unnoticed one-sided object of reification we call the "muted object." The spell renders the subject unaware of imagining, conceiving and experiencing her self-relation in objectifying ways. The subject as muted object might even think that she is right on a track approved by society's imaginations and by one's own imaginations as normal. But by wanting to give her life a form that is regarded as normal, the subject desires to establish a problematic unity between her own self-imaginaries and social imaginations. This unity is problematic because it is *only* fabricated by the subject's desire to conform to the imaginary-power nexus. Returning to the computer game *Perfect Mom*, we can state: becoming a supposedly perfect mom implies identifying with the allegedly normal social imaginations that cause one to fulfill all the imagined expectations imposed by society. Accordingly, the mother turns herself into the unnoticed one-sided object of her self-imaginations because she imagines the digitally imposed notions of normality as necessary. She can*not* question her own self-imaginaries.

UNDER THE SPELL OF PASSIVIZATION: "WHAT IS CALLED COMMUNICATION NOWADAYS IS BUT THE NOISE THAT DROWNS THE SILENCE OF THE SPELLBOUND" (ADORNO)

In contouring the third mode of the affective imaginary type of reification, we now examine how the subject falls *under the spell of passivization*, a mode in and through which the subject desires to imagine herself as a passivized object of the dominant imaginations of life formation. The book at hand describes this entanglement as the specific type of subjectivity that here is called the silencing machine. For the subject under the spell of passivization silences in an unnoticed way all those voices that are not part of the dominant voice, that is, the call for perfection.

This reading is grounded in a so far underestimated interpretation of Foucault's notion of the dispositive, conceived here as the dispositive of desire. Translated to the field of digital enhancement, the dispositive of desire yields a passivized subject because the subject is *not* able to *actively* question his blind imaginations because he considers the given logic of perfection as the social facts that are valid in life. To recap, the present book calls this entanglement the imaginary–power nexus. The motif of passivization is part of Foucault's early work: since his inaugural lecture *The Order of Discourse* at the Collège de France in 1970 Foucault is interested in the knowledge-power nexus that we encounter here in its variant of the imaginary-power nexus.

In the context of the spell of passivizaion we deal with the specific object of analysis that investigates digital enhancement as a mode of passivization. Conventionally, digital enhancement is characterized by the dynamics of activity. So far, it is rather neglected to consider the facet of passivization that is, however, by itself a key part of digital enhancement. With Foucault we examine the dispositve of desire as capable of producing settings of affective imaginary reification that are characterized by a strange combination of passivity and activity. The dispositive of desire effects a reified subject that is quite active in peculiar ways—we could also say she is quite busy. The dispositive of desire merely causes her to imagine a busy lifestyle as *the* way to exist. With Foucault's "philosophy of the present" this activity, however, can only be labeled a *pseudo*-activity insofar as the subject acts *only* superficially. She *indeed* actively thinks, lives, and feels, *but* she does so by immediately giving her life a form in imagining the logic of perfection as the best possible way. Hence, the subject is *solely* active and busy at a level that we can see if we "open [our] eyes": the subject, like the protagonist Lacie of the Netflix episode "Nosedive," is quite busy taking the best selfies everywhere she is, constantly checking any update of her rating, making friends only with a high rating score. But in so doing, she passivizes in the area of her dispositive of desire, the potential to create a different perspective on life formation. The possibility to create a dereified ethos is thus blocked. The horizon, to return to the Nietzschean semantics, of another "daybreak" (D) vanishes. Subjects like Lacie passivize possibilities of imagining life otherwise.

This process of affective, imaginary reification needs to be situated in the line of the three aspects already contoured in the previous parts: firstly, the subject is historically passivized as an object. The historically given perspective of the logic of perfection is responsible for this dynamic. Secondly, this dynamic is structured by various layers that cannot be situated only on one common denominator. We deal with the multiple layers of the "axes," to use the decisive Foucaultian vocabulary, of imaginations, power, and self-constitution. Hence, it would be a misunderstanding to conceive the dynamics of passivization as *merely* leading to arrangements of subjugation or domination. Thirdly, the subject is passivized by the systems of imaginations and power but also passivizes herself. Hence, she is *actively* involved in her own passivization.

With Adorno's *Negative Dialectics*, we further problematize the dynamic of passivization by constellating it in the broader context of "cultural industry."[38] In *Dialectics of Enlightenment*, Adorno contextualizes the pair "active" and "passive" in the context of the cultural industry, which he characterizes as mere noise. In his view, the "cultural industry" numbs subjects instead of raising their critical consciousness and instead of enabling them to *really* listen to the sound of life. The attribute "numb" describes the subject with a reified attitude. In this sense, the subjects' potential to create a dereified ethos is reified by the noisy machine of what Adorno calls "culture." Its noise is kept alive by the subject's imaginations, rendering the subject an accomplice of these processes. In further translating Adorno's formulation of the *Negative Dialectics* into our digital today, we address the problem of our digital present as the specific noisy machine called digital culture: "what is called communication nowadays is but the noise that drowns the silence of the spellbound."[39]

From this perspective, we problematize the strange activity, or the *pseudo*-activity, we encounter in the dynamics of unification introduced earlier: subjects like Lacie *only* manage to fabricate a passivized unity between her desire to imagine digital life differently and the established imaginations to act as usual. In terms of "noise," this unity drowns out the subject's desire for another perspective on life. This unity is acquired at a high price because it is mostly determined by the subject's desire to *just* adapt and thus to merely listen to the existing imagined social expectations. As a consequence, the given perspective of the logic of perfection passivizes the potential to perceive different perspectives on life formation, and the subject turns into the accomplice of this process. With Adorno, we thus note that the process of unification results in silencing the subject's desire for a dereified attitude. The superficially active, but still muted subject does indeed hear a sound, but this sound is only the "noise" that reaches and intensifies her desire to unify her different imaginations of life formation with the social expectations of perfection *one-dimensionally*: she imagines the "noise" of the cultural machine as *necessarily* gaining power *over* her imaginations of how to live differently, and we need to add: this subject is *not* aware of these dynamics of affective imaginary reification.

To illustrate the point, consider the following situation: various social media programs become so self-evident in the subjects' lived realities that their desire for a dereified attitude becomes muted, and the desire to listen to the noise causes the subject to turn herself into a muted object, which here we call the silencing machine; it silences in a quasi-automatic way different perspectives on life formation. The unifying process between the subject's perspective on different ways of life formation and the given one thus only considers the perspective of, in this context, the social media programs; the subject's desire for a dereified attitude is gambled away. Wanting to be a muted object and thus advancing a one-sided unification between possible different imaginations and the stereotype variants of how to imagine normally life causes the subject to be open to passivize her own life. The muted and passivized subjects in the noise of digitalization are *not* able to actively listen to the voice of their self-imaginaries and correspondingly to actively create a unity in which they fit. But at this point, the subjects are not able to undertake such a unification because the sound of the normal self-imaginaries balance out the subjects' subtle desire to give life a form that is open to other possibilities. Hence, the potential to create a different sound of, as it were, is muted. The unification imagined by the subjects as necessary to activate generates *at first* a great deal of noise.

With a differentiation between "active" and "passive," we can further understand that subjects engulfed in digital noise tend to turn themselves into the silently passivized object of, for example, personalized advertising strategies. Although it is well known that digital communication devices can be observed and instrumentalized for commercial purposes, the subjects' self-imaginaries are often so affected by the "noise" of a digitally communicating culture that it becomes part of their imagined self-understanding to feel the urge to utilize these channels, be it WhatsApp or Snapchat. In this sense, the subject is a pseudo-active one that does *not* realize how quasi-automatically or normally the digital culture could lead her to give her life a digital form. As a result she constantly takes selfies not only in famous places such as Berlin, Paris or New York, but also during the performance of ordinary tasks like cooking or playing with a cat. We could even say that the subject becomes so much a one-sided object of digital culture that she even *likes* the desire to give her life a form in which she continually presents and "expose[s]"[40] herself within the frame of the normally habitualized cultural norms.

It is no accident that the active subject turns into the addressee of the affective imaginary type of reification, especially amidst loud digital "noise." Above all, the active subject gives her life an imagined form in the terms of the very same digital "noise." Via her own activity, however, she is initially prompting herself to strengthen her desire to imagine herself as a one-sided object. Ironically, the passive subject, which is in fact more at risk of losing her desire for freedom, has better chances of resisting the affective imaginary type of reification developed here, obviously *without* being aware of doing so. The passive subject desires to lead a life adapted to the normal and accustomed ways of imagining life. Thus, this subject has the will to dispense parts of her active participation in life relatively freely and thus the parts that are responsible for producing the silencing machine. For it is the active subject that explicitly tries to develop new self-imaginaries while questioning her form of life. But in so doing, she silently activates the affective imaginary type of reification. The subject actually conceives of herself as a dereified subject, but she cannot perceive her own blind spots. The affective imaginary type of reification does not provoke a breakdown in the subject's desire to imagine a certain way of forming her life. Such a violent clash would *at least* become perceivable to the subject. However, the affective imaginary type of reification does not make this clear to the subject, but instead operates in "silence." Instead of prompting any easily perceivable kind of rupture, it just causes the subject to continue her life formation, only with the slight silent caveat that this life formation passivizes the subject's desire for a dereified attitude. Life goes on, so to speak, but at what "price," to say it with Foucault, remains shrouded in "silence."

GENEALOGICAL CRITIQUE AND THE CRITICAL CONSTRUCTION OF A DE-REIFIED AGENCY

In what follows, we stress the freedom-empowering side of the affect- and power-sensitive Janus face of subjectivity in digital enhancement in terms of de-reification. In general, the term "de-reification" is derived from Foucault's notion "desubjugation"[41] which he introduced in his lecture "What is critique?" toward the end of the 1970s. "Critique," as he says there, has the function of "desubjugation" and thus of *un*doing the patterns in and through which a subject becomes who she is in living, thinking and feeling in *un*noticed ways. We transfer this formulation to the field of digital enhancement in terms of affect-centered, power-based reification. Correspondingly, "de-reification" is understood as critique that can theoretically conceptualize how the subject can *un*do the patterns of reification that made her hitherto *blindly* think, live and feel as the one-sided object of the digital times.

In the previous chapter we outlined the specific patterns of affective imaginary reification. Undoing these patterns would involve an understanding of how a different intrasubjective world of images could be nourished in the subject, namely such a world that also depicts the problems that come with digital enhancement. For example, one could think of the different images of technology transported by Plato's myth *Protagoras*, in which Plato narrates how the Greek mythological figure Prometheus stole fire from the gods and brought it to human beings. As a result, Prometheus was punished by the gods and bound to a rock, where an eagle picked at his liver.[42] By such narrations, different images could emerge in the subject in and through which he could become aware of the hybrid striving of human beings that seek to play God by technological means.[43] As a consequence, the subject's hitherto unquestioned intrasubjective world of images that transported views of technology as rendering an alleged perfect world possible could become porous and could be opened up for new images that show the downsides of technology, too. In what follows, we, however, do not go further down this path.[44]

For we outline the *basic model* of a de-reified agency in the field of affect-centered power relations. It promises to theoretically conceptualize how the subject can undo patterns of affect-centered reification. Here, the present book analyzes how the critical methodology of genealogical critique and the understanding of enlightenment are the key components of this basic model of a de-reified agency. This chapter describes these critical procedures as the *de-reifying test*, which continues the enlightenment test outlined earlier. The de-reifying test is a specific kind of enlightenment test. It also consists of two aspects. Firstly, its task is to verify the kind of affect-based power relations at stake in the concrete contexts of digital enhancement. Here it needs to check the specific relations of power in terms of reification. Secondly, if the first step concludes that this is a case of affective, imaginary reification, then this shows that digital enhancement has become a problem. According to this diagnosis, the use of digital enhancement in these specific contexts is not justifiable in terms of affective, imaginary reification. This chapter's architecture presents the de-reifying test as follows: The freedom-endangering side shows, in terms of affective, imaginary reification, how digital enhancement can become a problem. The freedom-empowering side then shows, in terms of de-reification, that digital enhancement does not *have* to become a problem.

Against this background, the argument of this chapter is divided into three steps. In the first step, we introduce the basic grammar of the de-reified ethos with the motif "shifting attitudes in counterpoint." The second describes the basic understanding of genealogical critique and enlightenment by way of the following three, core aspects: the "history of ourselves," constructing a de-reified agency and the stylistic form of the experiment/essay. Thirdly, we specify two key modes of the subject's de-reified agency, analyzing the art of the parrhesiastic cry, and the art of the playful. With respect to the corpus of text, we deal in particular with the late Foucault of *The History of Sexuality* and its neighbored smaller texts and lectures, but also with the earlier texts that center on the *topoi* of knowledge, power and ethics.

The Art of Shifting Attitudes in Counterpoint

The basic grammar of the *de*-reified ethos consists of *the art of shifting attitudes in counterpoint.* In general, it encompasses the theoretical conceptualization of how the subject can become aware of his hitherto silently thought, lived and felt reified ethos as the muted object of the logic of perfection. In becoming aware of this, the subject can shift her hitherto silently reified attitude and can *construct a de-reified attitude* in and through which she can gain a voice that is *not* silently dominated by the logic of perfection. Thus, she *can transcend the contexts* of being the muted object of the logic of perfection that does not have a voice.

"Counterpoint" (in Latin: *punctum contra punctum*) derives from musicology, where it describes a technique of composition, used especially by Johann Sebastian Bach, for example, when composing his fugues. It means a web (*Gespinst*) of voices that relate to each other, but that are *nonetheless* independent. "Punctum contra punctum" therefore describes a dialectic. We transfer this concept to the field of digital enhancement, where we understand the relationship of "punctum contra punctum" as the relationship between the subject and the logic of perfection. In analyzing this, we show how the two things can relate to each other, *while nonetheless* remaining *in*dependent. In the context of the field of digital enhancement, we need to differentiate between global and local independence. As the chapter at hand seeks to show, the subject can gain local independence from the logic of perfection, but not global independence, as the logic of perfection is everywhere. To show this, we develop the motif of "shifting attitudes in counterpoint."

If we translate this picture to the field of digital enhancement, then the motif of the "counterpoint" specifies the motif of the de-reified attitude as follows: It explains that the subject is not able to *fully* transcend the contexts of the logic of perfection; for the logic of perfection belongs to the dominant, global perspective of our present day. But the subject *can transcend "punctum contra punctum" local contexts* that have hitherto been silently dominated by the ubiquitous, global logic of perfection. This means that she can *partially* transcend the global perspective of the logic of perfection. She can do so *by resisting point by point against the dominant points of the logic of perfection*. In this sense, the subject shifts "punctum contra punctum" from the muted object of the logic of perfection to the subject that can have a *partially* dissident voice.

The motif of shifting attitudes was already introduced in the chapter on Nietzsche, from whom the term derives. There, it was spelled out as the motif of affect-centered, power-based changes of perspectives or, in Nietzschean terminology, of "perspectivism."[45] The term "perspectivism" does not feature prominently in recent French philosophy and contemporary affect theory. But an implicit reading of related terms nevertheless shows its relevance. From this perspective, we conceive the motif with Foucault as shifting attitudes,[46] a term related to Eribon's understanding of "cultural transmission"[47] or to Massumi's approach of "navigating movements."[48] All of these formulations provide the tools to conceptualize the subject's potential to develop a de-reified attitude that empowers her to shift, transmit, navigate or wander *from* the given perspectives that structurally makes her into a muted object *to* possible new, different and dissident perspectives by which she is able to resist her status as a muted object. In order to develop such a motif, we do *not* follow the dominant discussions on perspectivism. For they mainly seek to clear the epistemic ground that perspectivism provides and thus ask to what extent perspectivism makes a true perspective possible. On the contrary, we refer in a new way to the notion of perspectivism and place it in relation to the method of genealogical critique and the notion of enlightenment that is specified here in its elements of de-reification. *Against* the common accusations that perspectivism involves relativism, perspectivism by contrast as it is understood here is *not* relativistic. For it shows the hitherto *wrongly* justified patterns of affect-centered domination that restricted the emergence of different, plural voices that dissent from the logic of perfection.

In the previous chapter, we already described the method of genealogical critique and the notion of enlightenment in its basic structure. In general, to cite Foucault, genealogical critique is the "historical investigation into the events that have led us to constitute ourselves and to recognize ourselves as subjects of what we are doing, thinking, saying" (WE, 315). In this sense, genealogical critique sheds light on the "history of ourselves." In doing so, the subjects can become aware that the way they have hitherto blindly thought, lived and felt is not given and thus not an "anthropological constant" (Foucault), but is rather contingent. "Contingent," as in the previous chapters, means "not necessary." In this sense, genealogical critique, to further draw on Foucault, enables the subject "from the contingency that has made us what we are" (WE, 315p) to understand and experience "the possibility of no longer being, doing, or thinking what we are, do, or think" (WE, 315p). But this type of "contingency" is *not* to be misunderstood as a *total* contingency that would identify every arrangement as not necessary (possible after Lukrez),[49] and thus as liquefiable, and would, therefore, lead into relativistic pitfalls. By contrast, the understanding of contingency in the terms of genealogical critique developed here does lead to the understanding of the critical construction of an *enlightened agency* that in this chapter is specified as a *de-reified agency*.[50] Hence, genealogical critique does not address every "history of ourselves" in which subjects have so far thought, lived and felt, and thus declares everything as contingent and therefore as possibly fluid; *it only* uncovers the specific "history of ourselves" that *excludes* different, dissident "histories of ourselves." In this sense, genealogical critique is the critical method that problematizes only *the history of what was so far justified, but what is not justifiable*. From this perspective, patterns of affect-centered reification belong to this problem that is tackled by genealogical critique. For they make blindly justifiable what is not justifiable, as they make the subject *blindly* turn into the muted object of the given contexts, that is, in the context of digital enhancement, the context of the logic of perfection. As a consequence, they *exclude* dissident voices that deviate from the given context of the logic of perfection.

In this context, genealogical critique needs to be conceived of as a *continuous process* that does *not* reach a last validation (*Letztbegründung*). By contrast, it theoretically conceptualizes that the subject is continuously called to shed light anew on patterns of affect-centered reification. Correspondingly, we speak of the *de-reification of de-reification*. According to this motif, the subject is not fully liberated once and for all from being the muted object of the logic of perfection. As shown in Foucault's critique of the "repressive hypothesis" and thus in his critique of Herbert Marcuse's and Wilhelm Reich's theory of sexual liberation,[51] Foucault clarifies that it would be shortsighted to conceive of subjects as eternally liberated from being the muted object.

The Three Aspects of Genealogical Critique and Enlightenment

In what follows, we specify the motif of shifting attitudes in counterpoint in terms of genealogical critique and the notion of enlightenment by ways of the following three key aspects: firstly by way of the "history of ourselves," secondly by way of the critical construction of de-reified agency, and thirdly by way of the stylistic form of the experiment/essay.

"The History of Ourselves"

The first key aspect of the motif of shifting attitudes in counterpoint encompasses the understanding of "the history of ourselves." In this context, Nehamas is right to state that genealogy is first of all histories. Nonetheless, it is not any history; genealogical critique is *historical critique*. What kind of historical critique that is we learn from Geuss, who clarifies that genealogical critique is *not* a stabilizing history as one could conceive of it at first. For example, one could be inclined to think of the history of a pedigree, to refer to Geuss' example. A pedigree tells the history of people in order to justify the way they are in the here and now. Agamemnon, for instance, tells the story of his fathers and grandfathers *in order to legitimize his power* as the king of Mycenae. In this sense, a pedigree has a vindicatory function. *By contrast*, shifting attitudes "punctum contra punctum" entails a *de*bunking and thus a *de*stabilizing aspect. For it theoretically conceptualizes *what was so far justified, but what is not justifiable*; hence, ascribing a vindicatory function to the motif "shifting attitudes in counterpoint" would *merely* continue *the justification of what is not justifiable*. The subject would *still* remain in his reified ethos in and through which he is the muted object of the dominant logic of perfection. It is against this background that the motif of shifting attitudes needs to be theoretically conceptualized as starting by liquefying and thus by making fluid what was hitherto justified but is not justifiable.[52] It is only this movement that theoretically conceptualizes how the subject can make room for new, dissident attitudes.

Moreover, the object of this historical critique is not any object, but "the history of ourselves." It makes plausible that when a subject tells "the history of ourselves," this subject can understand that his life formation is not a constant, that it is not naturally given, but has a history; when one sees the historicity, one's own life formation loses its allegedly self-evident appearance and turns into *a changeable process*. For if someone can say that something was this or that way at a certain time, he can also say that it is this way in the present, which in turn makes room for further developments in the future.

In general, understanding history as critique goes back to Nietzsche. In his text *Beyond Good and Evil* and the corresponding commentary *On the Genealogy of Morals* Nietzsche argues for the need also to historicize areas such as morals or truth that were hitherto not understood as historically made, but as naturally given. At the very beginning of the *Genealogy of Morals*, Nietzsche clarifies this task as detecting "the descent of our moral prejudices"

(GM, preface 2, 1), a task that he calls the analysis of "[his] a priori" (GM, preface 3, 4), a term that Foucault will later change to the "historical a priori" (AK, 126). The Nietzschean "a priori" in the *Genealogy of Morals* deals with the question "of what origin our terms good and evil actually have" (GM, preface 3, 4). Nietzsche thus tries to challenge the supposedly natural character of morals and shows that also this area, which was previously taken for granted, is historically made. Accordingly, it is not based on natural, eternal or substantial foundations, but on practices.

To illustrate the point, let us consider a today's fictitious example and refer to the mother in the computer game "Perfect Mom" of the film *Her*. She might consider herself to be a morally justified mother, in parenthesis: she fulfills the necessary task of being a good mother in accordance with the social expectations of her time, by being perfect in appearance, education, job, etc. But with Nietzsche we see that what she imagines is not based on necessary foundations; on the contrary, they are *historical* products, their alleged truth has only a timely character, like the belief, associated with the Germany of the 1950s, that to be a perfect mother one had to satisfy the so-called three 'K's': "Küche, Kinder, Kirche" (*kitchen, children, church*).

The Foucaultian "historical a priori" broadens the scope Nietzsche had in mind when speaking of morals and truth and considers the field of social facts. Correspondingly, Foucault questions the "historical a priori," that lead the subject to value the alleged social facts as self-evident, and thus to believe in the allegedly self-evident divisions between, for example, the mad and the non-mad, or the delinquent and the non-delinquent (NGH). In drawing Foucault's line further to the case of digital enhancement, we are able to theoretically conceptualize the "historical a priori" of life formation in digital times and can thus question the allegedly self-evident division between perfect and non-perfect. This view helps us understand that a subject that becomes aware of the historically produced character of the division between perfect and non-perfect is then able to conceive and experience that his life formation is not an allegedly self-evident, unchangeable social fact, but is historically shaped by the logic of perfection and is therefore changeable.

In this way, *the practical character of one's own life formation* becomes visible. In understanding that life formation is not a constant, but a process, the subject can learn to understand and experience that this process is shaped by several different practices. Accordingly, the subject can see that his very own life formation also consists of all the "self-practices" (Foucault) that one performs blindly day by day; this is the case, for example, when Snowden prepared his talk in Tokyo and had to examine the history of spying technology. By describing the different histories of the cultural, social and political practices that assume different forms over time, such as the practices of spying technology in Snowden's case and the histories of the practices of the prison, of the clinic, of punishment, of juridical forms or of life formation in Foucault's case, the subject can be empowered to question the alleged natural status of his life formation. As a consequence, the given perspective, which ensured that the subject tacitly denied this practical character, can become fluid.

However, understanding the practical character of one's own life formation requires time, as Snowden shows—in Snowden's case, three years. In Tokyo while delivering his talk on spying technology, Snowden *only haphazardly* realized that not only the Chinese government is spying on his people, but also the US government. Nonetheless, he did not realize at this point what this fact meant for his own life formation. He only tells us in his autobiography that after his talk in Tokyo he was so troubled that he could not really sleep. Three years later it occurred to him that, all of a sudden, he understood and experienced that he could no longer work for the NSA because of its spying practices. We can compare this process with the work of the mole. Analogously, this process signals that the subject who develops a de-reified attitude is able to delve down into his affective attachment to the logic of perfection, which he previously developed blindly. Just as the mole works invisibly beneath the earth, so too the subject works in similar ways beneath the surface of his visible behavior, and thus digs down into his affective life. But this invisible work is not transparent to the subject right from the beginning. The subject does so, without being *fully* conscious of what he is in fact doing. Snowden's key formulation for describing this process is in this sense: "it struck me."[53] Nonetheless, one day, like the creation of a mole hill, the process can reach the light of the day and the subject is able to highlight his hitherto blind attachment to the logic of perfection and, in so doing, create a new perspective on his life.

In other words, the subject can realize that one's own life formation consists of different layers of practices that make one think, live and feel. Woolf provides an ironic picture of how these different layers could be outlined when she writes in her feminist classic *Orlando* that life formation is composed of "selves [. . .], one on top of another, as plates are piled on a waiter's hand."[54] Just as the mole excavates the different layers of the earth, so too the subject excavates the different layers of his practiced life formation that are stapled on top of each other like plates over time. And, to further draw on Woolf, these different practices that shape the different layers of the subject's life cannot be understood as mere "amalgamates"[55] that stick together due to the subject's "Captain self."[56] This would mean that they would only be put together due to the subject's perspective that wants to make her immediately open to blindly pursuing the path paved by the perspective that is dominant in a specific time and that determines what one just needs to understand and experience as valuable. In today's times of digital enhancement, the subject's perspective of the "Captain self" would thus be marked by the logic of perfection. In Orlando's time, the "Captain self" represents the dominant perspective that makes the subject think and experience that leading an alleged valuable life would entail marrying a rich husband, in the case of women, or for men, pursuing a career, and in our time using digital means as much as possible. But the lens provided by understanding that the "history of ourselves" is shaped by *various* practices over time enables us to perceive that the subject's "Captain self" is also historically shaped. Thus, it only has power *over* Orlando's different perspectives on life due to its historically produced value. Hence, such an understanding can make the subject realize that in a different time there could be different "Captain selves." In seeing this, the contemporary practiced "Captain self" loses its value and Orlando can come to see a possible "great variety of selves"[57] that are not determined by her "Captain self."

To conclude: A subject that can gain a critical understanding of the "history of ourselves" is able to see that his own life formation must not be a mere continuation of the past life formation. Accordingly, this subject is able to distance herself from the claims of the past that she hitherto blindly assumed to be necessary. Now she can understand and experience that hitherto she has only deduced from past values what is valuable in her life formation in the here and now and thus also in the future. In this way, the subject can see that there is space for different perspectives on life formation that were hitherto covered immediately.

Critically Constructing a De-Reified Agency

Now we turn to the second key feature, "constructing a de-reified agency." It specifies the self-reflexive procedure in and through which the subject's dissident, new perspective on life formation is constructed by criticizing the "history of ourselves" in and through which the subject previously thought and experienced himself as the muted object of the given order. Thus far, we know that de-reified agency entails the ability to place the natural character of the given perspective on life formation in question, as Snowden was able to do, and thus it encompasses the ability to see the history of the given perspective. We also know that, in so doing, the subject opens up new spaces for developing a de-reified attitude, a dynamic that we described in terms of the motif of shifting attitudes, of switching from a reified to a de-reified attitude. However, we do not yet know how this process works in concrete terms.

To this end, we specify the understanding of the de-reified attitude as the type of subjectivity in and through which the subject can see the plural ways of attaching to the given order and thus undo the ways in and through which she was the one-sided object of that order. Translated to the case of digital enhancement, the given order is another name for the logic of perfection. Accordingly, constructing a de-reified agency entails the subject's ability to criticize her hitherto one-dimensional attachment to the order of perfection. In doing so, she can see the plural range of placing herself in relation to the order of perfection. Hence, this subject is able to suspend the logic of perfection in areas of her life formation. But this also means that the subject can attach to the logic of perfection, but only in a self-reflexive manner—that is, if she chooses to do so *in a free manner*.

The theoretical register that conceptualizes the potential plural range of the subject's attachments to the given order is social ontology already introduced in the previous parts. Basically, "social ontology" describes the specific circumstance that everything *that is*—i.e., subjects, animals, institutions, practices, things, etc.—is *inter*related and thus *inter*dependent in and through relations of power, that is, of a web of forces (*Kräfte*) that has emerged through causes and effects.[58] Foucault's perspective on social ontology entails the understanding of the "critical ontology of ourselves" (WE, 316).[59] It can theoretically conceptualize that the subject can become aware of the potential different ways of becoming who she is in the historically given web of forces (*Kräfte*). Power understood this way does not have a mere dominating function as described in the feature "power *over*"; it can also have a freedom-*enabling* aspect as described in the expression "power *to*."[60] In other words, social ontology is the theoretical register that shows that the subject can become aware of the *wide range of power relations* in which they are in the web of forces (*Kräfte*) generated in the here and now—ranging from patterns of domination to patterns of *em*powerment and thus of liberation. Affect-centered reification is part of patterns of domination. For it makes the subject become, in a blind manner, the one-sided object of the given contexts and in this sense it has power over her. Translated to the case of digital enhancement, the social ontological register can theoretically conceptualize that the subjects can understand and experience that they have the *power to* distinguish, in the ancient Greek sense of κρίνειν (*krínein*) and thus of "making distinctions," between how they *are* in the present as the muted object of the given contexts and possible ways of *being* a different subject that they were not able to develop until now. This subject does *not* emerge by mere relations of cause and effect like a billiard ball that is hit at some point and that then has the effect of knocking further billiard balls in different directions. In this scenario, there would not be the possibility to see different ways of becoming who one is; one just is the way one is hit by the billiard ball. In the case of digital enhancement, the social-ontological view clarifies that the subject can become aware that she is not only the object that immediately attaches to the given perspective of the logic of perfection—in other words, an affectively dominated subject—but potentially can be a different, dissident subject *by* *questioning* her hitherto blindly formed perspective on forms of life. Hence, the logic of perfection *no longer has power over* the subject. Moreover, such a view of power in terms of social ontology needs to be understood as "voluntaristic." For the subject, having a de-reified attitude is *not* the accident of the logic of perfection; she is *not* its muted object, but a subject that has a voice in virtue of which *she can choose* of how to relate to the logic of perfection.[61]

With Foucault's understanding of the social-ontological register, therefore, we see that the subjects' being in the present, i.e., their thinking, experiencing and living altogether, stands in the context of the "history of ourselves." It is no accident that Foucault also speaks of the "*historical* ontology of ourselves" (WE, 316, emphasis by the author). Hence, the subjects' potentiality to be different emerges, to translate it in Sabine Hark's words, *not* "from a point zero,"[62] but against the background of the "history of ourselves" that until now affectively dominated her being. But in conceiving and experiencing that their being is related to history, the subjects can become aware that their being could also be different, namely in a way that is not blindly dominated by the logic of perfection. Applied to the case of being the muted object of digital enhancement, we see that the subject and her momentous status as a muted object are *not* identical, i.e., the subjects as such are not the muted object, only their *current* attitude makes them immediately imagine, behave and feel as a muted object of the logic of perfection. This difference is important. For it shows that subjects have *a double affective disposition*: *on the one hand* to be a subject that is open for having her will formed and *on the other hand* the potential to generate a dereified attitude, and thus the specific potential here outlined as shifting the subject's attitude from being the muted object to becoming a subject who can raise her voice. Accordingly, the subject with such a voice is able to "to recognize [herself] as subject[] of what [she is] doing, thinking, saying" (WE, 315).

On this basis, the social-ontological register enables us to conceptualize the difference between affect-centered power relations and relations of affective domination in terms of *symmetry and asymmetry*. Usually, patterns of domination are understood by way of asymmetry and patterns of freedom by way of symmetry. Affect-centered power relations, following Foucault, however, are asymmetric, *but not necessarily* patterns of domination. The social-ontological register allows us to understand why this is *not* a contradiction. To this end, we refer to the case of our digital present. In the field of digital enhancement, the subject is *structurally* confronted with asymmetric affect-centered power relations from early on. The historically given perspective of the logic of perfection represents such an asymmetric power relation. The subject does not have a symmetric position, since she does not have the means to simultaneously and symmetrically initiate "an equilibrium" (SP,

218) between her perspective on life formation and the given perspective, to put it with Foucault. From an early stage, the subject needs therefore to be conceptualized as the muted object of the historically formed perspective, just as she is prompted to imagine her own perspective as necessarily subsumed under the given perspective. Hence, we do not perceive *a* symmetry between the subject's perspective on life formation and the perspective prescribed by the logic of perfection, but instead an *a*symmetry.

But the social-ontological character shows that the asymmetric power relations, though problematic in the here and now, have the potential to be changed, too. For it makes clear that affect-centered power relations consist of *different fields of forces (Kräfte)* that do *not* assume a substantive form, but instead a *relational one* that is open to change. In other words the social-ontological register enables us to conceptualize that the subjects' situation in digital enhancement should not be confused, to put it in Foucault's drastic terms, with "slavery" (SP, 221). On the contrary, we can see that the subject has the affective disposition to be free, by which we mean that these potentially free subjects, to quote Foucault again, "are faced with a field of possibilities in which several ways of behaving, several reactions and diverse comportments may be realized" (SP, 221). Thus, the social-ontological perspective provides us with the tool we need to shed light on the circumstance that the historically given perspective of the logic of perfection does not have the mere affect-centered *power over* the subject's life formation. By contrast, the social-ontological register shows that the subject can also acquire the affect-centered *power to* relate *differently* to the perspective of the logic of perfection, and thus to become aware of her affective disposition to creating new and dissident perspectives on life formation.

Seen from this angle, asymmetric affect-centered power relations *as such* are not problematic because the subject, far from being the *eternal* mute object of the pre-given perspective, can also begin to raise his voice, and thus become the subject of the given perspective, i.e., become a subject who, like Snowden, is able to pluralize his attachment to the given perspective. Hence, the subject does not act like an automaton that quasi-automatically deduces how to constitute himself from the given perspective in a causal manner. He can interrupt the mono-causal attachment and develop different ways of attaching to the given perspective. Snowden's case shows that he was able to interrupt his hitherto mono-causal life formation by pluralizing his perspective on digital enhancement. In so doing, he was able to create a new perspective on his life by detaching himself from his assumption that a valuable life is necessarily based on digital enhancement. From this perspective we see that life formation is *not* a continuous, merely harmonious endeavor; rather, it is an ongoing battlefield in which the different forces struggle for shifts, changes and turning points; its protagonists are called to *continuously* question, contest and rewrite the hitherto established history of life formation thus far determined by the logic of perfection. By questioning its dominant value, the subject can open up a whole new field of life formation and thus shift from a reified to a de-reified attitude. To recap, this book does not understand any contestation in freedom-empowering terms, only those, that pass the enlightenment test (see the previous chapters of this book).[63]

On this critical understanding of the specific "history of ourselves" in terms of affect-centered power relations, the subject is able to imagine, perceive, and experience that it is no longer necessary to imagine her attachment to the field of digital enhancement as "situations or states of domination" (ECSPF, 283, *Herrschaftstatsache*, Foucault, "Die Ethik der Sorge um sich als Praxis der Freiheit," 255). Only now is the subject able to recognize that she no longer needs to attach to the given perspective in ways that Foucault describes as "blocked, frozen" (ECSPF, 283), but in ways that account for the historical character of one's own life formation. In thus perceiving the history of life formation, the given perspective on life formation turns into an interpretable signifier that calls, as Foucault puts it, for a "hermeneutics of the subject" (HS) and thus *for further and further interpreting* the ways in and through which subjects in history so far were dominated: instead of being blindly determined by the dominant signifier of the logic of perfection, the subject can detect the potential for interpreting the signifier differently and thus for pluralizing the perspectives on life formation. Such a subject therefore has the de-reified attitude that he can perceive, understand and experience that his own life formation does not exist in determined, one-dimensional patterns, but on the contrary in pluralizable ways that we call, in analogy to Derrida, life formation *à venir*, a life formation which only be conceptualized as one that comes into being.[64]

The Stylistic Form of the Experiment/Essay

Thirdly, the motif of shifting attitudes in counterpoint in terms of genealogical critique and enlightenment entails the feature of *the stylistic form of the experiment/essay*. Obviously, in speaking of the stylistic form of the essay, we do *not* want to force Foucault into a corset of concepts. Rather, we are referring to a stylistic form that *only alludes* to an essayistic style. Hence, we are not claiming to develop a poetological conception of the essay with reference to Foucault. That would be too ambitious a project. Although questions of style are usually considered to be a mere luxury or an ornament, we consider them in line with Foucault to be an integral part of the argumentative method. In stressing the argumentative character of style, we are *not* favoring the stylistic form over the argument.[65]

Basically, the form of the experimental style developed here shares one core feature with the form of the essay: It promises to provide *the freedom of methodology* that is able to consider the subjective perspective. Since we are not carrying out a poetological analysis, we are not talking about the stylistic form of the essay in the narrow sense used in literary theory, but in a broader sense. This broad sense derives from the literal meaning of "essay." In French, and thus in the language in which the form of the essay was invented by the French Renaissance philosopher Michel de Montaigne,[66] "l'essai" means "attempt." The experiment is also characterized as an attempt. The understanding of "experimental" here, however, does *not* refer to just any attempt, it is *not* relativistic, but to an attempt *to search for enlightenment*. Foucault's term for this view is "experimental analysis" (HS 1, preface to the German edition, 7, trans. by the author; *Probebohrung*). In this sense, we describe the experimental style as conceptualized by the motif of "trying and searching."[67] This understanding benefits from the contribution not only of Foucault's short writings, such as his interviews or his smaller texts gathered in the standard Foucault reader *Dits et Écrits*, but also, if not in particular, by his longer major works, such as *Discipline and Punish*, and his lectures also form part of this line of essayistic thought.

Usually, the form of the essay is understood in opposition to philosophy. Whereas the essay is considered the form that is able to represent the subjective perspective due to its freedom in techniques and thus its formal mobility, philosophy is accused of restricting the subjective perspective due to its systematic character.[68] Here, we draw on Foucault in showing that philosophy in the sense of genealogical critique operates in a style that can be compared *in a broad way* to that of an essay *without* losing its inherent systematic character.[69] Genealogical critique, as developed here, is a methodology that can theoretically conceptualize how the subject, in the field of digital enhancement, can become aware of the affect-centered power relations of which she was previously unaware. This process of awareness-raising, as we have seen, is highly subjective. For it is only the subject herself that can undertake this process; as described in Snowden's case, it cannot be undertaken by someone else. So one can*not* expect, on the formal, stylistic level, a methodology that will be able to describe such a process *in general terms* that would fit every subject's process of awareness-raising. Instead, what is needed is a *flexible and mobilizable form* of style that is able to account *for particular cases by way of momentous snapshots*, but in such a way that the narration of these cases *could be generalizable* and thus could also be understood by other subjects as relevant to them. It is for this reason that genealogical critique is conceptualized as an experimental, essayistic style.

To demonstrate this experimental, essayistic understanding, we stress just one line of thought. It derives from the understanding of the essay as the small or minor form. Drawing on Deleuze and Guattari, "minor literature" is understood as the form that is able to account for the *non*-standard literature that is not dominated by the standard perspective on literature. It is this view that makes the philosophical methodology of genealogical critique experimental and in this sense essayistic. For the philosophical methodology of genealogical critique is the small form, so to speak, that liquefies the grand meta-narratives that seek to show how to account for life formation *as such*. By contrast, genealogical critique is minor in the sense of Deleuze and Guattari in that it *only* narrates *partial* "histor[ies] of ourselves" that take place in several contexts. Foucault describes the various partial "histor[ies] of ourselves" in the cultural, social and political contexts of the prison, the clinic or ancient and early Christian life formation, and this book aims to extend Foucault's line to contemporary digital life formation.

More specifically, the partial "histor[ies] of ourselves" set out from the particular, from possibly intimate situations, such as the sexual relationships of *The History of Sexuality*; they might begin at some random point that at first is not so obvious to the reader, such as Foucault's point of departure, the random year 1639, which he chooses to tell the story of the oppressed rebellion of the *Nu-pieds*. Or they might glimpse at what one only cares for *by the by*. From this perspective, the formal representation of the partial "histor[ies] of ourselves" assumes the guise of an assemblage. Assemblages, in formal terms, are experimental ways of putting together everyday things. In narrating the partial "histor[ies] of ourselves," the narrator puts together experiences and the memories from the subject's everyday life. From the subject's subjective perspective, these experiences and memories make sense as the object of the narration, which needs to be told this way and no other way. While the conception of the experiment is usually part of natural science, for example experimental physics, the conception of assemblage shows that we are dealing with a different kind of experiment. In natural science, experiments are undertaken to falsify results. Foucault, however, uses the understanding of experiments in a *non*-standard reading. He shares with the standard reading *the gesture of proving*. But he then refers to this method *not* in order to prove a result, but in order *to prove formally the plurality of forms of life*. Furthermore, the "experimental" aspect entails an understanding of assemblages of life formation in momentous snapshots, because these histories experiment with the object of analysis in the here and now. The author can choose different experiences in the moments of life formation as the object that he considers worth telling as a contribution to the "history of ourselves." It is this way of narrating the "history of ourselves" that mobilizes the form. As a consequence, such a mobilized way of narrating the partial histories can leave the room required for writing *different partial critical histories*. In this sense, genealogical critique is essayistic in the broad sense because it does not contribute to writing a dominant perspective of a so-called supposedly successful "history of ourselves" (*Erfolgsgeschichte*). *By contrast*, genealogical critique makes room for all the other partial histories that were excluded from the narratives told hitherto. It is in this sense that genealogical critique has the formal mobility to represent the plural character of life formation: For the narration of the partial histories shows the contingency of the "histor[ies] of ourselves." Hence, the way, in and through which these histories are told, makes room for further narrations of these "histor[ies] of ourselves," because the subject reading the histories can understand that different histories could also be told. In this sense, the experimental, essayistic style has a *pluralizing character*. In and through the way they are told, these narratives pluralize the space for the knowledge of the "histor[ies] of ourselves."[70]

And due to the experimental, to wit essayistic understanding developed here, the plural understanding of the critical "histor[ies] of ourselves" is not eclectic. One could argue that by following Foucault we end up speaking of a whole range of different critical "histories of ourselves": the critical history that questions the costs of the subject in prison, the different project that questions the costs of the subject in clinics, or of the subject of pleasure, to which we add the further project of the desiring subject in digital life formation. However, in spite of their heterogeneous and plural character, the formal style of the essays clarifies that these critical histories can*not* be told otherwise. For they are *necessarily* partial critical histories that only in doing so do *not* fall into the pitfall of writing the grand meta-narrative of an alleged successful life formation *as such*.

Furthermore, they are tied together by the search for enlightenment. For the experimental style of the essay is not to be misunderstood in relativistic terms, which would contribute to assumptions about its merely eclectic character. The essayistic representation of the subjective perspective entails a generalizable, enlightened perspective, which avoids this pitfall. It is a project of enlightenment because the various partial "histor[ies] of ourselves" represent (*abbilden*), in ways that are accurate to their specific context, the splinter in the eye of society. Here "enlightenment" means making this previously invisible splinter visible to everyone. Here, for example, we are dealing with the splinter in the subject's gaze that society thinks for example that, as Foucault puts it, it "must be defended" (SMD) against the alleged abnormal, against the alleged mad or against the alleged delinquent. It is in this sense that Foucault does *not* write *any* partial history; there is an *implicit normative common denominator*. He *only* writes the partial history that uncovers *what was hitherto justified but not justifiable*. The histories of the delinquent or the mad are thus the partial histories that tell the story of the hitherto justified, but not justifiable divisions into delinquent and non-delinquent or into mad and non-mad. We continue Foucault's line of thought and write the history of the hitherto justified but not justifiable division of life formation into perfect and non-perfect. The experimental character of the assemblages, however, does *not* mean that what Foucault says would not have much value. Here, we must be more specific: The critique of digital enhancement is valid, one can state in line with Foucault, as long as the form of digital enhancement exists. When it no longer exists, then one could say it has a historic character. The experimental character *only* signals that the critique of digital enhancement is *not the only* critique that is valuable for criticizing the current life formation.

It is in this sense that the experimental, essayistic style theoretically conceptualizes how one *tries to search for truth.* Hence, this search does *not* claim to have the final word, which would correspond to the type of an absolute, substantial or essential truth; this would pave the way for the type of the grand meta-narrative. Instead, what the essayistic style claims to have is *a partial word*, so to speak.[71] So, the experimental, essayistic style claims to search for the truth in the partial history of ourselves, for example in the contexts of the prison, the madhouse, digital life formation and so on. This kind of truth is thus *not* characterized by any type of "absolutes." As the types of partial "histor[ies] of ourselves" vary so much in the different local contexts, it also becomes clear why we can*not* expect to find a fixed principle to give an account of them.[72] This makes one think of a search that does not follow a linear, straight line "from A to B" (FWL, 136), but formally constructs the critical space to catch sight of what has so far been neglected, and to keep this space *continuously* open.[73]

This *also* makes it clear that the experimental, essayistic style is *not* to be understood in the sense of a metaphorical style. The role of metaphor, particularly in the work of the French philosopher Sarah Kofman, is considered as a key part of the French tradition of deconstruction, a line of thought whose origins are associated with Derrida.[74] Metaphors, however, are part of analogous reasoning, which stands *in opposition* to conceptual reasoning. Whereas conceptual reasoning claims to represent reality, analogous reasoning claims to represent reality *in merely relative ways*; in so doing, it *only* offers analogies that are characterized by an *as if* modus: They are used *as if* they represented reality. And Kofman also admits that the understanding of metaphor she develops comes close to that of poetry.[75] But with this closeness she simultaneously accepts a *mere relative relation to truth*, although she might *only* want to criticize a metaphysical understanding of truth.[76]

If Nietzsche speaks of concepts that are nothing more than metaphors, he does *not* do so in order to provide an entire metaphorology. *On the contrary*, what can be seen in the French tradition from Nietzsche to Foucault, as we attempt to show here, is the *practice-based* understanding of concepts. Concepts are not merely there, they are developed, arranged and re-arranged by human beings over time *with the aim of searching for truth*. Furthermore, they are not only "invented" or "found,"[77] but, as conceptualized here, have *a history*, namely the "history of ourselves," in and through which their development emerges. The essayistic style is the form that aims to represent this practice-based understanding of concepts and to highlight the mobilized, that is *experimental mediation* (*Vermittlung*) between concepts and reality. We thereby see that reality is *not* a mere positive fact, which can be objectively described by concepts. Rather, we seek to make plausible that the way concepts describe reality is already the result of *historically shaped* practices. These historically shaped practices are *not relativistic* because their value needs to be interpreted in the scope of their specific context and time. They are justified in case they pass the enlightenment test (see the previous chapters). Nietzsche, for example, gives plenty of illustrations of such a practice-based approach to the value of concepts. One of the most instructive is his formulation of the "columbarium of concepts" (WL, 1, 882, trans. by the author) in *On Truth and Lies in an Extra-Moral Sense*. The term derives from Latin and means "dovecote." Only this meaning is of relevance here; the contemporary meaning of "columbarium" in the sense of an "urn house" that stores the ashes of the dead is irrelevant. The "columbarium of concepts" shows that the value of concepts does not have an eternal character, but is volatile, like doves that fly away. In other words, the value of concepts is changeable. If the value of the concepts is no longer justified, the doves, or the content of the concepts, are enclosed in the "columbarium." Accordingly, subjects are called to continuously search for the enlightened justification of concepts.

This view promises to have the advantage of being *sensitive to practices of epistemic domination*, that is, the exclusion of dissident voices that is produced by history-blind forms of knowledge, such as prejudices in the context of madness or delinquency, to refer to Foucault's examples. The expectation now is that the experimental, essayistic style will be able to *flexibly* relate the specific local contexts of "the history of ourselves" (e.g., the history of the mad, of the delinquent and so on, in other words the content of concepts) to the concepts already given in society, in a manner that is sensitive to practices of epistemic domination. Here, the experimental style theoretically conceptualizes the enlightened search for the truth, which tells the story of the previously silenced partial history of the mad or delinquent. Practices of epistemic domination would silence these partial histories and would only care for the bigger picture of the alleged history *as such*, that is the usual understanding of madness or delinquency. But in explicitly being sensitive to the local contexts, the experimental style promises to counter patterns of epistemic domination, by adding further partial "histor[ies] of ourselves," which are able to shed light on the concrete practices of domination that have so far been forgotten. In doing so, the essayistic style is linked to the hope of *constructing an enlightened space for a pluralized use of concepts*, which can counter prejudice-based practices of concepts.

With the understanding of the experimental, essayistic style developed here, Foucault belongs to a larger stream of thought that is developed in nineteenth-century and twentieth-century critical philosophy in the tradition of Nietzsche and Adorno. If Nietzsche scholars mainly distinguish between the Nietzsche of the aphorism, that is, of the nuanced short passages that describe a subjective, but generalizable situation in life, be it a private, scientific or religious context, and the Nietzsche of the longer, supposedly more systematic works, such as *On the Genealogy of Morals*,[78] then the understanding of the essay as it is conceived of here considers Nietzsche's works as *a relative formal unity*. For their common formal denominator lies in the experimental search for representing the partial histories of life formation in their plurality without excluding possible different "histor[ies] of ourselves."

Later, in twentieth-century philosophy, it is especially Adorno who stands in this line, too. Adorno also places himself in line with Nietzsche when he states at the beginning of his major work *Negative Dialectics* that it is with Nietzsche that the "turning point"[79] (*Kehre*) in philosophy started. In particular, Adorno's shorter works *Eingriffe* or *Minima Moralia* can be called essays in the sense described above. They are essayistic insofar as they make room for de-identifying thinking and thus for a thinking in constellations. De-identifying thinking resembles the understanding of the partial "histor[ies] of ourselves" that opposes identifying, totalizing thinking. By contrast, identifying thinking is the thinking that is dominated by the perspective of how to lead life *as such*, namely in identifying with that which is given in the "administered world,"[80] for example in the culture industry. This identifying thinking, however, restricts the subject's different perspectives on life formation. For Adorno, grand meta-narratives would be the genre that represents the perspective on identifying thinking. It is in this sense that Adorno writes the variety of the "histor[ies] of ourselves," for example in the *Minima Moralia*, that incidentally consider themes like the subject's allotment garden or the surface of a sea. Writing these partial histories formally makes room, then, for writing different partial histories.[81]

The Parrhesiastic Cry and the Playful

In what follows, we concretize the art of the de-reifying attitude by sketching the art of the parrhesiastic cry and the art of the playful. The two modes of the parrhesiastic cry and the playful are related to each other as follows: While the parrhesiastic cry designates the outburst of the de-reified ethos, and thus the moment when the subject becomes aware of the new attitude, the art of the playful shows how the de-reified attitude can be kept alive by the subject's activated, mobilized and flexibilized attitude, here described in terms of the mode of the playful. In the previous chapters we already compared the process of consciousness raising, and thus of becoming aware of the de-reified attitude, with the work of a mole. To remain with this image, we can reformulate the relation between the two de-reified attitudes as follows: The moment of the parrhesiastic cry enters the scene when the mole suddenly creates his mole hill. We compare its creation to the moment when the subject gains an enlightened gaze and sees the new perspective in self-reflexive manners that was previously not clear to her. The playful characterizes the mole's continuous work of digging the tunnel without being determined to dig the tunnel exactly at a certain point, instead doing so flexibly.

The Art of the Parrhesiastic Cry

It is now time to outline the art of the parrhesiastic cry as a de-reifying ethos. Etymologically speaking, parrhesia comes from the ancient Greek verb "parrhesiazesthai," which means "to say everything"[82] (GSO, 43). The field of digital enhancement suggests the following interpretation, which takes as its starting point Owen's interpretation of Foucault;[83] according to this interpretation "parrhesia" is the mode in and through which the subject can raise her voice.[84] In digital enhancement, "parrhesia" provides the theoretical register required to understand how the subject is able to speak in her name that is *not* over-determined and thus reified by the dominant perspective of the logic of perfection, and thus can resist her status of the muted object of the logic of perfection.

The proposed interpretation is based on the following texts. The notion of parrhesia plays a crucial role not only in Foucault's late work,[85] but also in his earlier works.[86] In particular, it features in Foucault's late lectures *The Government of Self and Others*, where it plays a role in political, social or intimate contexts. We interpret "parrhesia" as a mode of the subject's de-reified ethos in and through which she can question her hitherto blindly reified attachment to the knowledge-power nexus. Thus it was also necessary to refer to the earlier Foucault who dealt with the themes of knowledge, power and ethics.

Whereas we already problematized the subject's blind attachment to the logic of perfection in and through which the subject, like Snowden, *slowly* gains a new perspective on his life, with the de-reified ethos of the parrhesiastic cry we encounter the sudden, unforeseen outburst of the subject's de-reified attitude. Here, we shift to the field of, in Foucault's terminology, "truth-telling."[87] The parrhesiastic cry contributes to the freedom-empowering side of the affect-centered, power-based Janus face of digital enhancement insofar as it conceptualizes how the subject is able to detach from his previously unquestioned blind belief in the "truth-telling" (Foucault) produced by the dominant knowledge of a certain time and society. In so doing, it can make plausible that the subject can question his hitherto blindly accepted status as the muted object of the dominant knowledge, and thus perceive that this knowledge is *not* necessary, although she regarded it until now as *the* truth that one just accepts. As a result, the subject is no longer the mere object of the dominant knowledge; to put it differently: she is no longer the mere receiver of what is supposedly usually or normally considered true, but she begins to actively conceive of and experience herself as the subject who also contributes to the shaping of knowledge. Thus, the subject begins to understand herself as having a voice in society that counts. In this sense, Foucault explicitly speaks of "truth-telling." Thus, it is *not* a self-evident social fact in whose production subjects are not involved. By contrast, the production of truth is the active process that emerges due to subjects that tell what is true and what is not true. This image resonates with Nietzsche's understanding of truth. Nietzsche formulates in similar ways that it is the subject that creates what is considered true or not but then over time forgets his involvement in the creation of truth (WL). Generally speaking, in this way we see that truth *is practiced*. The "parrheastic cry" is now one of the key practices, according to Foucault, in and through which truth is practiced.

To see this point let us examine how Foucault describes the parrhesiastic scene. To this end, we refer to the lectures *Government of the Self and Others* and single out the key scene that takes place between Creusa, a wealthy woman in Athens, and Apollo, the divinity of light. This scene goes back to the tragedy *Ion* written by Euripides around 418 BCE. It narrates the history of the secret intimate relationship between Creusa and Apollo that takes place in ancient Athens and from which the child Ion emerges. Creusa abandons her child on his birth, so she does not know that Ion is still alive. The problem begins at the moment when Creusa and her husband Xuthos come to the oracle of Apollo, and Creusa asks Apollo through the oracle where her son is. Apollo's words as conveyed by the oracle represents the perspective of truth that Creusa and Xuthos have to consider as the right "truth-telling." According to Apollo, the first man they will meet on leaving the temple will be their son. They meet Ion, who is *in fact* Creusa's son, but Creusa and Xuthos do not know that Ion is the child of the relationship between Creusa and Apollo. Hence, Apollo speaks the truth through his oracle, which must be recognized as valid. In the end, in Foucault's narration of the story, Apollo's truth turns Ion not into his own son, but into the alleged son of Xuthos from a liaison with a slave whom Creusa has to accept. Within the ancient normative orders, as Foucault observes, this scene belongs to the "worst that a

Greek noblewoman, a woman of high birth who has to continue the line of her ancestors, may face" (GSO, 118p).

Against this background the parrhesiastic cry begins. Creusa cries out her anger, fear and sorrow, and in so doing she destabilizes the eternal character of the truth that Apollo's "truth-telling" claims to have. Her cry shows the practical character of truth-making. For by resisting the truth as told by Apollo, Creusa questions its essential, natural or substantial character, which imposes on Creusa the status of a mute object that does not have a voice. This scene shows that, through the parrhesiastic cry, Creusa does not systematically argue for the necessity of a different truth in and through which she is not only the object; she just no longer tolerates the truth presented by Apollo, and thus she can no longer bear her status as an object. So, she is suddenly able to perceive, understand and experience truth as *not* eternally given. On the contrary, she recognizes that truth is socially constructed, produced and practiced by Apollo's oracle. In this way, Creusa becomes a subject who can raise her voice. Hence, Creusa no longer accepts the truth imposed on her and fights for a different truth-telling which makes her leave her status as the mere object who would only be subjugated to the history of being a betrayed wife who has to care for her husband's son. So, Creusa's rage prompts her to no longer believe and imagine that Apollo's truth-telling is the only valid truth. As a consequence, she can begin to adopt a self-reflexive attitude in and through which she develops a vivid, flexible and mobile "mode of relating" to the given perspective. Thus she can take her destiny in her own hands not as mere object, but as a subject, too, who can raise her voice. In this context, Creusa's perspective on truth-telling is *not* to be misunderstood in relativistic terms. She does not say that there are no truths; rather, what she grasps is the circumstance that the process of truth-making matters to her as she plays an active part in this process.

In the field of digital enhancement, we are obviously *not* dealing with the case of a divinity, such as Apollo, who tells what is supposed to be recognized as true. Nevertheless, the practices of digital enhancement also simply appear to the subject as the unquestioned truth, just as Creusa in the beginning was prompted to believe in the oracle given by Apollo. And also in digital enhancement subjects with a de-reified ethos are able to perceive the practical character of the truth that until now turned them into muted objects. With the episode "Arkangel" of the Netflix series *Black Mirror* we can illustrate a drastic variant of such a parrhesiastic cry. This episode problematizes the case of a mother and her child Sarah. Because the mother fears that she is not good enough at raising her child, she looks for help and finds the supposed remedy in digital media. Sarah becomes the muted object when she receives a digital implant in her head, without her consent, by which she can be controlled at all times by her mother. Her mother can even regulate what she can see. Sarah discovers by chance that her life is totally determined, and thus fully reified, by digital media and thus also by her mother who is in charge of the technology. She expresses her rage and sorrow in a parrhesiastic cry and starts an argument with her mother; in the ending she leaves her mother. At this point, Sarah's parrhesiastic cry also expresses, as in the scenario with Creusa sketched above, that the truth formulated by the dominant perspective of digital media is not essential, substantial or natural, but has a practical character and thus is changeable. In Sarah's case, the truth was practiced and even invented by some engineers who regarded technology as a remedy that might be better at raising children than parents. But the way Sarah's life unfolds reveals that it is not true that her life was perfected by digital media; on the contrary, she lives the nightmare of being fully controlled by digital media and her mother.

With respect to the three key features of the subject's de-reified ethos we can state the following: The first feature, the "history of ourselves," makes the subject aware of the object of history, and that is the social character that the history of her life formation has. Hence, Sarah comes to realize that she is not the muted object of a natural history against which resistance would not make sense. But because she is the object of a social history, she can begin to resist the perspective that presented itself as her destiny. With the second feature, that is, critically constructing a de-reified agency, we see that Sarah constructs her subjectivity in and through which she is no longer the muted object of digital media, but the subject that can criticize her life formation performed hitherto by crying out her anger. The third feature, the form of the experiment/essay, explains that the narration of Sarah's life formation is not the grand meta-narration of life formation as such, but a partial history that leaves the room required for telling life formation differently.

The Art of the Playful

After having sketched the mode of the "parrhesiastic cry," the present book now delineates the aesthetic category of the playful as a key variant of the subject's de-reified ethos. While the parrhesiastic cry shows that the subject's consciousness rises suddenly, the second motif of the aesthetic category of the playful in turn sketches *the activated manner* in and through which the subject develops her de-reified attitude. This motif thus takes up the thread from the previous chapter of the pacified and pseudo-active subject and shows how she can change the situation as her *real* participant. In this sense, the notion of the game should not be misunderstood as the situation familiar from Beckett's play *Endgame*. For Beckett's endgame does not leave any space for changes, which in our case means shifts in attitudes, and thus for the subject's specific shifts from a reified to a de-reified attitude. Different to

Adorno, we do not "try to understand Endgame";[88] we just leave it there by stating that the absurd theatre play *Endgame* is about the absurdity of gaining meaning in life. In the play, this theme is taken up by the dysfunctional communication between Hamm, an old man, and his servent Clov while Hamm's parents Nagg and Neil sit in two garbage cans. As there seems to be no way out of this meaningless situation, the title *Endgame* is taken from chess, where the endgame is the final part of the game where no return is possible. Transferred to the situation of shifting attitudes, Beckett's absurd theatre *Endgame* would not leave any room for shifting attitudes. For Beckett shows that the protagonists are caught in their attitudes. Clov, for example, remains Hamm's servant whether he wants to or not. His efforts to leave Hamm are in vain. From this perspective, it becomes obvious that the Foucaultian mode of access calls for a reconsideration of the understanding of the game. Here, "game" must be understood in such a different way that it can open up the space for what we can call, following Nietzsche "entr'actes" (*Zwischenspiele*, JGB, 5, heading of the 4th section, trans. by the author)—so that for Foucault, in contrast to Sartre, "Les jeux *ne* sont *pas* faits."[89] Another situation of the game could be possible.

The aesthetic category of the playful has a long history. In Western philosophy, it is usually traced back to Plato's essay *Ion*. There the category of the game has the function of showing that things go easily. With Aristotle, the playful is connected to children and thus acquires the negative connotation of childishness. The understanding of play underwent a reevaluation in German Idealism and German Classicism. Kant's and Schiller's understanding of the playful is instructive for the approach developed here insofar as they both theoretically conceptualize with the motif of the playful how the subject is activated to resist patterns of domination. Hence, they consider the motif of the playful a mode of creating freedom. In this sense, the play is linked to Kant's last critique, *The Critique of Judgment* of 1790, to the subject's forces of imaginations (*Einbildungskraft*). The free play of the forces of judgment, to use the Kantian expression, enables the subject to arrive at a judgment of taste that is *neither* subsumed under a specific end (*Zweck*) *nor* under a concept (*Begriff*).[90] Schiller situates himself closely in the Kantian tradition; his emphatic reevaluation of the notion of the playful even ascribes an anthropological value to the playful: "[Man] is only wholly Man when he is playing,"[91] to quote the famous sentences from Schiller's *Letters to An Aesthetic Education of Man*.[92] In twentieth-century philosophy the notion of the playful is considered further. In *Negative Dialectics*, Adorno ascribes to the playful the not "accidental"[93] function that is part of the subject's non-totalizing thinking and thus a mode of thinking in constellations. Here the playful is read as a critical resisting dynamic against the totalizing structures of thinking, of already always identifying something as something; it enables the subject—to return to the ancient Greek understanding of critique as κρίνειν (*krínein*) or making distinctions—to make a distinction to the once set identification of something as something. Hence, the playful on Adorno's conception shows how something could be different.

Foucault's understanding of the playful stands in this tradition as he takes up the thread of conceiving the playful as a mode of enabling freedom. Commentators rarely associate Foucault with the aesthetic category of the playful. If he is considered at all, then it is in reference to his understanding of play, and not, as it is explicitly put here, to the play*ful*, and this reference to the play is done *without* taking the aesthetic dimension seriously.[94] But in so doing, such studies, however instructive they may be, fail to overlook the freedom-empowering dimension developed in the present book along the lines of the ancient Greek conception of *aisthesis* and thus of sensory perception in terms of affect-centered power relations. Foucault addresses the notion of the playful in various contexts, as part of sexuality, of critique, as a mode of "truth-telling" or of playful spaces of freedom (*Spielräume der Freiheit*). The common denominator of these different contexts lies, as we try to sketch, in the freedom-enabling potential *to activate* the subject to adopt a different attitude toward the perspective of the logic of perfection in and through which the subject can see different perspectives on life.

From this perspective we consider the motif of the "art of *playfully* shifting attitudes." It encompasses the understanding of the subject as no longer perceiving her attachment to the logic of perfection, as Foucault puts it, in blindly "blocked, frozen" ways. The playful takes up the theoretical register of social ontology and theoretically conceptualizes the wide range of the subject's *possible* attachments to the logic of perfection. Hence, we deal with an understanding of games of power (*Machtspiele*). The motif of the playful thus theoretically conceptualizes that the subject can become aware "that what exists is far from filling all possible spaces" (FWL, 140). The point that we stress in this context follows Foucault in highlighting the subject's potential to continuously play with her initially dominated attachment to the logic of perfection, to keep the form of her life dynamized and thus vivid. Viewed through this lens, the notion of the playful conceptualizes that the subject can become aware that truth and knowledge does not possess timeless validity *despite of* its temporary dominant character in society; on the contrary, it makes plausible that the subject can understand and experience that also the context of truth and its related areas of knowledge consists, in Foucault's terms, of the "games of truth" (ECSPF, 281). Thus, as with Nietzsche, we also see with Foucault that in different historical epochs subjects invented what they valued and justified as true.[95] In this sense, the understanding of the "games of truth" encompasses an understanding of the games of justification. Accordingly, the motif of seriousness is *not* the alleged counter-part of the playful. For a justification is serious. On the contrary, the counterpart of the playful is the understanding of fixed and blocked relations of power. They would lead to the impasse of a last justification.

The understanding of the playful developed here thus conceptualizes that the subjects can be empowered to recall that their justifications and beliefs in truth and knowledge are the product of the struggles that were once won in society.[96] They do *not* mirror and represent the alleged eternal truth or any kind of last justification. On the contrary, they function as, to put it further with Foucault, a "regime of knowledge" (SP, 212, trans. by the author), in other words, as systems of knowledge and power that determine what is only usually or normally justified as true in a society in a specific time. But in this function the different and dissident forms of truth, knowledge and belifes were excluded, "disqualified" (SMD, 7) and "buried" (SMD, 8), as in today's digital enhancement the different and dissident forms of knowledge and beliefs that do *not* follow the logic of perfection.[97] Playfully shifting attitudes thus entails the motif of keeping alive the memory of the different "histor[ies] of ourselves," that is of the different "subjugated knowledge" (VG, 21, trans. by the author), as Foucault puts it, that were defeated by the dominant knowledge in a specific context and time. For memorizing *not only* the dominant "history of ourselves" along the lines of perfection can empower the subject to realize the mobility in relating oneself to oneself in the context of the logic of perfection.

To conclude, if we combine the motif "playfully shifting attitudes" to the three features of the de-reified ethos—i.e., the features of the "history of ourselves," of "critically constructing a de-reified ethos, and of the stylistic form of the essay—then we gain the following picture: the subject that playfully shifts his attitude from a reified to a de-reified ethos historicizes her given attachment to the logic of perfection. This subject is able to understand and experience that her life formation is not naturally given, but socially constructed by blindly attaching to the historically produced perspective, which in our case is represented by the logic of perfection. So, this subject is aware that her perspective on life formation is dominated by regimes of truth. The feature of the "history of ourselves" can make the subject aware that the regimes of truth are only the result of the "games of truth." This point leads us to the second feature. For this view can construct a de-reified subjectivity that does not understand and experience herself as the object of "games of truth," but as *an active co-player* of the "games of truth." Thus, this subject is able to pluralize her attachment to the logic of perfection and can generate a pluralized, flexible, activated, and as we put here, playful attachment to the logic of perfection. Thirdly, to keep the dice rolling on the formal level, too, the playful refers to the form of the essay in the understanding developed here. On the level of form, this view also leaves sufficient space so that there could be various “histor[ies] of ourselves” that narrate in plural ways and in manifold contexts the story of the co-player that plays an active part in the truth formation.

FIFTH CONCLUSION

In this section, we described the affect- and power-sensitive Janus face of subjectivity in digital enhancement in terms of reification. First of all, we outlined digital enhancement as an affect-centered training laboratory and thus continued to pave the way that we started to walk with Nietzsche. Furthermore, we analyzed the freedom-endangering side of the Janus face of subjectivity in digital enhancement by developing the type of affective, imaginary reification. In so doing, we outlined how the subject *already* imagines the pre-existing perspectives of the logic of perfection in immediate ways as a social fact that she simply values as necessary and thus turns blindly into the one-sided object of the logic of perfection. This entanglement is echoed by what we call the imaginary-power nexus in analogy to Foucault’s understanding of the knowledge–power nexus. The functioning mode of this specific type of reification was illustrated by the understanding of subjectivity in terms of the habitualizing machine, the normalizing machine, and the silencing machine. Moreover, we then continued to describe the freedom-empowering side of the affect- and power-sensitive Janus face of subjectivity in digital enhancement. To this end, we theoretically conceptualized how the subject can develop a de-reified ethos, a development we called the de-reifying test. Here we outlined the basic model of a subject’s de-reified ethos in “shifting attitudes in counterpoint,” a motif we delineated with reference to genealogical critique and the understanding of enlightenment. We continued the line of argumentation by further analyzing the motif of shifting attitudes in counterpoint in terms of its three key features the “history of ourselves,” “critically constructing a de-reified agency,” and the “stylistic form of the essay.” By describing the motif of the art of the parrhesiastic cry and the art of playfully shifting attitudes, we concretized the de-reified ethos.

NOTES

1. Foucault was not able to finish the manuscript of the fourth volume before his death in June 1984. Due to Didier Eribon’s comments Foucaultian scholars knew that the manuscript existed. But since Foucault stated “no posthumous publications,” it came as a surprise when the fourth volume *Les aveux de la chair* was in fact published in 2018. Didier Eribon, *Michel Foucault*, 2nd ed. (Paris: Éditions Flammarion, 2011

[1989]), loc. 7119, Kindle; Agustín Colombo and Edward McGushin, "Confessions of the Flesh–Guest Editors' Introduction," *Foucault Studies: Special Issue. Foucault's History of Sexuality Vol. 4, Confessions of the Flesh* 29 (2021): 1–5; Guillaume Le Blanc, "Why Read Foucault's 'Confessions of the Flesh' Today?," *Critique 13/13*, December 2, 2019, available at: http://blogs.law.columbia.edu/critique1313/guillaume-le-blanc-why-read-foucaults-confessions-of-the-flesh-today-english-version/.

2. HS 4, 366pp [French edition].

3. However, it would be a misunderstanding to think that Foucault discovered antiquity only in his late thought. Already in his first lecture, *Lectures on the Will to Know*, Foucault was concerned with the ancient context. Cf. WK, 99pp.

4. From this perspective, it is clear that the present book does *not* contribute to the discussion to what extent Foucault has interpreted the ancient and early Christian settings historically accurately. To these debates, see: Wolfgang Detel, *Foucault und die klassische Antike: Macht, Moral, Wissen* (Frankfurt a.M.: Suhrkamp, 1998).

5. To this debate: Daniel Zamora and Michael C. Behrent, eds., *Foucault and Neoliberalism* (Cambridge/Malden: Polity Press, 2016).

6. Such a perspective can be traced back to Geoffroy de Lagasnerie's reading of Foucault's text on neoliberalism: Geoffroy de Lagasnerie, *La dernière leçon de Michel Foucault? Sur le néoliberalisme, la théorie et la politique* (Paris: Fayard, 2012), 16pp.

7. DE 4/354, 861.

8. See Nehamas, *The Art of* Living; Wilhelm Schmid, *Philosophie der Lebenskunst: Eine Grundlegung* (Frankfurt a.M.: Suhrkamp, 1998); Hadot, *La philosophie comme manière de vivre*; Arnold I. Davidson, *The Emergence of Sexuality: Historical Epistemology and the Formation of Concepts* (Cambridge: Harvard University Press, 2004); and Saar, "Nachwort."

9. Lorenzini has shown that antiquity was structured by "moral perfectionism." Lorenzini, *Éthique et politique de soi*. Such a perspective is required in particular to understand how the neoliberal subject is already rooted in antiquity, as Harcourt highlights. Bernard E. Harcourt, "Foucaut's Keystone: Confessions of the Flesh. How the Fourth and Final Volume of 'The History of Sexuality' Completes Foucault's Critique of Modern Western Society," *Columbia Public Law Research Paper*, no. 14–647 (2019): 2pp. Recently, on the notion of neoliberalism Thomas Biebricher, *Die politische Idee des Neoliberalismus* (Berlin: Suhrkamp, 2021), 21pp.

10. Cf. in particular the second, third and fourth volume of *The History of Sexuality*.

11. Originally "l'homme de désir" (HS 2, loc. 168, Kindle); the German edition uses the term of "Begehrensmensch" (HS 2, 21).

12. If Foucault is often accused of only providing an elitist project through the motif of the art of living merely reserved for the wealthy and white man as in antiquity and early Christianity, it is clear from today's perspective that the project of the art of living encompasses everyone. From this perspective, it is also clear that Foucault does *not* conceive the ancient art of living as a role model (GE, 256).

13. Originally, Foucault had in mind to write a *different* history of sexuality. The back cover of the first volume announced his intention to write the following five further volumes: *The Flesh and the Body*; *The Children's Crusade*; *The Woman, The*

Mother, and the Hysteric; *The Perverse*; and *Population and Races*. These volumes were *not* published, although Foucault had already written first drafts of *The Flesh and the Body* and *The Crusade of Children.* Cf. Stuart Elden, "The Problem of Confession: The Productive Failure of Foucault's 'History of Sexuality,'" *Journal for Cultural Research* 9, no. 1 (2005): 23–41.

14. In current debates, there is talk of agonistic experimentation, too. Cf. Paxton, *Agonistic Democracy*, 99pp. Of course, this notion might be helpful. But since the notion of agonism might be misleading, at least in Mouffe's version (see part 1), this book does *not* speak of agonistic experimentation, but instead of enlightened experiments.

15. Paradigmatically: Maurice Blanchot, *Michel Foucault*, trans. Barbara Wahlster (Tübingen: edition diskord, 1987), 54.

16. Freud, *Triebe und Triebschicksale.*

17. Cf. The Foucault-handbook does not mention the term: Clemens Kammler et al., eds., *Foucault Handbuch: Leben—Werk—Wirkung* (Stuttgart: Metzler, 2014).

18. The term "alienation" is a related topic. Within this concept, we are dealing with a "relation of relationlessness," as Jaeggi puts it in *Entfremdung*, 19.

19. In discussing Karl Marx, Max Weber and Georg Simmel, Georg Lukács sketches the term reification in its economic aspects in *History and Class Consciousness.*

20. Axel Honneth, *Verdinglichung: Eine anerkennungstheoretische Studie* (Berlin: Suhrkamp, 2015).

21. Cf. In 1939, Freud founded the journal "American Imago." Without author, "American Imago," available at: https://www.press.jhu.edu/journals/american-imago. However, the present book does *not* offer a psychoanalytical analysis. On the relationship between critical theory and psychoanalysis, see: Martin Jay, *The Dialectical Imagination: The History of The Frankfurt School and the Institute of Social Research 1923–1950* (Boston and Toronto: Little, Brown and Company, 1973).

22. See Moira Gatens, *Imaginary Bodies: Ethics, Power, Corporeality* (London/New York: Routledge, 1996); Drucilla Cornell, *The Imaginary Domain: Abortion, Pornography, and Sexual Harassment* (London/New York: Routledge, 1995); Immanuel Kant, *Kritik der Urteilskraft.* AA V (Berlin/New York: de Grutyer, 1971 [1790]), 217pp.

23. Annie Ernaux, *The Years*, trans. Alison L. Strayer (London: Fitzcarraldo Editions, 2017 [2008]), emphasis by the author.

24. The general understanding of these three "axes" goes back to Saar, see: Martin Saar, "Analytik der Subjektivierung: Umrisse eines Theorieprogramms," in *Techniken der Subjektivierung*, ed. Andreas Gelhard et al. (Paderborn: Fink, 2013), 17–28.

25. The computer game depicted in the film is an allusion to the social credit system that is already applied in China. See Lily Kuo, "China Bans 23m From Buying Travel Tickets as Part of 'Social Credit' System," *The Guardian*, March 1, 2019, available at: https://www.theguardian.com/world/2019/mar/01/china-bans-23m-discredited-citizens-from-buying-travel-tickets-social-credit-system.

26. Jacques Lacan, *Das Spiegelstadium als Bildner der Ichfunktion, wie sie uns in der psychoanalytischen Erfahrung erscheint*, Schriften I, trans. Peter Stehlin (Weinheim/Berlin: Quadriga, 1986 [1936/1949]); Jacques Lacan, *Das Seminar. Buch II: Das*

Ich in der Theorie Freuds und in der Technik der Psychoanalyse, trans. Hans-Joachim Metzger (Weinheim/Berlin: Quadriga, 1991 [1954/1955]). In current debates, Cornelius Castoriadis's texts represent another fruitful resource for approaching the imaginary. Castoriadis draws less on psychoanalysis than on an ontological conception of the imaginary. He develops the notion of the imaginary to explain what society *is*, for the imaginary can show that a society is created upon the imaginations and forms, or institutions, that its members share. See Cornelius Castoriadis, *The Imaginary Institution of Society* (Cambridge: MIT Press, 1997 [1975]).

27. For an instructive overview of the general history of the imaginary, cf. Saar, *Immanenz der Macht*, 316pp.

28. Similarly, Foucault also speaks of the "machine" (GOS, 57) in and through which subjects value ordinary social roles as *the valuable* social role one *simply has* to adopt. This discussion is part of Foucault's analysis of Kant's understanding of enlightenment at the beginning of his lecture "Government of the Self and Others" in the early 1980s.

29. Joseph John Earl, ed., *Ferdinand de Saussure: Critical Assessment of Leading Linguists*. Vol. 3: *Saussure, Structuralism and Post-Structuralism* (London: Routledge, 2013).

30. In general on asceticism in Foucault cf. Arnold I. Davidson, "Ethics as Ascetics: Foucault, the History of Ethics, and Ancient Thought," in *The Cambridge Companion to Michel Foucault*, ed. Gary Gutting (Cambridge: Cambridge University Press, 1994), 115pp.

31. Sloterdijk, *Du musst dein Leben ändern*. Sloterdijk's title goes back to Rainer Maria Rilke's poem "Archaischer Torso Apollos." Needless to say, Rilke does *not* misunderstand the potential to change life as a reifying training camp. Rainer Maria Rilke, "Archaischer Torso Apollos," in *Der neue Gedichte anderer Teil*, ed. Ernst Zinn, 7th ed. (Frankfurt a.M.: Insel, 1995 [1957, 1908]), 503.

32. See the English edition: Peter Sloterdijk, *You Must Change Your Life: On Anthropotechnics*, trans. Wieland Hoban (Cambridge: Polity Press, 2013), 4 (emphasis by the author).

33. In contemporary discourse, Menke in particular shows the freedom-impairing effects of training, but without describing the affective dimension developed here: Menke, "Zweierlei Übung," 199–210.

34. From Deleuze we know that a series starts by doing something three times: Deleuze, *Abécédaire*.

35. As a side note: Foucault's interest in norms can be traced back to his teacher, Georges Canguilhem, among others: Georges Canguilhem, *Le normal et le pathologique* (Paris: PUF, 2009 [1966]). On the relationship between Foucault and Canguilhem see: Maria Muhle, *Eine Genealogie der Biopolitik: Zum Begriff des Lebens bei Foucault und Canguilhem* (München: Fink, 2013), 103pp.

36. Louis Althusser, "Ideologie und ideologische Staatsapparate (Notizen für eine Untersuchung)," in *Ideologie und ideologische Staatsapparate. 1. Halbband*, trans. Frieder O. Wolf and Peter Schöttler (Hamburg: VSA-Verlag, 2010 [1969]), 37–102.

37. Katja Diefenbach et al., ed., *Encountering Althusser: Politics and Materialism in Contemporary Radical Thought* (London: Bloomsbury, 2013); Warren Montag,

ed., *Althusser and his Contemporary: Philosophy's Perpetual War* (Durham: Duke University Press, 2020); Saar, "Analytik der Subjektivierung," 17–28.

38. Max Horkheimer and Theodor W. Adorno, *Dialektik der Aufklärung: Philosophische Fragmente*, 23rd ed. (Frankfurt a.M.: Fischer, 2017 [1947]), 128. The line of cultural critique goes back to Jean-Jacques Rousseau, *A Discourse on Inequality* (London/New York: Penguin Classics, 1984 [1755]).

39. Theodor W. Adorno, *Negative Dialectics* (New York/London: Continuum, 2007 [1966]), 348.

40. Harcourt, *Exposed*.

41. WC, 15, trans. by the author.

42. Plato, *Protagoras*, trans. Hans-Wolfgang Krautz (Stuttgart: Reclam, 1986).

43. Today, of course, we live in "a secular age," to put it with Charles Taylor, *A Secular Age* (Cambridge, MA/London: Harvard University Press, 2007). Accordingly, this perspective would need to be translated into *a secular language*. Here, this book proposes *the language of enlightenment* (see the enlightenment test). In general, on the relationship between socio-cultural and religious narratives see Willliam Schweiker, *Dust that Breathes: Christian Faith and the New Humanisms* (Oxford: Wiley-Blackwell, 2010).

44. This view on imaginations does *not* conflict with Adorno's ban on images. For Adorno *only* criticizes images that paint (*ausmalen*) a better future. This, however, is *not* the aim of the imaginations described here, which do not seek to envision a better future in and through digital media. On the contrary, these imaginations are conceptualized here as making room to imagine things *differently*. In this sense, they only provide *partial* pictures, which do not claim to give the entire picture. Plato's myth *Protagoras* is *one* way to make room to imagine the digital future differently, but in referring to this myth we are *not* saying that this is the *only* way to represent the digital future *as such*. Hence, imaginations as construed here are *not totalizing* images. This was the main problem that Adorno was afraid of. Adorno, *Negative Dialectics*, 207.

45. Warren demonstrates the affinity between the notion of perspectivism and the genealogical method. Mark Warren, *Nietzsche and Political Thought* (Cambridge, MA/London: MIT Press, 1991), 102pp.

46. Also François Ewald, "Foucault: Ein vagabundierendes Denken," in *Michel Foucault: Dispositive der Macht* (Berlin: Merve, 1978), 7.

47. Didier Eribon, *Returning to Reims* (London: Penguin Books, 2019 [2009]), 223.

48. Massumi, *Politics of Affect*, 18.

49. Lukrez, *Über die Natur der Dinge oder vom Wesen des Weltalls*, trans. Eva Marie Noller (Stuttgart: Reclam, 2021 [around 55 BCE]).

50. By this, it is also clear that this book does *not* refer to the understanding of contingency that is part of today's discussions in post-fundamentalism, as it *mis*trusts *every* norm. Cf. Oliver Marchart, *Das unmögliche Objekt: Eine postfundamentalistische Theorie der Gesellschaft* (Berlin: Suhrkamp, 2013).

51. Herbert Marcuse, *Triebstruktur und Gesellschaft. Ein philosophischer Beitrag zu Sigmund Freud* (Frankfurt a.M.: Suhrkamp, 1965 [1955]); Wilhelm Reich, *Die sexuelle Revolution* (Frankfurt a.M.: Fischer, 1993 [1936]).

52. A central historical example of what was justified, but is not justifiable is the anti-Semitic scandal of the Dreyfus Affair in France. In general, one understands by the Dreyfus Affair the *il*legitimate life sentence handed to Alfred Dreyfus, a Jewish officer, in France in the late nineteenth century. The scandal illustrates the *un*justifiable justification of anti-Semitic narratives: Sarah Bianchi, *Das deutsche Kaleidoskop: Die Dreyfus-Affäre in der wilhelminischen Öffentlichkeit zwischen 1898 und 1899* (Frankfurt a.M.: Peter Lang, 2012).

53. Snowden, *Permanent Record*, 214.

54. Virginia Woolf, *Orlando*, London: HarperCollins Publishers, 2014 [1928], 209.

55. Woolf, *Orlando*, 210.

56. Woolf, *Orlando*, 210.

57. Woolf, *Orlando*, 209.

58. Hence, we are not dealing with an individual ontology, as would be conceivable in the tradition of René Descartes, *Philosophische Schriften.* In einem Band. Mit einer Einführung von Rainer Specht (Hamburg: Meiner, 1996).

59. The view on ontology is not undisputed, see Simon Critchley, *Infinitely Demanding, Ethics of Commitment, Politics of Resistance* (London/New York: Verso, 2007).

60. The basic differentiation between "power over" and "power to" goes back to Hanna Fenichel Pitkin, *Wittgenstein and Justice: On the Significance of Ludwig Wittgenstein for Social and Political Thought* (Berkeley: University of California Press, 1972), 276.

61. Usually, the concept of social ontology is *not* understood in implicitly normative ways. It is, however, conceptualized in this way in the present book, due to the enlightened potential of genealogical critique. For the classic, non-normative reading of ontology, see Johanna Oksala, "Foucault's Politization of Ontology," *Continental Philosophy Review* 43, no. 4 (2010): 445–466. In general, the ontological tradition generally goes from Aristotle through Spinoza to Nietzsche and contemporary post-modern thinkers. Here too, the standard reading advances a non-normative view on ontological questions, see especially Étienne Balibar, *Spinoza et la politique* (Paris: PUF, 1985); Moira Gatens and Genevieve Lloyd, *Collective Imaginings: Spinoza, Past and Present* (London: Routledge, 1999) and William Connolly, "Spinoza and Us," *Political Theory* 29, no. 4 (2001): 583–94.

62. Sabine Hark, *Koalitionen des Überlebens: Queere Bündnispolitiken im 21. Jahrhundert* (Göttingen: Wallstein Verlag, 2017), 17, trans. by the author.

63. As shown in the chapter on affects (cf. part one of this book), this book does *not* speak of agonistic contestations, struggles and disputes. Mostly, agonistic aspects would consider *every* dispute and struggle as productive, see Chantal Mouffe, *Über das Politische: Wider die kosmopolitische Illusion* (Frankfurt a.M.: Suhrkamp, 2007 [2005]). For a nuanced understanding of agonistic dimensions that does *not* fall into Mouffe's pitfall see Owen, *Nietzsche, Politics, and Modernity*. Against this background, the present book *only* cares for *enlightened* contestations, disputes and struggles (see the enlightenment test).

64. Jacques Derrida, *Schurken: Zwei Essays über die Vernunft*, trans. Horst Brühmann (Frankfurt a.M.: Suhrkamp, 2003), 111. Also Jacques Derrida, *Politik der*

Freundschaft. Trans. Stefan Lorenzer, 5th ed. (Berlin: Suhrkamp, 2018) and Oliver Flügel, "Jenseits von Prozedur und Substanz: Jacques Derrida und die normative Demokratietheorie," in *Politische Philosophie und Dekonstruktion: Beiträge zur politischen Theorie im Anschluss an Jacques Derrida*, ed. Andreas Niederberger and Markus Wolf (Bielefeld: Transcript, 2007), 119–42.

65. For a general account of the "tension" between argument and style, see Didier Eribon, *Gesellschaft als Urteil*: *Klassen, Identitäten, Wege,* trans. Tobias Haberkorn (Frankfurt a.M.: Suhrkamp, 2017 [2013]), 159, trans. by the author.

66. The stylistic form of the essay goes back to the French Renaissance philosopher Montaigne, who created the form of the essay by reading ancient sources. Michel de Montaigne, *Essais*, ed. Ralph-Rainer Wuthenow (Frankfurt a.M.: Suhrkamp 2001 [1580]).

67. Thus, the stylistic form that can remind us of an essay is not to be confused with poetry (*Dichtung*). For while poetry (*Dichtung*) does not claim any kind of truth, the experimental, in some ways essayistic form developed here is characterized *by a search for enlightenment.* This also entails the search for truth, not for a substantial, essential or eternal understanding of truth, but for a practice-based truth, see later on in this chapter.

68. Georg Lukács, "Über Form und Wesen des Essay," in *Georg Lukács: Ästhetik, Marxismus, Ontologie. Ausgewählte Texte*, ed. Rüdiger Dannemann and Axel Honneth (Berlin: Suhrkamp, 2021 [1910]), 47pp.

69. Foucault is widely discussed in literature theory. These contemporary debates especially consider his understanding of the author or his understanding of discourse theory. Achim Geisenhanslüke, "Literatur und Diskursanalyse," in *Michel Foucault: Eine Einführung in sein Denken*, ed. Marcus S. Kleiner (Frankfurt.M./New York: Campus, 2001), 60–71. However, to date Foucault has only rarely been considered as contributing to the literary form of the essay, see Nina Hahne, *Essayistik als Selbsttechnik: Wahrheitspraxis im Zeitalter der Aufklärung* (Berlin/Boston: de Gruyter, 2015). *In passing*, Honneth speaks of Foucault's work in terms of essays. Here, he is thinking of Foucault's essays on Kant, see: Honneth, *Einleitung: Genealogie als Kritik*, 118.

70. Here, we deal with *a subject-based understanding of contingency*. This understanding does *not* entail a contingent understanding of the process of history *as such*.

71. Hence, the Nietzschean and Foucaultian tradition does not seek to contribute to the current discussions on truth, which center on questions of correspondence, coherence and consensus. In this sense this is not about the extent to which truth and facts correspond, assertions are free of contradictions, or opinions coincide. Concerning this debate, see Bianchi, *Einander nötig sein*, 55pp.

72. Though Bernd Magnus has a point in stating that Nietzsche considers an existential imperative, see Bernd Magnus, *Nietzsche's Existential Imperative* (Bloomington: Indiana University Press, 1978). If this perspective is transferred to the argument here, the existential imperative and thus to a certain extent the principle that ties Nietzsche's and Foucault's experimental, essayistic style together is the claim to uncover patterns of affective domination in telling the partial "histor[ies] of ourselves."

73. This attempt needs to be situated in the context of French philosophy in the Nietzschean tradition. From different angles, this tradition reflects the *crises of language* and the precarious relationship between language, concepts and their relationship to reality. In contrast to the French *de*constructive understanding according to which language and concepts are problematic *as such*, that is they *structurally fail* to adequately represent reality, we offer a *constructive* understanding; it hopes to show that *especially certain kinds of practices* of concepts are problematic. The experimental, essayistic style here promises to specify the kinds of practices that are problematic, namely those that depart from a *fixed* mediation between language, concepts and reality and thus that neglect the *flexible, history-based* mediation between them.

74. Jacques Derrida, "La mythologie blanche: La métaphore dans le texte philosophique," *Poétique* 5 (1971): 1–52.

75. See Paul Patton, "Nietzsche and Metaphor," in *Enigmas: Essays on Sarah Kofman*, ed. Penelope Deutscher and Kelly Oliver (Ithaca: Cornell University Press, 1999), 97p.

76. Here, we do not contribute to symptomatological investigations as conceived by Gilles Deleuze, *Nietzsche and Philosophy*, trans. Hugh Tomlinson (New York: Columbia University Press, 1983), 3. For the lesson of symptoms would be based on an understanding of illness. For example, flu is characterized by different symptoms, such as a cough. If one observes concrete symptoms, one can then draw conclusions about the bigger picture of the illness.

77. For the general discussion of "finding" and "inventing," see Richard Rorty, "Relativismus: Entdecken und Erfinden," *Information Philosophie*, no. 1 (1997): 5–23.

78. Nehamas, *Nietzsche: Life as Literature.*

79. Adorno, *Negative Dialectics*, 23.

80. Horkheimer and Adorno, *Dialektik der Aufklärung*, ix.

81. On the affinities between Foucault and Adorno see Harcourt, Saar and Bianchi, "The Critique and Politics of Identity"; Bianchi, "Unter dem und wider den 'Bann des Einheitsprinzips';" and Sarah Bianchi, "Zu einer aufklärerischen Mikrologie. Adorno, Foucault und die digitalisiert 'verwaltete Welt,'" in *Adorno und die Medien*, ed. Lioudmila Voropai and Judith-Frederike Popp (Berlin: Kadmos, 2023), 199–214.

82. Frédéric Gros, "Introduction." In *Michel Foucault. Discourse and Truth and Parrésia*, ed. Henri-Paul Fruchaud and Daniel Lorenzini (Chicago/London: Chicago University Press, 2019), xvii.

83. Owen, "Perfectionism, Parrhesia, and the Care of the Self."

84. On a critique of this gesture see Gayatri Chakravorty, "Can the Subaltern Speak?," in *Marxism and the Interpretation of Culture*, ed. Cary Nelson and Lawrence Grossberg (Urbana: University of Illinois Press, 1988), 271–313. According to Spivak, Foucault is not able to consider the position of the colonial subject; cf. Ina Kerner, "Gayatri Chakravorty Spivak, Can the Subaltern Speak?," in *Schlüsselwerke der Kulturwissenschaften*, ed. Claus Leggewie et al. (Bielefeld: transcript, 2012), 13. Basically, on the role of silence, cf. Nikita Dhawan, *Impossible Speech: On the Politics of Silence and Violence* (Sankt Augustin: Academia, 2007).

85. See GSO and GSO2.

86. WOE; GL.

87. Cf. Foucault's late lectures in Louvain: WDTT.

88. Theodor W. Adorno, "Trying to Understand Endgame," *New German Critique*, no. 26 (1982): 119–50 [1961].

89. The formulation "Les jeux sont faits" goes back to Sartre's play entitled the same. Jean-Paul Sartre, *Les jeux sont faits* (Paris: Gallimard, 1998 [1947]).

90. Kant, KU, AA 05: 218.21–34.

91. Schiller, *On the Aesthetic Education of Man*, 79.

92. In the cultural philosophy of the first half of the twentieth century, the anthropological dimension of the playful was taken up by Huizinga, who spoke of the "homo ludens." Johan Huizinga, *Homo ludens: Vom Ursprung der Kultur im Spiel* (Hamburg: Rowohlt, 2004 [1938]). Usually, "homo ludens" is considered as the counter-term of "homo faber," see Max Frisch, *Homo faber: Ein Bericht* (Berlin: Suhrkamp, 2011 [1957]). Foucault's understanding of the play, however, shows that "homo faber" is only the *alleged* counter-term as the play is enmeshed in contexts of production. Ruth Sonderegger criticizes the tradition of the play after Schiller for being anthropocentric, see: Ruth Sonderegger, *Für eine Ästhetik des Spiels: Hermeneutik, Dekonstruktion und der Eigensinn der Kunst* (Frankfurt a.M.: Suhrkamp, 2000).

93. Adorno, *Negative Dialectics*, 24.

94. Cf. Norbert Alexander Richter, *Grenzen der Ordnung: Bausteine einer Philosophie des politischen Handelns nach Plessner und Foucault* (Frankfurt a.M./New York: Campus, 2005).

95. Cf. TJF.

96. See SMD.

97. In similar ways, see Benjamin's mode of access in his theses on history. Walter Benjamin, "Über den Begriff der Geschichte." In *Walter Benjamin: Werke und Nachlaß. Kritische Gesamtausgabe, Vol. 19*, ed. Gérard Raulet (Berlin: Suhrkamp, 2010 [1942]).

PART III

When Scenes of Affect-Centered Power Relations Can Be the Beginning of the Subject's "Flight"

While the previous parts focused on the critical-theoretical microanalysis of the affect- and power-sensitive Janus face of subjectivity in digital enhancement, part three is interested in understanding how this Janus face can be activated. To this end, this book describes the external catalyst of such an activation in *scenes of affect-centered power relations that can turn into the beginning of the subject's "flight,"*[1] as Édouard Louis puts it. That is, the subject is flying into the new world where he can value his dissident perspective on life formation as a real "option."[2] Such an analysis promises to show that, when the subject stumbles haphazardly over contradictions in life, *an intrasubjective communication between the two sides of the Janus face of subjectivity in digital enhancement*—a channel of communication, which was previously blocked—can emerge. It is this intrasubjective communication that this book considers as the birth of the subject's critical, enlightened voice and thus as the beginning of the affect- and power-sensitive Janus face of subjectivity in digital enhancement. We need to note from the start that the two sides of the Janus face do not exist *in*dependently; there are *not* two distinct sides. Rather, there is the subject's desire to be part of digital enhancement, and then a critical, enlightened side can emerge which *criticizes this desire* to be part of digital enhancement; such a critical side can cause the subject to desire *differently*.[3]

Against this backdrop, part three consists of four steps. Firstly, it explores previously unexamined resources, which can help to understand the activation of intrasubjective communication. Here it brings into dialogue newer French philosophy and newer French literature, as well as Virginia Woolf's feminist classic, *Orlando*. Secondly, it specifies "activating" as the process by which an external catalyst can activate the affect- and power-sensitive Janus face of subjectivity in digital enhancement. Thirdly, with reference to Snowden, it analyzes the general model of activation as a procedure of translation. Fourthly, with reference to Louis, Ernaux and Woolf, it specifies the specific steps of such a procedure of translation.

1. Édouard Louis, *The End of Eddy*, trans. Michael Lucey (London: Harvill Secker, 2017 [2014]), 122.
2. Louis, *The End of Eddy*, 141.
3. We refer mainly to the English versions of the works of newer French literature, since we also quote from them. In cases where there is no English version available, the original French version is used.

Chapter 6

Updating the Resources

Part three starts with *an update of the resources.* While Foucault reads ancient Greek and early Christian literature in order to understand the emergence of the desiring subject, we propose a further focus on newer French literature and the feminist classic *Orlando.* Bringing into dialogue newer French philosophy with this corpus of texts promises to conceptualize the intrasubjective communication between the two sides of the affect-centered, power-based Janus face of subjectivity in digital enhancement. Here, part three refers to the works of the contemporary authors Ernaux and Louis, central voices of newer French literature.

Obviously, there is no homogenous group of texts referred to as "newer French literature." But these authors deal with common topics, which are also relevant to newer French philosophy. Both newer French literature and newer French philosophy use the subject's particular perspective to highlight the structural problems of society as a whole, for example the problem of social roles, i.e., the way society expects the subject to behave at home, at work, among friends and so on.[1] Here, we can see an *inter*dependence between newer French philosophy and newer French literature: they inform each other *mutually*; the firm boundaries between philosophy on the one hand and literature on the other hand become fluid, *though* each of them remains within their distinct disciplines. For the philosophical understanding of aesthetics developed here, this includes elaborating the general conceptions that enable us to theoretically conceptualize the conditions of the "history of ourselves" in our digital today, conditions that are socially constructed by affect-centered power relations. For the understanding of newer French literature, this includes the motif of the "ethnologist of herself,"[2] to which Ernaux often refers, and which also features in Louis's work. Just as ethnology is interested in deciphering the plurality of forms of life,[3] the "ethnology of ourselves" is concerned with understanding the specific plural histories that have shaped the each author's past. The authors gather these histories into assemblages, telling the story of how they became who they are, stories that could be relevant for everyone. It is in this sense that Ernaux, for example, tells the story of herself as a girl in the book entitled *A Girl's Story*.[4] Hence, this is anything but a personal story about the particular girl that Ernaux was when she grew up in a working class family in French society in the second half of the twentieth century. Rather, it is a theoretical reflection on the structural problems of French society, which made Ernaux become who she is, and think, live and feel the way she does. In the case of the "girl" in the book's title, this meant becoming a girl who slipped blindly into relations of power abuse with a man in a French summer camp in the second half of the twentieth century.

Although Louis and Ernaux describe how they climbed the social ladder, from the working class to the highest institutions of society (earning Ernaux the Nobel Prize in literature in 2022, and turning Louis into a famous author, whose writings have been adapted for theatres around the world), they are *not* to be confused with social climbers in the sense of *parvenus*. So they are *not* part of the classic literature on social climbing (*Aufstiegsliteratur*), whose well-known exponents include Guy de Maupassant and Stendhal. In *Bel-Ami*, for example, Maupassant describes how the protagonist Georges Duroy, a poor former NCO, does everything he can to become rich and famous in nineteenth-century Paris.[5] In *The Red and the Black*, Stendhal portrays the ambivalences of social climbing.[6] The protagonist Julien Sorel, who comes from a brutal family of carpenters in nineteenth-century France, is a social climber who uses people to ascend the social hierarchy. However, Stendhal also shows that Sorel is emotionally affected by the people, that he originally intended to use. *In contrast*, the works of Ernaux and Louis are *not* part of this *Aufstiegsliteratur*. Such an interpretation would fail to see that their academic rise is *not* a means to achieve external goals, such as becoming rich or famous. *On the contrary*, their academic climbing is *part of their self-reflexive transformation*. Hence, reading, studying and writing have self-transformative effects on both Louis and Ernaux: these activities empower them to find their own voices in a society that does not usually listen to such voices, and normally silences what they have to say. As we learn from Ernaux and Louis, Ernaux's voice as a woman is not usually heard in the patriarchal French society of the mid-twentieth century, and Louis's voice as a gay man is not heard in the world of his childhood in homophobic, rural Northern France in the late twentieth century.

Woolf's work does not belong to newer French literature, of course. First and foremost, Woolf is a notable British feminist author, who wrote in the early twentieth century, and her works belong to the feminist canon, along with those of writers such as Simone de Beauvoir, bell hooks, or George Sand.[7] With respect to its themes, however, her writing is *not entirely dissimilar* to the newer French literature that refers to her work.[8] Woolf also produced more or less autobiographical texts, describing structural struggles in society that are *not* hers alone. She depicts in particular structural struggles with gender roles, that is, with the specific social roles in the domain of gender, i.e., the alleged division between male and female social behavior. Especially in *Orlando*, Woolf shows how the societal order assigned men and women different places in society, which they had to accept.[9] The theme of gender roles is important throughout Woolf's works. In *A Room of One's Own*, she sketches the difficult working conditions of female writers, who deviated from stereotypical social ideas at a time when women were not seen as writers.[10] *To the Lighthouse* problematizes gender roles in the context of ordinary familial bonds.[11]

It goes without saying that newer French literature and the feminist classic *Orlando* are not *the* key literature one would expect in a critical-theoretical microanalysis of digital enhancement. However, they correspond to the aesthetic approach developed here with reference to Foucault and Kracauer: With this aesthetic approach, we can see that their texts are *"by the by" reflections* on our digital age. They do *not explicitly* discuss digital enhancement, the topic under scrutiny here, but their indirect approach promises to inform the reader of "the substance of things" (Kracauer), that is, the glue that *subtly* binds society, more effectively than the *self*-narrations that society *itself* provides. For the self-narrations of society run the risk of *only* justifying that society, and thereby hiding the structural problems that people might face in the time they live in. On the contrary, the indirect approach can highlight structural societal problems that are also relevant in times of digital enhancement.

Although Ernaux, Louis and Woolf describe different contexts than those discussed in this book with reference to the field of digital enhancement, these authors give a particularly nuanced theoretical reflection on the "substance of things," to use Kracauer's term. The "substance of things," which is similar in Ernaux, Louis and Woolf, *as well as* in the case of digital enhancement described here, consists in an understanding of *the effects of the logic of perfection on the subject's life formation.* The little-discussed common denominator for understanding the logic of perfection in Ernaux, Louis and Woolf is the author's (self-)description of *having the will to fit perfectly into the society of the time.*

And their microanalyses offer a drastic narrative of what it means when subjects fail. Ernaux, for example, describes how she did not fit perfectly into the society of her time. She had an abortion at a time when abortion was illegal;[12] she had a romantic relationship with a man who was half her age,[13] at a time when such relationships were only socially acceptable for men; and she did not follow the path that her parents from a working class background had wanted her to take.[14] Louis also depicts how he did not fit perfectly into society. He did not fit in because he was gay at a time when it was not acceptable to be gay, to have a high voice or to like ballet.[15] In *Method* (trans. by the author), Louis describes in an intimate manner all the "method[s] (trans. by the author)"[16] that he tried out—unsuccessfully—in order to be accepted, first in Amiens where he went to college and then in Paris where he studied. He wanted to succeed in the world of literature, but *at first* he did not have a place in this world. The places were reserved for people from academic backgrounds, who were used to navigating in a social language that was foreign to Louis, who came from homophobic, rural, working-class Northern France. To fit into this academic society nonetheless, he tried *not* to be gay, tried to imitate so-called tough, straight guys, had body surgery to be thin (like the stereotype of the academic), and became a prostitute, since he did not have sufficient money to survive in Paris.

In *Orlando*, Woolf depicts how the fictitious character Orlando did not fit perfectly into the society of his/her time. Unlike the works of Louis and Ernaux, *Orlando* is not an autobiographical narrative. Yet, it has been suggested that it can be regarded as a somewhat autobiographical text, in which Woolf implicitly writes about her romantic relationship with "her, [SB] honoured Lady Sackville."[17] Lesbian relationships were not represented in the romantic order of early-Twentieth-century society in Great Britain, and thus were not commonly accepted as possible. Be that as it may, in general terms *Orlando* describes the "gender trouble" (to borrow a term from Butler's contemporary feminist classic) faced by both men and women.[18] In this sense, Woolf, like Louis and Ernaux, portrays people who have trouble finding their place in society, although they are expected to fit in perfectly. In Woolf's novel, somewhat mysteriously, the protagonist Orlando falls into a deep sleep every time his/her problems become too heavy, and wakes up in a different society at a different time.

Moreover, the texts by Woolf, Louis and Ernaux describe how subjects (or Louis or Ernaux themselves) can learn to want to be part of society *in a different way*. They also describe how subjects (or they themselves) are able to become who they are *despite* the potential misrecognition by society. Here, Woolf, Ernaux and Louis depict the different ways in which they learned, *by the by*, not to care about the logic of liking they had once blindly shared. This "by-the-by" approach can uncover a subtle understanding of the "substance of things." This is what makes Ernaux and Louis's autobiographical texts and Woolf's quasi-autobiographical *Orlando* paradigmatic case studies, which are highly instructive for a critical-theoretical microanalysis of the "substance of things" in digital enhancement—where the subject's likeability is one of the most important currencies. This is not to say that other media, such as films, TV shows, or pictures, cannot enable us to undertake such a microanalysis of the "substance of things." Unlike Aristotle or Walter Benjamin, this book does *not* conceive of specific media as being able to show the "substance of things." Aristotle, for example, regarded the theater as the place from which he expected to prompt a cathartic transformation in the audience, that is, a self-reflexive change for the better, capable of purging the audience of their alleged sins.[19] Benjamin saw Russian films in particular as having the potential to provoke a self-reflexive transformation in the spectator. He elaborated this view in his influential work of critical theory, *The Work of Art in the Age of Mechanical Reproduction* (*Das Kunstwerk im Zeitalter seiner technischen Reproduzierbarkeit*). Here, Benjamin argues that a canvas *only* leads the viewer into "contemplation" (Benjamin, *The Work of Art*, 31), causing him "to give himself up to his chains of associations" (Benjamin, *The Work of Art*, 32). *On the contrary*, Russian films (in particular) *can interrupt* these "chains of associations." In the context of digital enhancement, however, we are dealing *with a range of media* that can show the "substance of things," as we have already seen in the previous chapters. The common denominator of all these possible media is that they do *not* provide a self-narration of how contemporary society wants to be seen. What is required for a critical-theoretical micro-analysis of the "substance of things" is a narrative that shows the structures that *subtly* bind society, but *only* does so *by the by*.

NOTES

1. Didier Eribon also belongs to this stream of thought. Didier Eribon, *Returning to Reims*, trans. Michael Lucey (London: Penguin Books, 2019 [2009]).
2. The precise term, that is used by Ernaux, is "an ethnological study of myself," see Ernaux, *Shame*, 19. From a philosophical perspective, see Raymond Geuss, "Nietzsche's Philosophical Ethnology," *A Journal of Humanities and the Classics* 24, no. 3 (2017): 89–116.
3. Ethnology derives from the ancient Greek word "ethnos" which means "tribe, nation, people." Cf. Geuss, "Nietzsche's Philosophical Ethnology."
4. Annie Ernaux, *A Girl's Story*, trans. Alison L. Strayer (London: Fitzcarraldo Editions, 2020 [2016]).
5. Guy de Maupassant, *Bel-Ami*, trans. Douglas Parmee (London/New York: Penguin Classics, 1975 [1885]).
6. Stendhal, *The Red and the Black*, trans. Roger Gard (London/New York: Penguin Classics, [1830]).
7. Obviously, the understanding of a canonical literature is problematic insofar as it produces exclusion, and mainly promotes the reading of well-known authors. Nonetheless, it is helpful for a first take on the topic.
8. Ernaux, *A Girl's Story*.
9. Woolf, *Orlando*.

10. Virginia Woolf, *A Room of One's Own* (Chicago: Albatross Publishers, 2015 [1929]).

11. Virginia Woolf, *To the Lighthouse* (London/New York: Penguin, 2018 [1927]).

12. Annie Ernaux, *Happening*, trans. Tanya Leslie (London: Fitzcarraldo Editions, 2019 [2000]).

13. Annie Ernaux, *Le jeune homme* (Paris: Gallimard, 2022).

14. Annie Ernaux, *A Woman's Story*, trans. Tanya Leslie (New York: Seven Stories Press, 2003 [1987]); Annie Ernaux, *A man's place*, trans. Tanya Leslie (London: Fitzcarraldo Editions, 2012 [1983]).

15. Louis, *The End of Eddy*.

16. Édouard Louis, *Changer: méthode* (Paris: Seuil, 2021).

17. Virginia Woolf, *The Diary of Virginia Woolf*. Vol. 3: *1925–1930* (New York: Mariner Books, 1981), 180.

18. Judith Butler, *Gender Trouble: Feminism and the Subversion of Identity* (London: Routledge, 2006).

19. Aristotle, *Poetik*, ed. Manfred Fuhrmann, trans. Manfred Fuhrmann (Stuttgart: Reclam, 1994 [335 BCE]).

Chapter 7

The Basic Method

How Enlightened Stumbling Blocks Can Activate an Intrasubjective Communication

How were the real figures Ernaux and Louis and the fictitious figure of Orlando able to turn the tide and start to critically reflect on how their life could be *different*? The basic method is "activating." The present book outlines the basic method as an activation of the subject's intrasubjective communication. This communication takes place between the subject's *un*critical voice and his critical, enlightened voice. It is this view that this book considers as the *tertium comparationis* between newer French literature, the feminist classic *Orlando* and the today's field of digital enhancement.

We learn from Ernaux, Louis, and Woolf, however, that such an activation is not easy to undertake. It is not comparable to the situation of Münchhausen, who lifted himself out of the mud by pulling his hair.[1] For unlike Münchhausen, today's subjects in the field of digital enhancement, as well as Louis, Ernaux, and Woolf, are *initially* unable to perceive that they are stuck in the mud. They are blind *at first,* which makes them just want to live, feel and act in ways that are similar to the usual patterns of how to think, live, and feel. So, in contrast to Münchhausen, who can in fact see, feel and experience the mud around his legs and the legs of his horse, the subjects in digital enhancement like Woolf, Ernaux and Louis cannot so easily see the mud that fill their life. In other words: the social order, in which Orlando, Ernaux and Louis are raised, and the field of digital enhancement are so smart that they cause subjects to want to put their mind on standby, if not turn it off entirely, and this means the subjects are initially unable to see how they can update their minds, and thus they cannot see the subtle effects of the logic of perfection on their life formation.

Nonetheless, as Louis, Woolf, Ernaux and the field of digital enhancement show, there are two sides to this issue. It would thus be reductive to conceptualize the subject exclusively as subjugated. Thus far, however, we have not specified the external catalyst that can make the subject's updating his mind and thus activate the intrasubjective communication. From the earlier chapters it should be clear that the subject can*not* be conceived as being capable of doing this alone; the subject *is dependent* on an external catalyst that can make the activation possible. Now this chapter proposes that one such external catalyst is *scenes of affect-centered power relations.* These can become an external catalyst, as the chapter aims to show, because they *can turn, by the by, into stumbling blocks*, that is, contradictions in life that are perceivable by the previously blind subjects. Originally, the theme of scenes derives from the context of theater. From this context, we are familiar with actors that play their role on a stage, as for example in presenting Bertold Brecht's *Three Penny Opera*. We approach the topic of scenes from a non-standard perspective, which focuses on everyday life. The term "scene" is used here to describe the specific expressions of affect-centered power relations, which can turn into stumbling blocks. Here, contradictions in the subject's everyday life are considered as stumbling blocks. It will be remembered that affect-centered power relations are those specific relations that make the subject open to being *blindly* formed in his will to think, live and feel in a certain manner, *without* being aware of this.

Yet, from the scene of theater, this non-standard view on scenes borrows the relation between scene and its audience. In the theatrical scene, actors play scenes and the audience is the recipient that sees and hears those stories unfold. In this sense, a relationship develops between the scene and its audience. This relational aspect is also relevant in our digital present, *although* we obviously do not observe actions in theatrical scenes. The notion of scene here draws on William Shakespeare's well-known phrase "all the world is a stage."[2] In the debate on social roles this perspective is adopted in particular by Erving Goffman in his far-reaching analysis *The Presentation of Self in Everyday Life*.[3] Especially the German translation of Goffman's book takes up Shakespeare's point already in its title *Wir alle spielen Theater*. Accordingly, it is not only the more or less professional actor who performs in a scene, as one might think, but the world *as a whole* becomes scenes "and all the men and women [. . .] players," as Shakespeare puts it (Shakespeare, *As You Like It*, 42). In this sense, the scenes with which we are confronted in digital times is not the literal scene found in the context of theater. Rather, we are confronted in our digital present with everyday life, which can have a relational effect on the subject, too.

In this book, the relational aspect is conceptualized as *stumbling blocks*. Certain scenes in life can make the subject stumble, *by the by*. From the analysis of the essayistic form, we see that *every* scene here can potentially turn into such a stumbling block. Since it is the subjective perspective in the form of the essay that designates a context as relevant for analysis, it is not possible to specify the content of the scenes. Instead it is the subtle effect of the structural problems on the subject's own life formation that marks the common denominator connecting the different possible scenes. The image of the stumbling block shows us that the structural problem affects the subject in a relatively unpredictable manner. The scenes of affect-centered power relations cause unforeseen contradictions, like someone stumbling haphazardly over a stone in the street. This is explicitly not intended by the subject. Here, we need to specify that *not every* contradiction can be understood as such a stumbling block, only those contradictions that can be called *enlightened contradiction*; they draw attention to contexts of domination that were so far rather overlooked, put differently: they pass the enlightenment test (see the earlier chapters). In this sense, the present book speaks of *enlightened stumbling blocks*.

Such an enlightened stumbling block is comparable to what has been called "intersectionality" in recent feminist debates. The term "intersectionality" was coined by Kimberlé Crenshaw about 40 years ago.[4] Since she realized that there was a gap in the United States' legal system to the effect that the experiences of black women were thus far not structurally represented in the law, she sought to fill in this gap with the conception of intersectionality. With this term she showed how the different social categories of race, class and gender can intersect, like streets at a crossroads. The gap Crenshaw saw in the United States' legal system meant that people could not articulate experiences of intersecting discrimination. This gap became obvious to Crenshaw in the case of Black women filing a suit against the automobile company General Motors in the late 1970s. Several Black women filed this suit due to the company's "last hired, first fired" policy.[5] Compared to other social groups, Black women were hired relatively late, and were the first to be fired. However, the women lost the case based on the justification that there were Black men working in the company as mechanics and there were also women working as secretaries. Thus, according to the court's justification, *neither* women *nor* Black people were being discriminated against. But US law was unable to deal with *intersecting* categories, such as the consideration of subjects that are *both* Black *and* female. Therefore, there were no theoretical tools to express the specific case of, for example, being confronted with Black women who did not want to be the first to be fired by a car company.[6]

The notion of the enlightened stumbling block is analogous to the intersecting streets of race, class, and gender, which cross at a particular point. Yet, the enlightened stumbling block produced by scenes of affect-centered power relations present in digital enhancement might be different. Here the focus is not on the intersecting scenes of race, class and gender. Instead, the understanding of this stumbling block entails the scenes of affect-centered power relations, that is, in our case, the subject setting herself in relation to the logic of perfection. Nonetheless, the crucial *tertium comparationis* lies in the fact that they, too, intersect like streets at a crossroads, to return to Crenshaw's crucial image. In the language of the present book, the crossroads and thus the intersecting streets correspond to the small surface-level expressions, to use Kracauer's phrase.

Newer French literature, for example, is conducive for a theoretical reflection on such "small surface-level expressions." In very concrete terms, Louis for example describes such surface-level phenomena when he depicts in his book *Who killed my father?* the reduction of the housing subsidies by five euro per month,[7] introduced by president Emmanuel Macron.[8] In general terms, *Who killed my father?* is an analysis of *the social death* of his father. In Louis's autobiographical narration, this "death" represents the place of the silent, poor, working class families, which are not killed by nature or by biology, but, according to Louis's provocative analysis, by the structural patterns that make them silent. They become silent, as Louis describes, by not being able to participate in social life.

It is here that the description of the surface level enters the scene. For it is through the concrete surface-level phenomenon of a 5-euro reduction in the housing subsidies that Louis shows the importance of this seemingly small amount of money in the lives of this segment of society: it was a further impediment for his father and his family to expand their social way of life. The reduction of five euro had such significant consequences for his family that they could no longer go to the beach in summer, for example. This concrete surface level phenomenon intersects with the global logic of perfection, according to which people are socially expected to have the will to fit perfectly into society. Here, the intersection mirrors the structural problem that Louis's family was *structurally* unable to fulfill social expectations, because they were excluded from society by society, in this case by Macron's reduction of the housing subsidies. As a consequence, the conceptualization of this surface phenomenon shows that it is not a personal problem faced only by Louis's family, but a problem confronting the entire society.

And it is these intersecting points, i.e., the intersection of the surface-level phenomena with the global perspective of society as a whole, that we see as the *external catalyst that can activate the Janus face of subjectivity* in digital enhancement. In and through them, the subject can become aware of the *structurally* intersecting scenes with which he is subtly confronted in his life formation, that is, the logic of perfection. When such scenes intersect, the subject can stumble over these intersections. Seemingly small intersections can thereby become meaningful to the subject. They can empower him to see the larger problems of the macro-level, that is, the logic of perfection, that were not previously visible to him. Now the subject can start to see them *as happening in front of his own eyes*. As a consequence, the subject can be activated to develop the enlightened gaze required to see the blind routinization of his life formation so far, which has made him develop his uncritical voice, in conformity with societal expectations, and has thus silenced his enlightened voice. It is this activation that can empower the subject to realize that he is actually, day by day, unconsciously silencing his enlightened voice.

In newer French literature we can find many examples of such small, but meaningful surface phenomena. Louis also describes the surface-level phenomena of having "a high-pitched voice"[9] or of his "way of walking,"[10] while Ernaux mentions the surface-level phenomena of, for instance, "the reality of others, their way of speaking, of crossing their legs, of lighting cigarettes,"[11] and thus towards all the seemingly small phenomena that might be so negligible, however, which express what silently holds society together. In our digital present, people are seduced in similar ways into silencing their critical voice, for example by the small surface-level phenomena of the sound of a new WhatsApp message, tweet or Facebook message. As the whistleblower Wylie writes in his autobiography, social media is all about these gimmicks that create an immediate desire to hold the supposedly fancy new tools in one's hands.[12] In other words, these superficial phenomena can make subjects react quasi-automatically to whatever has just beeped.

NOTES

1. Described as the "baron of lies," Münchhausen is a literary figure based on a historical figure. He became well-known for his extraordinary tales of adventure, such as riding a cannonball or tying his horse to the top of a church tower. See, for example, Irene Altenmüller, "Wie Baron von Münchhausen zum 'Lügenbaron' wurde," *NDR*, May 11, 2020, https://www.ndr.de/geschichte/koepfe/Baron-von-Muenchhausen-Vom-Pagen-zum-Luegenbaron, muenchhausen280.html. Some of his expressions, such as lifting oneself out of the mud by pulling one's own hair, became well-known proverbs. See Marco von Münchhausen, *Das Münchhausen-Prinzip: Wie Sie sich am eigenen Schopf aus dem Sumpf ziehen* (München: Kösel, 2011).
2. William Shakespeare, *As You Like It* (Leipzig: Amazon Distributions, without year [1603]), 42.
3. Erving Goffman, *The Presentation of Self in Everyday Life* (New York: Doubleday, 1959); Erving Goffman, *Wir alle spielen Theater: Die Selbstdarstellung im Alltag*, trans. Peter Weber-Schäfer (München: Piper, 2003).

4. Kimberlé Crenshaw, "Demarginalizing the Intersection of Race and Sex: A Black Feminist Critique of Antidiscrimination Doctrine, Feminist Theory and Anti-racist Politics," in *University of Chicago Legal Forum* 140, no. 1 (1989): 139–67.

5. Cf. Maisha-Maureen Auma, "Kimberlé Crenshaws Einfluss auf mein gerechtigkeitsstrategisches Denken: Fangen wir noch einmal bei der Critical Race Theory an," in *Heinrich Böll Stiftung*, April 15, 2019, https://www.gwi-boell.de/de/2019/04/15/kimberle-crenshaws-einfluss-auf-mein-gerechtigkeitsstrategisches-denken.

6. See Without author, "Intersektionalität: Eine kurze Einführung," in *Heinrich Böll Stiftung*, April 12, 2019, https://www.gwi-boell.de/de/2019/04/12/intersektionalitaet-eine-kurze-einfuehrung.

7. Édouard Louis, *Who killed my father?*, trans. Lorin Stein (London: Harvill Secker, 2019 [2018]).

8. Louis, *Who killed my father?*, 78.

9. Louis, *The End of Eddy*, 15.

10. Louis, *The End of Eddy*, 15.

11. Ernaux, *A Girl's Story*.

12. Wylie, *Mind*ck*, 108pp.

Chapter 8

Activating in General

A Procedure of Translation

In what follows, we specify *the procedure of translation* by which the affect- and power-sensitive Janus face of subjectivity in digital enhancement can be activated. Here, we refer to the specific procedure of the intrasubjective communication that can create the subject's enlightened voice, which was previously silenced by his uncritical voice. The motif of the uncritical voice represents the subject's blind will to fit perfectly into society, whereas, as this book goes on to show, the theme of the enlightened voice encompasses the subject's self-reflexive ability to think, live and feel *differently*, that is, the self-reflexive ability to want to become who one is *in opposition* to a blind will.

We use the case of Snowden to outline the basic model that makes the procedure of translation plausible. Generally speaking, the basic model shows how the *intrasubjective communication between the two sides of the Janus face of subjectivity in digital enhancement can emerge*. To analyze this model, we single out the moment when Snowden was preparing for his talk in Tokyo. Here, Snowden only came to develop a critical voice *by the by*, as he himself writes in his autobiography. He had *actually* planned to do something slightly different, namely to show that the Chinese were the bad cops in surveillance technology, while he was supposedly one of the good cops, so to speak. In short, he had a fairly strong uncritical voice, fed by the clichés about his digital today. But it turns out that it was this preparation for his talk that made Snowden become different, and more aware of what he was doing in his work for the NSA. He realized that it was not only China that was using digital media to spy on people, but also the US—as shown, for example, by the PRISM scandal revealed by Snowden. While Snowden might have a feeling of being implicitly addressed by his preparation of his talk in Tokyo, we need to differentiate between different steps in this procedure.

Initially, Snowden may just have read historical stories about spying processes in China. He may have read the passages that describe how subjects are spied upon by the social credit system. He may have gone on to read and learn about how people are blocked from web pages in case they lose too many credit points due to social behavior qualified as "bad" by the Chinese authorities, for example protesting against the Chinese government. Snowden may have seen pictures of people protesting against the Chinese government. Thoughts like "Ah, that's what it's like in China; how different it is in America" may have crossed his mind. In a second step, he might react strongly to what he has read. For example he might feel sorry for the Chinese people in their digital today, who might be receptive to the disciplinary mechanisms provoked by the social credit system and even want to conform to the social expectations that come with this social credit system. At this point, Snowden may also be terrified by what he has discovered. He might wonder how the Chinese people can become so submissive that they even desire to live, feel and act as the object that is subjugated to what the social credit system wants them to be. Who would want to live in such a digital setting? This and similar thoughts might occur to him while preparing his talk.

Next, Snowden might begin a process of translating what he has read about the historical events in China into the context of his own life. What seemed to be far removed from his current life now draws close to his eyes. At this point he might perceive the scenes as if they are happening in front of his own eyes. In this crucial step Snowden might start to think: Ah, are there also situations in the today's US in which one can be observed and pursued by digital means?

And with this key question Snowden might begin to realize that all these stories that were seemingly so far removed from his daily life can also be found in what he sees when he "open [his] eyes" (Foucault) and see the territory of digital enhancement. Through this process of translation Snowden can become aware that these historical narratives are also stories about his own life in his digital present. This process describes the beginning of the path to develop a critical voice. Snowden begins to question his habitualized status as the object of digital change and becomes the subject of digital change. In other words: Snowden is able to perceive the form of his own life from a new, dissident perspective. This new perspective allows him to see that his life in the US is also structurally confronted with today's version of surveillance technologies. He in the digital age of the US might then see that the digital surveillance technology becomes the dominant perspective that might prompt him to only see his life through the lens of the logic of perfection, which, in Snowden's case, includes an understanding that digital surveillance technology makes life perfect and does not spy on ordinary people. Although this passage might suggest that the procedure of translation takes place within minutes, Snowden clarifies the contrary: he states that it took him over three years to gain a new understanding of what he was *really* doing.

From Snowden's case, we learn the following lesson: The general model of activation consists of three interwoven, translating processes. *Firstly*, it consists of doing something "by the by." *Secondly*, it consists of processes of questioning that can emerge when the subject haphazardly stumbles over what he is doing. *Thirdly*, it involves processes of self-reflexive realization of the things one has previously done silently.

Chapter 9

Activating Specifically

Realizing That "Flight Is an Option"

The fourth step specifies the general model of activation. Here, the three steps mentioned above—including doing something by the by, questioning one's own actions, self-reflexively realizing one's silent actions—are discussed with reference to the film *Her*, which we already introduced in the previous chapters, and to the works of Louis, Ernaux and Woolf. This enables us to outline the specific mode of activation that shows, in Louis's words, that "flight is an option."[1]

Starting with the first point, we see that the subject does something *by the by*, like Snowden's preparation for the talk in Tokyo. "By the by" includes, for example, processes of habitualization or of normalization. The common denominator of these processes is that the subject does something *without* being aware of what he does. In other words, "by the by" is characterized by a *non*-self-reflexive process. Here, we are dealing with the subject's *un*critical voice and thus with the freedom-endangering side of the Janus face of subjectivity in digital enhancement. At this point, the subject does not see that "flight could be an option," in Louis's words. This has consequences for the starting point of intrasubjective communication. This shows us that the subject does *not intend* to start the process; rather, it happens when he does something different. At this point, then, the intrasubjective communication is "*on mute*."

Here we need to be more specific. For the crucial point is that this subject is *not* aware that his intrasubjective communication is on mute. In this sense, subjects can be conceptualized as resembling what Deleuze, referring to Foucault, conceived as the problem horizon in his time:[2] the subject as the "passenger par excellence" (Deleuze, *Foucault*, 136, trans. by the author). The subject is an exemplary passenger, he is travelling through his life, but this travel is silently forced in ways that are *not* noticed by the subject: In this sense he "is a prisoner" (Deleuze, *Foucault*, 136, trans. by the author) who does *not* know he is in the prison of his time. In other words, this "passenger par excellence" (Deleuze) is characterized by the fact that he himself is *not* governing his life, but his routines govern the passenger in ways that are *not* transparent to the subject. This chapter considers these intransparent ways of being governed as the reason for the muted intrasubjective communication. Prior to being able to generate an enlightened voice, this passenger has not the slightest clue that he is living a life that he does not hold the reins himself. Thus, different to the subject's self-understanding of serving as the captain of his own life, his routines have taken over the helm and blindly direct the course of his life. As a consequence, it is not necessary to force the subject to put his life in the chains of routines and to mute his intrasubjective communication. Far from it—this subject wants to do so more or less of his own accord.

This point becomes even more consequential for people like Theodore in the film *Her*, to which we referred to earlier. Theodore explicitly does not want to lead a life in a muted intrasubjective communication. Yet, he ends up in doing so *nonetheless*, and worse: without being aware of this fact. To recall, Theodore falls in love with his digital assistant, who calls herself Samantha. While the spectator learns from the ending that the entire society is in fact having love affairs with their smartphones, Theodore thinks that he is the only one involved in such an unusual romantic relationship. Despite Theodore's ambitions to *not* live his romantic life in ways that are merely similar to the accustomed way of loving and thus to not simply adapt to the usual social expectations of what a romantic life should look like—a handsome couple, beautiful children, a house on the outskirts of the city amidst green meadows, respectable hobbies, and so on—Theodore is *not* able to *really* achieve a different form of life, in ways that are *dis*similar to what he sees. For it is Theodore's uncritical voice that silently puts his enlightened voice on mute. And his uncritical voice is clever enough to make Theodore believe that his life is governed by his presumed enlightened voice.

This understanding of Theodore's peculiar voices about romantic life is important to note, because it makes the specific muted intrasubjective communication plausible. On a superficial level, i.e., the level of Theodore's ordinary behavior, Theodore is quite willing to have a conversation with himself about his unusual love life. He constantly tells himself that he is fine with his unusual romantic relationship with Samantha, that it is Samantha who cares for him and in fact knows more about him than he himself could ever know. But still, Theodore is caught in the thrall of his uncritical voice, which blindly excludes his enlightened voice.

The silent lack of the dissident voice in Theodore's intrasubjective communication becomes obvious when we consider the key scenes where his ex-wife's social expectations have such power over Theodore that he damages his relationship with Samantha. *Instead* of being able to be *in*different to what his ex says and thus to be able to listen to his enlightened voice, he does *not* notice that he is silently following the calls of his uncritical voice. For example, after his ex makes fun of Theodore by saying that he is "dating his laptop" (*Her*), he starts an argument with his beloved AI Samantha without noticing that he is silently submitting to the social expectations imposed upon him by his ex-wife. So, Theodore's ex-wife subtly nourishes in Theodore the feeling of no longer wanting to discover different perspectives on how romantic life might look in ways that are *dis*similar to what one usually sees. In other words, his ex gets him back on the supposedly "right track." And this track is paved in similarities that silence his possible enlightened voice. In this way, Theodore's mindset follows his uncritical voice and by this he just wants to be similar to the other couples in the streets, similar to all those couples that can hold hands and do not have difficulty feeling each other and similar to all those couples that can simply spend time together without others observing their behavior, wondering why, for example, a human being feels at home while sitting next to a smartphone. The list of similarities that silently conducts Theodore's programmed mindset could easily go on. These aspects now cause Theodore to value the existing social expectations as able to represent his own life fairly well. So, according to Theodore's perspective, if a subject feels that they do not fit into the patterns of social expectations, then it is up to this subject to change his life. It is not the social expectations that are problematic, but rather the form of the subject's life becomes a problem that he needs to fix. As a consequence, one must return to the alleged safe space that is provided by society's expectations. At the end of the day, these expectations might not allow Theodore to discover other ways of living, but the promise with which they make Theodore conform is too seductive to him: *he is liked by society*. And in the world of digital enhancement, likeability is among the most valuable currencies. So, why should Theodore want to resist?

Secondly, to make plausible how the subject *can start* to *un*mute the muted intrasubjective communication between the two sides of the Janus face of subjectivity in digital enhancement, we learn from Snowden that there is an external catalyst, such as the preparation for his talk in Tokyo, that can start the conversation with oneself. The book at hand considers scenes of affect-centered power relations as the theoretical tool that can conceptualize such an external catalyst. They show that it is not the subject himself that starts this process, but *the subtle contradictions in life* that can make the subject stumble over what he is doing, by the by and thus blindly. "Blindly" revisits the motif of affect-centered power relations, which make the subject open to having his will formed in ways that he does not notice. This book considers the external catalyst as the subtle effects on the subject's life formation that can make the subject *realize what is really happening in front of his eyes*. In terms of intrasubjective communication, we can say that the subject can, at this point, start to see that there could be potential for intrasubjective communication, and thus for his enlightened voice, which has previously remained silent.

This incipient intrasubjective communication takes the guise of a *question*. It is not that the subject begins an intrasubjective dialogue where he tells himself what to do, what he has been doing wrong so far, and so on. Rather, the subject begins to wonder what he has been doing so far. This resembles Snowden's situation when he was starting to rethink his assumption that only Chinese surveillance technology was a means for spying on ordinary people. Hence, to put it in the language of the affect- and power-sensitive Janus face of subjectivity in digital enhancement, it is now the subject's blind, uncritical voice that is on mute. By questioning his thinking, doing, and feeling, the subject's enlightened voice can be heard. The subject can see that there could be an alternative to the way he has lived so far, following a more or less blind routine. But at this point the subject is not able to really listen to his enlightened voice. The enlightened voice is still in its infancy, and the subject is not yet able to give credit to this voice.

We specify the second step of the activation of intrasubjective communication with reference to Louis's autobiographical text *En finir avec Eddy Bellegueule* (*The End of Eddy*), with the motif of "questioning life, shifting roles." In general terms, *The End of Eddy* describes Louis's self-reflexive transformation from "failure to flight."[3] He starts off as Eddy Bellegueule (Louis's original name), who conceives and experiences himself as "failure" because he does *not* fit into the provincial world of his hometown, Hallencourt in Northern France, with its stereotyped, homophobic expectations of what boys should be like—that is, straight and not gay. He then shifts to "flight" by "[b]ecoming"[4] Édouard Louis, that is, someone who is able to start to resist these stereotypes and who can fly into the new world, where Louis's gay voice has a chance of being recognized by Louis himself. By changing his name from Eddy Bellegueule to Édouard Louis, he made this self-reflexive transformation explicitly visible.

If we refer to the motif of "questioning life, changing roles," we are tapping into a widespread, international social-theoretical debate on social roles, which has been part of the fields of sociology, social psychology and philosophy since the 1950s and 1960s. This debate became especially well-known with the works of Goffman and Ralf Dahrendorf.[5] To recall, social roles are usually understood as patterns of behavior in the context of work, family or friends, which society expects us to fulfill. For example, working in a bank usually requires one to conform to a certain dress code, for instance to wear a suit. Social roles are commonly understood either as restricting the subject's life formation (Dahrendorf) or as providing the possibility for social transformation (Jaeggi).[6] We contribute to the latter strand. In this domain, *an ethical critique* of social roles is dominant, which considers the *topos* of social roles in the ethical context of life formation.[7] The book at hand provides *an aesthetic critique*, which considers the *topos* of the subject's self-reflexive transformation. It shows how Louis's "flight" into the new world, where he recognizes himself in his gay role, begins with his subjective perspective on how to perceive his own thinking, living and feeling *differently*, and how to evaluate it *positively*. The motif of "questioning life, changing roles" is developed here as a means *to open up a space to transcend the general context* of how to behave in social relations, which has previously restricted the subject's creation of his subjective perspective on social roles.

The motif "questioning life, changing roles" starts from the point where the subject does *not* see *any* possibility of a dissident shift in life—what Louis calls "the flight." As Louis states in *The End of Eddy*: "I had to get away. But early on it doesn't occur to you to get away [. . .]. You don't even understand that flight is an option."[8] At this point, the subject's intrasubjective communication is on mute. *Though* Louis already had a sense of being different from the rural, homophobic world where he was raised, his intrasubjective communication was *still* dominated by his uncritical voice, which had silenced his enlightened voice. Up till this point, an understanding and experience of possible different perspectives on how to think, live and feel had *not* been available to Louis. At the time, Louis perceived himself as a "failure," unable to fulfill the social norms of this rural world. This understanding echoes Louis's uncritical voice, with which he blindly validates his own sense of failing to meet perceived social standards.

We will now turn to Louis's "theatrical experience,"[9] as a central surface-level phenomenon that helped him, *while doing something differently, to get onto the self-reflexive path* to developing his enlightened voice, which differs from the voice of society, and which he can validate as a real option in life. A key part of Louis's "theatrical experience" as a surface-level phenomenon is the fact that, as Louis writes, he did not get involved in theatre to get onto this self-reflexive path, but *rather* "to get people to like me."[10] Since he was good at acting, people even said: "Maybe you can be the next Brad Pitt."[11]

Here, two points are relevant: Firstly, Louis already had a sense of being different from the rural world of his hometown in Picardy. Secondly, however, he was *not* able to perceive his difference as a good way of thinking, living and feeling. Thus far, Louis *de*valued his difference. For example, he put a lot of effort into imitating others in order to eliminate this unwelcome difference, such as standing in front of his mirror every morning, practicing how to behave like an alleged "tough guy."[12] In this sense, we see that Louis was strongly on the side of his uncritical voice, which made him blindly wish to be liked by others and put his critical voice on mute.

If we understand Louis's stage acting as a small surface-level phenomenon, this can explain how it was acting, ironically, that empowered Louis to begin to value his difference. Since Eddy was quite good at acting, he was given the chance to enroll at a lycée in Amiens that was known for its theatre program. Upon meeting other people at the lycée who had a different family background to his own, he started to question his earlier life. This is the moment when his intrasubjective communication began. There, he started "to say to [him]self"[13] things that were *not* dominated by the general perspective of his hometown. This was generated especially by all the new "bodily habits"[14] he saw in Amiens. People dressed differently, they behaved differently, they said hello differently; men even kissed each other on the cheeks. It was this visual difference that made Louis start to question his former life in Hallencourt, and to question the general perspective of his hometown, which he had previously blindly validated as the right way to live.

So, it was in Amiens that Louis came to realize that "flight could be an option," to come back to his key formulation. Questioning his previous way of thinking, living, and feeling opened up a possible space for a *different* way of living, thinking and feeling, although at that time he was not yet able to turn into Louis who could live his gay life as his normal way of being. But this was the moment when Louis *began* to question life, and to shift his social role. He began the "flight" from Eddy, who felt like a "failure" because he did not fit into the general pattern of life in his hometown, to Édouard Louis, now a well-known author who is not ashamed to tell people his story and talk about his difficulties in (as he himself puts it) "becoming" gay. This is the "flight" into someone who starts to develop his different, dissident voice.

Thirdly, the theoretical conceptualization of intrasubjective communication involves the subject's realized self-reflexive transformation. In this step, the subject can perceive his own hitherto unrealized thinking, doing and living as *a real option for thinking, living and feeling*, which is valuable to the subject on a cognitive level, too. Hence, in addition to the second step, the third step entails the understanding and experience of *really* flying (to put it in Louis's words once again) into the dissident world that one has not previously seen as an option. To illustrate this step, we refer to Woolf's classic feminist novel *Orlando*. With reference to Orlando, we specify the third step as the theoretical conceptualization of the subject's enlightened voice, which enables him/her criticize his/her former uncritical voice—the voice that had previously made him/her blindly silence his/her critical voice. Orlando's enlightened voice expresses the following point: Orlando "need[s] neither fight her age, nor submit to it; she was of it, yet remained herself."[15]

Here, we consider Orlando's critical voice in the field of gender roles. Orlando is paradigmatic in this context because, as Woolf describes it, he/she is able to switch between the assumed gender roles of male and female over the centuries. Usually, gender roles are understood as the specific social roles that unfold in the domain of the presumed binary division between male and female. This is explicitly *not* about the category of sex, which refers to biological characteristics such as chromosomes. Gender is about the *social* understanding of what it means to be male, female or diverse, such as one's habits.[16] In the widespread, current feminist debates, there is a tripartite view on gender roles.[17] *On the one hand*, difference feminists stress the differences between the gender roles that are ascribed to men and women.[18] *On the other hand*, equality feminists highlight the similarities between the gender roles of men and women.[19] *Different to these two positions*, pluralizing feminists opt for the *pluralizing of gender*.[20] It is to this third strand that Orlando's enlightened voice contributes a central perspective, as discussed here: Orlando's enlightened voice shows that the problem is *the alleged binary division between female and male*. According to this understanding, the pluralizing of gender roles needs to start at the division machine that blindly produces the presumed binary division between female and male.

With the paradigmatic case of Orlando, we see how Orlando's enlightened voice includes the ability to *be in opposition* to the division machine that divides gender roles into the supposedly correct categories of male and female. Orlando is able to be in opposition to the "spirit of her age";[21] she is able to distance herself, in a self-reflexive manner, from what is usually and normally understood as right or wrong in the context of gender. With a sense of irony, Woolf paradigmatically satirizes these clichés by having Orlando say: "The man has his hands free to seize his sword, the woman must use hers to keep the satins from slipping from her shoulders."[22]

Given his/her distance from such common understanding of gender roles, Orlando is able to raise the doubtful question: "Is this [. . .] what people call life?"[23] And the mere fact that he/she raises the question tells the reader that Orlando is aware of the possibility of transcending the given contexts of gender roles. For him/her, gender roles and life in general involve more than "what people call life."[24] One central way to understand what Orlando's new and dissident life could look like is her need to write. This is not a one-off occurrence. "She wrote. She wrote. She wrote."[25] Moreover, she does not write just any story, but what we have called (with reference to Foucault and Tully) "the history of ourselves." Orlando puts this in the following words: "Life? Literature? One to be made into the other?"[26] Writing the "history of ourselves" becomes the surface-level phenomenon (to borrow Kracauer's term) that makes her do something by the by, and, in the process, do something differently, namely start to change her view of gender roles. In writing "the history of herself," Orlando begins "to look about her for what she had come in search of—that is to say, life [. . .]."[27] And this means she becomes aware that the formation of gender roles and life formation in general is based on contingent conditions, i.e., conditions that are not necessary and could be different. To borrow Woolf's formulation from her feminist text *A Room of One's Own*, we see that there is no single "truth"[28] about how to be a man or a woman, but only "opinions"[29] about how to practice gender roles.[30] As a result, Woolf shows that there is not just one true gender role that describes Orlando, but a whole set of plural options that can describe her. Thus, Orlando could be called "the boy who sat on the hill; the boy who saw the poet; [. . .] the Gypsy; the Fine Lady; [. . .] the girl in love with life [. . .]."[31] Moreover, Woolf conceptualizes Orlando's critical, enlightened, dissident voice as a voice in and through which Orlando is able to switch between her different roles "as quickly as she drove."[32]

We will continue to describe the subject's enlightened voice, now stressing its risky aspect, with reference to Ernaux's autobiographical text *Happening*. The main "happening" described here is Ernaux's abortion. Ernaux tells the reader how she had an abortion at the age of 23, at a time when it was illegal in France (1963). Here we are not concerned with the legal or illegal status of abortion, but only with Ernaux's enlightened voice. As will be highlighted here, expressing one's enlightened voice can even mean risking one's own life. Ernaux describes how she nearly died because when she had an illegal abortion in Rouen she did not receive the medicine she needed: although the medicine was available, it was illegal to give it to her. As a consequence, she came close to bleeding to death. In the beginning of the text, the reader learns about Ernaux's critical voice. This voice encompasses her ability to oppose the common view, in France at the beginning of the second half of the twentieth century, that a pregnancy must be continued *whatever* circumstances accompanied the potential birth. Ernaux's circumstances were such that she could not have the baby: the pregnancy was the result of a random encounter with a friend from university, and having a baby as an unmarried woman would have prevented Ernaux from continuing her academic path at university. She had had to fight hard for this, since she came from a non-academic background where it was not common to go to university.

It is in this sense that Ernaux writes: "Sex had caught up with me, and I saw the thing growing inside me as the stigma of social failure."[33] The German translation is more specific; it specifies that sex not only caught up with "me," that is with Ernaux, but with Ernaux's "social background."[34] Here, then, the problem is already connected to the French society in which Ernaux grew up and more specifically to her family background, to her father and mother who were laborers and shopkeepers and did not want her to go to university. According to Ernaux, there were two major problems that kept people "in the legacy of poverty": either the "pregnant girl"[35] or "the alcoholic."[36] And it was this potential social loss of her academic life that made Ernaux so strong in her critical voice that she took the risk of nearly dying due to the abortion. As we learn from the autobiography, Ernaux's critical voice involves an understanding of and a self-reflexive distance from *the structure of the legal judgment* that makes abortion illegal. This understanding enables her, as she puts it, to "judge the law" instead of only judging "according to the law."[37] And this meant that she—unlike most people at that time—was able to *not* perceive abortion as "wrong"[38] because it was "banned,"[39] *nor* to perceive it as "banned" because it was "wrong."

SIXTH CONCLUSION

Part three has analyzed how scenes of affect-centered power relations can be the beginning of the subject's "flight" (Louis) into a different world, where the subject is able to generate and listen to his enlightened voice and value it as a real and positive "option" in life. To show this, part three brought newer French philosophy into dialogue with newer French literature and the feminist classic *Orlando*. In referring to Louis, Ernaux and Woolf, this book aimed to show how scenes of affect-centered power relations can turn into an enlightened stumbling block, i.e., a contradiction in life, which can, *by the by*, make the subject aware of his previously uncritical voice. Up to this point, the subject's uncritical voice has silenced his enlightened voice, and made him open to being formed in his will by the logic of perfection. So up till now, the subject has *wanted to fit perfectly into the societal order of his time*, which gives him his status and thus his place in society. But when the subject stumbles over scenes of affect-centered power relations, his enlightened voice can be activated. With reference to Snowden, this book outlines the basic model of such an activation as a *procedure of translation*. With reference to newer French literature and Woolf's novel *Orlando*, this book further analyzed the specific procedures that characterize this process of translation. Here, the book at hand discussed the motif of "questioning life, changing roles." This is when the subject begins to shift, by the by, from his uncritical voice, in which he understood and experienced himself as a "failure," to his enlightened voice, which makes him to see the beginning of his "flight" and the positive value of his difference. With reference to Woolf, the book at hand sketched the motif of "pluralizing gender roles" and hoped to show how the subject can learn to value his different gender role. With reference to Ernaux, this book then described the risks that the subject might take when the two sides of the Janus face of subjectivity start to be activated.

NOTES

1. Louis, *The End of Eddy*, 142.
2. Deleuze, *Foucault*.
3. Louis, *The End of Eddy*, 122.
4. Louis, *The End of Eddy*, 141.
5. For a helpful summary, see Dieter Claessens, *Rolle und Macht* (München: Juventa, 1974).
6. Ralf Dahrendorf, *Homo sociologicus: Ein Versuch zur Geschichte, Bedeutung und Kritik der Kategorie der sozialen Rolle*, 5th ed. (Köln/Opladen: Westdeutscher Verlag, 1965), 27p; Jaeggi, *Entfremdung*.
7. Jaeggi, *Entfremdung*, 91pp; Rahel Jaeggi, *Kritik von Lebensformen* (Berlin: Suhrkamp, 2014).
8. Louis, *The End of Eddy*, 141.
9. Louis, *The End of Eddy*, 175.
10. Louis, *The End of Eddy*, 175.
11. Louis, *The End of Eddy*, 175.
12. Louis, *The End of Eddy*, 14.
13. Louis, *The End of Eddy*, 190.
14. Louis, *The End of Eddy*, 190.
15. Woolf, *Orlando*, 179.
16. In general, cf. Sabine Hark, *Dissidente Partizipation: Eine Diskursgeschichte des Feminismus* (Frankfurt a.M.: Suhrkamp, 2005); Ina Kerner, "Geschlecht," in *Politische Theorie: 25 umkämpfte Begriffe zur Einführung*, ed. Gerhard Göhler et al., 2nd ed. (Wiesbaden: VS Verlag für Sozialwissenschaften, 2011), 126–41; Hanna Meißner, *Jenseits des autonomen Subjekts: Zur gesellschaftlichen Konstitution von Handlungsfähigkeit im Anschluss an Butler, Foucault und Marx* (Bielefeld: transcript, 2010).
17. This discussion takes place against the bigger picture of the women's movement. In general, the women's movement is described as consisting of three waves. Obviously, the movement is more heterogeneous than this clear-cut structure might suggest; however, this division is helpful for a first impression. The first wave of feminism in the second half of the nineteenth century was particularly concerned with the struggle for women's right to vote. Second-wave feminism, which emerged in the second half of the twentieth century, showed that although women now had the right to vote, there were still areas where they did not have equal opportunities. Second-wave feminism was especially concerned with self-determination and the right to abortion. Third-wave feminism began in the 1990s. One of its main concerns is to show that feminism cannot only include white women, but also needs to address Black women's rights. Another aim is to show that the debate over gender should include LGBTQ* rights, and not only the binary scheme of men and women. See Ina Kerner, "Konstruktion und Dekonstruktion von Geschlecht: Perspektiven für einen neuen Feminismus," *gender . . . politik . . . online* (2007), available at: https://www.fu-berlin.de/sites/gpo/pol_theorie/ Zeitgenoessische_ ansaetze/KernerKonstruktion_ und_ Dekonstruktion/kerner.pdf.

18. Basically: Sylviane Agacinski, *Politique des Sexes* (Paris: Seuil, 1998); Carol Gilligan, *In a Different Voice: Psychological Theory and Women's Development* (Cambridge, MA: Harvard University Press, 1982).

19. Frigga Haug, "Sozialistischer Feminismus: Eine Verbindung im Streit," in *Handbuch Frauen- und Geschlechterforschung. Theorie, Methoden, Empirie*, ed. Ruth Becker and Beate Kortendiek (Wiesbaden: VS Verlag für Sozialwissenschaften, 2004), 49–55. Susanne Weingarten and Marianne Wellershoff, eds., *Die widerspenstigen Töchter: Für eine neue Frauenbewegung* (Köln: Kiepenheuer und Witsch, 1999).

20. Patricia Hill Collins, *Black Feminist Thought: Knowledge, Consciousness, and the Politics of Empowerment*, 2nd ed. (New York/London: Routledge, 2000).

21. Woolf, *Orlando*, 179.

22. Woolf, *Orlando*, 123.

23. Woolf, *Orlando*, 128.

24. Usually, Western political thought in accordance to Marx considers contradictions as principal and secondary contradictions. In this sense, questions of gender roles would be a secondary contradiction, whereas economic contradictions would belong to the principal contradiction. The understanding of subtle contradictions developed here seeks to dissolve the differentiation between principal and secondary contradictions. This means that contradictions in the context of gender roles are no longer understood as secondary. Jaeggi's and Fraser's discussion inspired this perspective: Nancy Fraser and Rahel Jaeggi, *Capitalism: A Conversation in Critical Theory*, ed. Brian Milstein (Cambridge: Polity, 2018).

25. Woolf, *Orlando*, 179. As shown in the chapters on Nietzsche and Foucault, repetition starts when a subject does something three times.

26. Woolf, *Orlando*, 192.

27. Woolf, *Orlando*, 125.

28. Woolf, *A Room of One's Own*, 6.

29. Woolf, *A Room of One's Own*, 6.

30. Today, of course, the understanding of LGBTQ*+ people must be included here. But at the time, when Woolf was writing, in the early nineteenth century, this view was not yet developed.

31. Woolf, *Orlando*, 209.

32. Woolf, *Orlando*, 210.

33. Ernaux, *Happening*, loc. 201, Kindle.

34. Ernaux, *Das Ereignis*, 16, trans. by the author.

35. Ernaux, *Happening*, loc. 191, Kindle.

36. Ernaux, *Happening*, loc. 191, Kindle.

37. Ernaux, *Happening*, loc. 298, Kindle.

38. Ernaux, *Happening*, loc. 298, Kindle.

39. Ernaux, *Happening*, loc. 298, Kindle.

Conclusion

AESTHETIC CRITIQUE OF OURSELVES

Relationality, Structural Epistemic Discrimination, and Structural Epistemic Anti-Discrimination

Instead of summarizing the main points advanced in this book,[1] the final section will raise further considerations and address the relational character of aesthetic critique. While the previous sections may have given the impression that aesthetic critique is *only* concerned with the subject, this would be a misunderstanding, and would give rise to traditional interpretations of Nietzsche and Foucault as *merely* individualistic.[2] Nietzsche would then be misunderstood as advancing an orthodox existentialist reading of elite aristocracy,[3] and Foucault as elaborating the view of the *a*political artist of living, best represented by the figure of the *flâneur* in Baudelaire, who *only* cares for his singularity. Further exploration of the topic of digital enhancement, the theme whose heartbeat follows the rhythm of perfection, would fuel this misunderstanding. *However*, such a reading would miss the relational point of the line of argument developed here. The previous section dealt *implicitly* with the relational aspect; this conclusion now deals with it *explicitly*.

To see this, let us reconsider how one becomes who one is in the context of digital enhancement, based on the grammar of Nietzsche and Foucault. Such a becoming includes the areas of knowledge, affect-centered power relations and self-constitution. Thus, one becomes oneself as a subject structured by the dominant knowledge of one's time. *At first* the subject immediately attaches herself to this knowledge, and constitutes herself *in conformity with* this immediate attachment to the dominant knowledge. In the field of digital enhancement, we are dealing with the specific dominant knowledge, which holds up the logic of perfection as the only valid way to lead one's life. In line with Foucault, the present book calls this entanglement the knowledge-power nexus. This view includes *implicitly* relational aspects. These are not to be misunderstood as doing something alone or as being lonely. *Rather*, the relational aspects need to be seen *on the structural level of becoming oneself*. The word "structural" highlights the fact that the subject becomes who she is *in relation to the structures of knowledge and power* that she also reiterates in her self-constitution. Hence, the relational aspects entail an *anonymous* relationality, which this book has examined more closely with reference to the theoretical register of social ontology. This showed us *not only* that the World Wide Web is a net, but also that structures are the specific network through which *everyone is related to everything*, that is, other subjects, animals, institutions and so on. So, the understanding of this structure-based, relational web does *not only* entail a person-based relationality, but *an anonymous relationality* that is generated by the structures of knowledge and power.

In what follows, we broaden the understanding of the anonymous relationality to the relational aspect as part of the enlightened gaze. Here, we quasi-define the relational aspect by connecting the motif of the aesthetic critique of ourselves to the motif of seeing the other in his contingency and plurality. In the previous chapters we showed that aesthetic critique stands in the tradition of the political philosophy advanced by Tully. Tully, with reference to Foucault, developed the conception of the "history of ourselves." This outlines the conditions in and through which the subject's way of life is shaped. According to this view, the way subjects *are* in the here and now is dependent on historically shaped contexts. By conceptualizing this, the notion of the "history of ourselves" can explain how subjects can become aware that their *current* way of life could also be different.

We transferred Tully's conception of the "history of ourselves" to the understanding of aesthetic critique in the Aristotelian tradition of *aisthesis*; this enabled us to highlight the fact that the subject's sensory perception, such as his gaze, is *already* part of the "history of ourselves." The present book thus proposed an understanding of the affect-centered, power-based conditions that subtly form the subject's sensory perception, i.e., his gaze, telling him how to think, live and feel. The book at hand argues that an awareness of these conditions can allow the subject to see the contingency of life formation, and thus understand that life does not have to be formed as it has been so far, but can be formed in multiple ways. The book at hand referred to this self-reflexive development as the subject's enlightened gaze.

This enlightened gaze does *not* operate in relativistic terms, and thus does not see the need to pluralize everything. Here, the present book showed that the historically shaped contexts are *not* problematic *as such*, and are therefore not the object of the enlightened gaze; the enlightened gaze only focuses on subtle patterns of domination, that is, patterns that narrow the subject's view on life *without* him being aware of this. We analyzed this view in terms of affect-centered power relations and thus in the context of affective domination. "Affective domination" describes the subtle effects of the logic of perfection on the subject, which make her open to being blindly formed in her will to see her life from the lens of perfection. The previous sections concentrated on a microanalysis of the freedom-endangering side, in terms of a quasi-automated subjectivity that has the will to adapt quasi-automatically to the logic of perfection. We then analyzed the freedom-empowering side, explaining that the subject should not be understood as always being caught in this quasi-automated subjectivity, but can develop an enlightened subjectivity, which is able to see the affect-centered power relations that are at stake and thus can transcend the quasi-automated contexts. We conceptualized this aspect with reference to genealogical critique and the notion of enlightenment.

Now we connect the understanding of the subject's enlightened gaze with the conception of *explicit relationality*. We aim to show that the way a subject's becoming is structured has effects on the way another subject becomes who she is. This will be discussed with reference to the motif of "seeing the other in her contingency and plurality." To illustrate this point, we refer to two contemporary examples: the storming of the Capitol on January 6, 2021 and the Black Lives Matter movement in the context of George Floyd's death. We refer to the storming of the Capitol to discuss the police strategy of facial recognition as a type of structural epistemic discrimination, and we use the case of the Black Lives Matter movement to analyze the motif of structural epistemic anti-discrimination.

Basically, the theme of "seeing the other in her contingency and plurality" is based on the relational understanding of the aesthetic critique of ourselves that has been developed here. The relational aspect involves the understanding that the subject's awareness of the plurality of life formation *has effects on other subjects' ways of becoming*. For the subject who becomes aware of the plural way of living broadens the space for thinking, living and feeling—not only for himself, but also for other subjects. There is no need to assume that the subject has good or bad intentions here; this is simply the way it is, since the subject is a subject among other subjects, who is *structurally* related to everything that is. So, one person's actions have structural effects on those of others. As a consequence, the broadened space for thinking, living and feeling also has consequences for the way that other subjects become themselves. For as we saw, becoming oneself involves the subject's relation to the systems of knowledge and power that he initially perpetuates in his self-constitution. If the space of knowledge is broadened, then the other subject's way of becoming can also be understood as broadened. For this subject then relates to a broadened space of knowledge, which he then perpetuates in his life formation.[4]

We illustrate this point with the case of gender roles. The space of knowledge includes all the knowledge available at a certain time about being (for example) a good woman. In earlier times, e.g., the 1950s, this space of knowledge was narrowed because of stereotypes that were, at the time, not commonly identified as an epistemic prejudice, that is, the allegedly justified understanding of something that is not justifiable. At that time, becoming a supposedly perfect woman was commonly understood to involve what is referred to in German as the three Ks, *Küche, Kinder, Kirche* (*kitchen, children, church*). A girl growing up in that period initially learned that she too must become a woman who embraces this stereotype of perfect womanhood. Educational institutions, such as kindergarten, school or university, would have stabilized this stereotyped knowledge. In schoolbooks, for instance, children might have seen pictures or read stories that reproduced this view. However, if subjects can become aware that this stereotyped knowledge does *not* represent the *only* way to be a woman, and can thus see the plurality of forms of womanhood, then there could also be different, dissident kinds of knowledge circulating in society. Today, for example, it is common knowledge that the three Ks, *Küche, Kinder, Kirche,* are nothing more than clichés. As a consequence, a girl raised in our time is no longer confronted with these epistemic prejudices, which include the fulfillment of the three Ks as the supposed evidence of being a "good woman." This means that she can see, from early on, more possible ways to become a woman.

To illustrate this point we now refer to two contemporary cases: the storming of the Capitol on January 6, 2021 and the Black Lives Matter movement in the aftermath of George Floyd's death. We have already referred to the storming of the Capitol in earlier sections. To begin with, we need to briefly restate what happened on January 6, 2021. January 6 was meant to be the day when Congress confirmed the Electoral College and thus Biden's victory over Trump. Normally, this is just a formality. But during this ceremony a mob incited by Trump, made up of mainly right-wing voters coming to Washington from all over the United States, stormed the Capitol. This mob was stimulated by Trump, who had constantly denied Biden's victory and continuously called on his supporters to stand by him, in particular via his favorite social media, Twitter. And even during his speech on January 6, 2021 Trump encouraged his voters to come to the capitol by saying, as *The New York Times* reports: "You have to show strength" (Trump).[5] What the Trump supporters understood by this call became clear the very same day. They gathered in Washington, violently entered the Capitol, broke through barricades and killed six people.[6]

In the introduction and the first part of the book we highlighted the seductive role of social media, which Trump used to manipulate his followers—not only those in the virtual net, but also those in the real streets. We will now consider another thread. Here we want to draw attention to the use of surveillance technologies, such as facial recognition, which the FBI used to find the Trump supporters. Evan Greer, director of the digital rights advocacy group "Fight for the Future," told the *Washington Post* that he is worried about people's privacy due to the use of surveillance technologies: "Whenever you see this technology used on someone you don't like, remember it's also being used on a social movement you support."[7]

Now, we set out to criticize digital facial recognition by advancing a structural argument against its use. To this end, we develop the motif of "structural epistemic discrimination." Criticism of facial recognition is widespread in today's debates. First of all, there is criticism of the way the technology works, the main claim being that it is dysfunctional. Facial recognition belongs to the field of machine learning. Here, we are dealing with the type of artificial intelligence that is fed by human beings, with a set of data consisting of pictures of people's faces. Upon this basis, the artificial intelligence is able to learn, by itself, to recognize certain recurring facial patterns.[8] The common criticism is that the AI is biased because it depends on the data with which it is fed. So far, for example, the AI has had problems recognizing pictures of Black women. Here, the rate of misrecognizing and thus misidentifying a person is about 30 percent. The rate of misrecognition for Black men is about 6 percent. On the contrary, this rate is only about 0.3 percent for white men.[9] Hence, there is a gender and racial bias.[10]

So far, the theme of discrimination is widely spread.[11] But the notion of discrimination *itself* is less analyzed. In what follows, we thus analyze the notion of discrimination *itself* and criticize the practice of facial recognition as producing *structural epistemic discrimination*. To this end we ask: What is structural epistemic discrimination? Here, we first need to clarify the term "discrimination." "Discrimination" derives from the Latin verb *discriminare*, meaning "to divide." Seen in this way, discrimination is a tool that divides people. From the perspective provided by Foucault, we can say that discrimination is the problematic dividing tool that produces the allegedly justified divisions between the mad and the non-mad, the delinquent and the non-delinquent. Epistemic discrimination specifies where such allegedly justified divisions take place, namely in the field of the dominant knowledge. It is therefore the dominant knowledge that is the source of the supposedly justified divisions between the mad and the non-mad, the delinquent and the non-delinquent. For example, in the field of digital enhancement the dominant knowledge includes the view that a supposedly perfect life involves using as many digital tools as possible. This dominant knowledge, then, is the source that produces the allegedly justified division between supposedly perfect and imperfect life formations. "Structural" revisits the relational thread discussed above. It signals that epistemic discrimination takes place in the field of the systems of knowledge and power. This shows that we are *not* dealing with *a problematic situation*, for example with the situation-based problem that a person has bad intentions, but *with a problematic structure*. And this structure matters not only to one subject, but to subjects in relation to other subjects. In this sense, epistemic discrimination has the problematic structure that makes people *blindly* justify the dominant knowledge, which is, however, not justifiable. And understanding the structure of this problem shows that it not only concerns one single subject, but subjects *in a structure-based relation to* other subjects. For the relational thread makes it clear that the subject, with such a biased, epistemically discriminating gaze, not only has a narrowed view of his own life formation, directed by the dominant knowledge of his time, but *also restricts the commonly shared space of knowledge* that is available in society for everyone. As a consequence, other subjects are also more likely to develop a biased, epistemically discriminating gaze, since this is the alleged normal societal understanding of what is good. So, "relationality" does not only entail face-to-face relationships in and through which people are related to each other in a direct way. This type of discrimination reaches a broader scope and has effects on people that are not in a direct relationship. This can be seen, for example, if we consider familial contexts and the raising of children. If a child is raised in an area that is dominated by a biased gaze, then he will have more trouble finding his way (see section two of the book at hand). The word "blindly" evokes the quasi-automatic subjectivity that is socially constructed by affect-centered power relations. As a consequence, the possibility of seeing the contingency and the plurality of the other is squandered.

Seen from this perspective, facial recognition is problematic because it reproduces patterns of structural epistemic discrimination, since the artificial machine is fed with data that mainly represents the dominant knowledge of a particular time. The contemporary criticism of facial recognition reveals that data including pictures of Black women is neglected, as shown by the failure rate of 30 percent. In other words, the artificial machine does indeed learn to recognize patterns of pictures of faces by itself, *but it is not able to learn by itself to correct the prejudices* with which it is fed. Hence, if the engineers have a biased, epistemically discriminating gaze, the machine does, too. The present book has shown that people are not condemned to be trapped within their biased gaze, and that the critical methods of genealogical critique and enlightenment theoretically conceptualize how such a biased gaze can be self-reflexively overcome and replaced by an enlightened gaze. But self-reflexivity is *not* achievable by machines, which are automated. This is already clear from the etymology. An automaton, from the ancient Greek word "αὐτόματον," is something that *can move by itself.* Moving by itself, however, *only* means being a self-operating machine. This makes it clear that the critical methods of genealogical critique and enlightenment are *not* translatable into the impoverished language of artificial machines, which only operate by the binary scheme of 0 and 1. Hence, they become a dividing machine, which divides life not into mad and non-mad or delinquent and non-delinquent, as Foucault analyzed, but into the binary scheme of 0 and 1. This, however, prevents them from seeing the contingency and plurality of the other. For them, the other is the result of a probability-based calculation around these binary schemes of 0 and 1. The understanding of the historicity of these patterns thus remains invisible. But becoming aware of the historicity of the conditions in and through which one becomes oneself is the necessary condition to create the space that allows people to see not only their own differences, but also those of others. On their own, artificial machines are *not* able to capture this point, and thus can only perpetuate the allegedly justified divisions, which are, however, not justifiable. It is in this sense that facial recognition *can* become the realization of what Kafka ironically depicted in *The Trial*.[12] In brief, *The Trial* tells the story of Joseph K., who is sentenced for a crime that he did not commit. People who are wrongly identified by facial recognition might feel like a contemporary Joseph K., for example the Afro-American Jacky Alcine and his friend, both Black,[13] who recently realized that they had been tagged as gorillas by Google facial recognition.[14]

Nonetheless, it would be reductive to *only* conceive of digital tools in terms of their freedom-endangering aspects. In what follows, we refer to the Black Lives Matter movement in order to show that digital media can also empower subjects to gain greater self-reflexive freedom. We analyze this point by discussing the motif of "structural epistemic anti-discrimination."

In brief, the Black Lives Matter movement was founded in 2013 by Alicia Garza, Patrisse Cullors and Opal Tometi in reaction to the death of the Afro-American Trayvon Martin in Sanford, Florida.[15] Trayvon Martin was shot dead by the security guard George Zimmermann—in self-defense, he claimed. Zimmermann was acquitted. Black Lives Matter was founded to shed light on the problem of racism in Northern America but also globally; in total, about 30 million people participated in their protests in the US and on a global level. The Black Lives Matter movement stands in the long tradition of the Black civil rights movement in the US. The American Civil War officially ended the long history of the enslavement of Afro-Americans in the US, but did not end the unequal treatment of Afro-Americans, which remained visible in police controls, in employment or at university. In general terms, the Black civil rights movement (not just Black Lives Matters, but also the Black Panther Party and the non-violent demonstrations led by Martin

Luther King in the second half of the twentieth century) seeks to highlight and abolish the unequal treatment of Black people. The Black Lives Matter movement became particularly well known after the death of George Floyd in 2020—after a policeman had knelt on his neck until he died. George Floyd's last words, "I can't breathe," which the police officer ignored, were circulated worldwide.

Digital media played a crucial role in the constitution of the Black Lives Matter movement from early on. It was the #BlackLivesMatter hashtag on Twitter that made this social movement well known. We will now analyze the role of social media within the Black Lives Matter movement in terms of structural epistemic anti-discrimination. The present book develops the understanding of structural epistemic anti-discrimination as follows: Basically, structural epistemic anti-discrimination analyzes how patterns of structural epistemic discrimination can be *un*done. Here, the book at hand considers the motif of structural epistemic anti-discrimination as contributing to the understanding of aesthetic critique of ourselves. Thus, structural epistemic anti-discrimination can enable us to theoretically conceptualize the possibility of "seeing the other in her contingency and plurality." It can do so because it promises to highlight the historicity of the supposedly valid division between mad and non-mad, between delinquent and non-delinquent, and, in today's case of digital enhancement, between perfect and non-perfect. Seeing the historicity of today's division between perfect and non-perfect includes seeing that this historicity is co-produced by digital technologies, such as facial recognition. For they feed the subject's blind belief that the world would become a more perfect place if it were made safer by using digital technologies such as facial recognition. But in doing so, they also exacerbate the subject's blindness to the fact that the digital devices that are supposed to make us perfect can have flaws.

Hashtags such as #BlackLivesMatter can empower subjects to *un*settle their blind belief in this division between perfect and non-perfect. Collecting the dissident histories of the people who have been excluded by this division and putting them online can empower subjects to gain pluralized views on life formation that go beyond binary divisions. This can enable subjects to become aware of the multiple views on how to think, live and feel, in ways that transcend these binary divisions. As a consequence, the space of knowledge is also pluralized. While the space of knowledge was once mainly occupied by the dominant knowledge, there is now a chance to have a more pluralized knowledge available here. This knowledge is sensitive to the different and plural views that were previously excluded. This sensitivity to difference is *not* relativistic. For it means considering the difference that was previously excluded by the patterns of affective domination, which justified what is *not* justifiable. The justification of racial prejudices belongs to the scope of what is justified but is not justifiable. These prejudices are also reproduced by digital technologies such as facial recognition, which is more likely to *mis*identify Black female faces. The same prejudices make Black people more likely to be stopped and searched in the streets by police officers. This is because the set of data that was originally stored was already biased. For if there is the prejudice that Black people are more likely to sell drugs and so on, then this prejudice leads to more controls of Black people than of white, which then supposedly provides evidence to support the prejudice. But if hashtags like #BlackLivesMatter can make such prejudices public, then they can help to pluralize the space of knowledge. This can then allow plural, dissident voices to emerge, namely all those voices that deviate from the assumed binary division between perfect and non-perfect and its accompanying prejudices. Now these voices can be heard in society.

This was the case for George Floyd, whose last words—while a police officer refused to remove his knee from Floyd's neck—were: "I can't breath." Due to social media, his cry was recorded; the video with the murder spread around the world, and people were shocked by the brutality. So, it is partly due to social media that Black people have gained such a power in the digital world, and also in the streets, that in April 2021 the policeman, Derek Chauvin, who killed the Afro-American George Floyd was found guilty.[16] This verdict might seem self-evident,[17] but in a world that is still imprinted with structural racism it is anything but self-evident.[18]

NOTES

1. For an overview of the main sections, see the summaries that follow each section.

2. Such as advanced by Undine Eberlein, *Einzigartigkeit: Das romantische Individualitätskonzept der Moderne* (Frankfurt a.M./New York: Campus, 2000), 42pp, 48pp, 57pp.

3. For an existentialist reading of Nietzsche that does *not* neglect the relational character of his work, cf. Bianchi, *Einander nötig sein*. Unlike the individualist readings, "existential" is understood here in a broad sense, centering on the subject's various contexts of living.

4. This view thus stresses the seldom-seen relationship between *self-reflexive freedom* on the one hand and *social freedom* on the other. "Social freedom" is a prominent *topos* in current social and political philosophy, see Axel Honneth, *Das Recht der Freiheit: Grundriss einer demokratischen Sittlichkeit* (Berlin: Suhrkamp, 2011). Usually, the understanding of social freedom entails a relational understanding of the other as a necessary part of one's own becoming. Here, the other can both enable and endanger freedom, see Bianchi, *Einander nötig sein*. This book has explored self-reflexive freedom with reference to the critical method of genealogical critique and enlightenment. It defined this freedom as the subject's potential to become aware

of the historicity of her life formation. Social freedom and self-reflexive freedom *overlap* in the relational understanding outlined in this conclusion.

5. Lauren Leatherby et al., “How a Presidential Rally Turned into a Capitol Rampage,” *New York Times*, January 12, 2021, available at: https://www.nytimes.com/interactive/2021/01/12/us/capitol-mob-timeline.html.

6. Rachel Chason and Samantha Schmidt, “The Freedom To Assemble in Two Acts: Lafayette Square, Capitol Rallies Met Starkly Different Policing Response,” *Washington Post*, January 14, 2021, available at: https://www.washingtonpost.com/dc-md-va/interactive/2021/blm-protest-capitol-riot-police-comparison/.

7. Drew Harwell and Craig Timberg, “How America’s Surveillance Networks Helped the FBI Catch the Capitol Mob,” *Washington Post*, April 2, 2021, available at: https://www.washingtonpost.com/technology/ 2021/04/02/capitol-siege-arrests-technology-fbi-privacy/.

8. Obviously, *not every type* of machine learning is discriminatory in the sense described above. There are also types that are *not* a problem. These include, for example, machine learning to organize infrastructure faster, such as planning where to build fiber optic cables to improve internet access. Cf. Without author, “Von der Straße bis zum Internet: Infrastruktur schneller planen dank Machine Learning,” *Forschung Kompakt*, May 8, 2019, available at: https:// www.fraunhofer.de/de/presse/presseinformationen/2019/mai/von-der-strasse-bis-zum-internet.html.

9. Cf. Frederike Kaltheuner and Nele Obermüller, “Diskriminierende Gesichtserkennung: Ich sehe was, was du nicht siehst,” *Netzpolitik.org*, available at: https://netzpolitik.org/2018/diskriminierende-gesichtserkennung-ich-sehe-was-was-du-nicht-bist/; Karen Hao, “This is How AI Bias Really Happens–And Why It’s So Hard to Fix,” *MIT Technology Review*, February 4, 2019, available at: https://www.technologyreview.com/2019/02/04/137602/this-is-how-ai-bias-really-happensand-why-its-so-hard-to-fix/.

10. For a general account of the gender data gap, see Caroline Criado-Perez, *Unsichtbare Frauen: Wie eine von Daten beherrschte Welt die Hälfte der Bevölkerung ignoriert*, trans. Stephanie Singh (München: btb, 2020).

11. Davide Castelvecchi, “Is Facial Recognition too Biased to be let Loose? The Technology is Improving–but the Bigger Issue is How it’s Used,” *Nature* 587 (2020): 347–49.

12. Franz Kafka, *Der Prozess und ausgewählte Parabeln* (Paderborn: Schöningh, 2001 [1925]).

13. Jethro Mullen, “Google Rushes to Fix Software that Tagged Photo with Racial Slur,” *CNN*, July 2, 2015, available at: https://edition.cnn.com/2015/07/02/tech/google-image-recognition-gorillas-tag/index.html.

14. For the problem of structural racism in general see Sarah Bianchi, “Färben und Wissen: Wie die Kulturtechnik des Malens weltgestaltend wirkt. Über: Ludger Schwarte. ‘Denken in Farbe,’” *Deutsche Zeitschrift für Philosophie* 70, no. 3 (2022), 559pp.

15. Barbara Ransby, *Making All Black Lives Matter: Reimagining Freedom in the Twenty-First Century* (Berkeley: University of California Press, 2018).

16. John Eligon, "Derek Chauvin Verdict Brings a Rare Rebuke of Police Misconduct," *New York Times*, April 20, 20021, available at: https://www.nytimes.com/2021/04/20/us/george-floyd-chauvin-verdict.html.

17. To recall, only a few days before the verdict on Floyd's murder, there was another murder of a Black man in Minnesota. Another police officer killed an Afro-American during an ordinary traffic stop close to Minneapolis, and thus only a few miles away from the place where the jury was discussing Floyd's case. This police officer killed the man simply because, as she said, she confused her teaser with her pistol. Bogel-Burroughs et al., "Minnesota Officer Who Shot Daunte Wright Meant to Fire Teaser, Chief Says."

18. Even president Biden admitted institutional racism as a problem in the US. See without author, "Joe Biden: There is Institutional Racism in America," *CNBC*, October 22, 2020, available at: https://www.cnbc.com/video/2020/10/22/joe-biden-there-is-institutional-racism-in-america.html.

Bibliography

Abramson, Ashley. "Burnout and Stress Are Everywhere." *Monitor on Psychology* 53, no. 1 (2022): 72, available at: https://www.apa.org/monitor/2022/01/special-burnout-stress.

Adorno, Theodor W. "Trying to Understand Endgame." *New German Critique*, no. 26 (1982): 119–50 [1961].

Adorno, Theodor W. *Minima Moralia: Reflexionen aus dem beschädigten Leben.* Frankfurt a.M.: Suhrkamp, 2003 [1951].

Adorno, Theodor W. *Negative Dialectics*. New York/London: Continuum, 2007 [1966].

Adorno, Theodor W. *Noten zur Literatur II.* Frankfurt a.M.: Suhrkamp, 1961.

Agacinski, Sylviane. *Politique des Sexes*. Paris: Seuil, 1998.

Agamben, Giorgio. *Homo sacer: Die souveräne Macht und das nackte Leben*. Trans. G. Roßler. Frankfurt a.M.: Suhrkamp, 2002 [1995].

Ahmed, Sara. *The Cultural Politics of Emotion.* 2nd ed. Edinburgh: Edinburgh University Press, 2014.

Allen, Amy. "Dripping with Blood and Dirt from Head to Toe: Marx's Genealogy of Capitalism in Capital, Volume 1." *The Monist* 105, no. 4 (2022): 470–86.

Allen, Amy. "Foucault and Enlightenment: A Critical Reappraisal." *Constellations* 10, no. 2 (2003): 180–98.

Allen, Amy. *The Politics of Our Selves: Power, Autonomy, and Gender in Contemporary Critical Theory.* New York: Columbia University Press, 2013.

Altenmüller, Irene. "Wie Baron von Münchhausen zum 'Lügenbaron' wurde." *NDR*, May 11, 2020, available at: https://www.ndr.de/geschichte/koepfe/Baron-von-Muenchhausen-Vom-Pagen-zum-Luegenbaron,muenchhausen280.html.

Althusser, Louis. "Ideologie und ideologische Staatsapparate (Notizen für eine Untersuchung)." In *Ideologie und ideologische Staatsapparate. 1. Halbband.* Trans. F. O. Wolf and P. Schöttler. 37–102. Hamburg: VSA-Verlag, 2010 [1969].

Althusser, Louis. *Freud und Lacan: Die Psychoanalyse im historischen Materialismus.* Berlin: Merve, 1976 [1970].

"American Imago." available at: https://www.press.jhu.edu/journals/american-imago.

Aristotle. *Metaphysik*. Trans. H. Bonitz. Hamburg: Meiner, 1995 [348–322 BCE].

Aristotle. *Poetik*. Trans. M. Fuhrmann. Stuttgart: Reclam, 1994 [335 BCE].

Ashenden, Samantha and David Owen. "Introduction. Foucault, Habermas and the Politics of Critique." In *Foucault contra Habermas: Recasting the Dialogue between Genealogy and Critical Theory*. Ed. S. Ashenden and D. Owen. 1–20. London: Sage et al., 1999.

"Audio Guide Museum of Modern Art." Available at: https://www.moma.org/collection/works/79251.

Auma, Maisha-Maureen. "Kimberlé Crenshaws Einfluss auf mein gerechtigkeitsstrategisches Denken; Fangen wir noch einmal bei der Critical Race Theory an." In *Heinrich Böll Stiftung*, April 15, 2019, available at: https://www.gwi-boell.de/de/2019/04/15/kimberle-crenshaws-einfluss-auf-mein-gerechtigkeitsstrategisches-denken.

Balibar, Étienne. *Spinoza et la politique*. Paris: PUF, 1985.

Balke, Friedrich. "Selbstsorge/Selbsttechnologie." In *Foucault Handbuch. Leben—Werk—Wirkung*, ed. C. Kammler, R. Parr and U. J. Schneider, 284–90. Stuttgart: Metzler, 2014.

Bargetz, Brigitte. *Ambivalenzen des Alltags: Neuorientierungen für eine Theorie des Politischen*. Bielefeld: Transcript, 2016.

Barthes, Roland. *Mythen des Alltags*. Frankfurt a.M.: Suhrkamp, 1964 [1957].

Bayle, Pierre. *Toleranz: Ein philosophischer Kommentar*. Ed. E. Buddeberg and R. Forst. Trans. E. Buddeberg. Berlin: Suhrkamp, 2016.

Beckett, Samuel. *Endgame*. Ed. R. McDonald. London: Faber and Faber, 2009.

Benhabib, Seyla et al. *Feminist Contentions: A Philosophical Exchange*. New York: Routledge, 1995.

Benhabib, Seyla. *Selbst im Kontext: Kommunikative Ethik im Spannungsfeld von Feminismus, Kommunitarismus und Postmoderne*. Trans. I. König. Frankfurt a.M.: Suhrkamp, 1995 [1992].

Benjamin, Walter. "Über den Begriff der Geschichte." In *Walter Benjamin: Werke und Nachlaß. Kritische Gesamtausgabe, Vol. 19*. Berlin: Suhrkamp, 2010 [1942].

Benjamin, Walter. *The Work of Art in the Age of Mechanical Reproduction*. Trans. J. A. Underwood. London: Penguin Books, 2008 [1935].

Berger, Alois. "Billiges Fleisch, unhaltbare Zustände." *Deutschlandfunk*, May 5, 2021, online: https://www.deutschlandfunk.de/schlachthoefe-in-europa-billiges-fleisch-unhaltbare.724.de.html?dram:article_id=481030.

Berlant, Lauren: *Cruel Optimism*. Durham: Duke University Press, 2011.

Bernstein, Richard J. "Hilary Putnam: The Pragmatist Enlightenment." *Graduate Faculty Philosophy Journal* 38, no. 2 (2017): 373–83.

Bianchi, Sarah and N.N., eds. "Subjektivierung und Aufklärung: Von Nietzsche zu kritischen Theorien und feministischen Theorien." *Nietzscheforschung* 33 (2026), forthcoming.

Bianchi, Sarah, ed. "Schwerpunkt: Zu Kants 300. Geburtstag: Aufklärung, Freiheit und Nicht-Beherrschung.'" *Zeitschrift für Praktische Philosophie* 11, no. 1 (2024), forthcoming.

Bianchi, Sarah "Zu einer aufklärerischen Mikrologie: Adorno, Foucault und die digitalisiert 'verwaltete Welt.'" In *Adorno und die Medien*. Ed. L. Voropai and J.-F. Popp. 199–214. Berlin: Kadmos, 2023.

Bianchi, Sarah. "Unter dem und wider den 'Bann des Einheitsprinzips': Identifizierungsmacht, Subjekt und Affekt nach Adorno." *Zeitschrift für Praktische Philosophie* 9, no. 2 (2022): 15–48.

Bianchi, Sarah. "Mikropolitiken und Aufklärung: Sozialontologische Perspektiven nach Nietzsche und Foucault." In *Nietzsches Perspektiven des Politischen*. Ed. C. Schubert and M. Ruehl. 57–71. Berlin/Boston: de Gruyter, 2022.

Bianchi, Sarah. "Färben und Wissen: Wie die Kulturtechnik des Malens weltgestaltend wirkt. Über: Ludger Schwarte. 'Denken in Farbe,'" *Deutsche Zeitschrift für Philosophie* 70, no. 3 (2022), 559–65.

Bianchi, Sarah. "Verletzlichkeit und Leben: Nietzsche im Kontext der Anerkennungsdebatte." In *Oldenburger Jahrbuch für Philosophie*. Ed. M. Bauer and N. Baratella. 111–38. Oldenburg: BIS-Verlag, 2019.

Bianchi, Sarah. "'Maschinen-Cultur' und Anerkennung: Über Nietzsches Verständnis von sozialer Individualität." *Nietzscheforschung* 25 (2018): 435–48.

Bianchi, Sarah. "'Aus seinem Leben ein Kunstwerk machen.' Ästhetische Selbstgestaltung nach Nietzsche und Foucault." In *Kongress-Akten*, Vol. 4: *Das ist Ästhetik!*, ed. J. Rebentisch, online: http://www.dgae.de/wp-content/uploads/2017/06/Aus-seinem-Leben-ein-Kunstwerk-machen_Bianchi_DGÄ.pdf, 2018.

Bianchi, Sarah. "Vom Scham-ersparen: Über Nietzsches fragiles Verständnis von Anerkennung in Abgrenzung zur französischen Verkennungslinie." *Zeitschrift für praktische Philosophie* 4, no. 2 (2017): 15–40.

Bianchi, Sarah. "'Oh Voltaire! Oh Humanität! Oh Blödsinn!' Über den Zusammenhang von Anerkennung und menschlichem Selbstverständnis bei Nietzsche." *Zeitschrift für praktische Philosophie* 4, no. 1 (2017): 15–36.

Bianchi, Sarah. *Einander nötig sein: Existentielle Anerkennung bei Nietzsche*. Paderborn: Wilhelm Fink, 2016.

Bianchi, Sarah and Christopher Th. Scott. "Enhancing Human Embryos: Restraint or Full Speed Ahead?." *The Enhancing Life Project*, February 25, 2016, available at: http://enhancinglife.uchicago.edu/blog/enhancing-human-embryos.

Bianchi, Sarah. *Das deutsche Kaleidoskop: Die Dreyfus-Affäre in der wilhelminischen Öffentlichkeit zwischen 1898 und 1899*. Frankfurt a.M.: Peter Lang, 2012.

Biebricher, Thomas. *Die politische Idee des Neoliberalismus*. Berlin: Suhrkamp, 2021.

Binoche, Juliette. *Clouds of Sils Maria*. DVD. Dir. O. Assayas. London: Artificial Eye, 2014.

Blanchot, Maurice. *Michel Foucault*. Trans. B. Wahlster. Tübingen: edition diskord, 1987.

Bogel-Burroughs, Nicholas et al. "Minnesota Officer Who Shot Daunte Wright Meant to Fire Teaser, Chief Says." *New York Times*, April 12, 2021, available at: https://www.nytimes.com/2021/04/12/us/brooklyn-center-police-shooting-minnesota.html.

Böhme, Gernot. *Aisthetik: Vorlesungen über Ästhetik als allgemeine Wahrnehmungslehre*. München: Fink, 2001.

Bostrom, Nick and Eliezer Yudkowsky. "The Ethics of Artificial Intelligence." In *The Cambridge Handbook of Artificial Intelligence*. Ed. K. Frankish and W. M. Ramsy. 316–34. Cambridge: Cambridge University Press, 2014.

Bostrom, Nick. "Why I Want to Be a Posthuman When I Grow Up." In *Medical Enhancement and Posthumanity*. Ed. B. Gordijn and R. Chadwick. 107–36. Berlin: Springer, 2008.

boyd, danah. *It's Complicated: The Social Lives of Networked Teens*. New Haven, CT/London: Yale University Press, 2014.

Bröckling, Ulrich. *Gute Hirten führen sanft: Über Menschenregierungskünste*. 3rd ed. Berlin: Suhrkamp, 2019.

Bröckling, Ulrich. *Das unternehmerische Selbst: Soziologie einer Subjektivierungsform*. 7th ed. Frankfurt a.M.: Suhrkamp, 2007.

Bröckling, Ulrich. "Das demokratisierte Panopticon: Subjektivierung und Kontrolle im 360 Grad Feedback." In *Michel Foucault: Zwischenbilanz einer Rezeption: Frankfurter Foucault-Konferenz 2001*. Ed. A. Honneth and M. Saar. 77–93. Frankfurt a.M.: Suhrkamp, 2003.

Brown, Wendy. *Politics out of History*. Princeton/Oxford: Princeton University Press, 2001.

Brühl, Jannis et al. "Was ist eigentlich bei Facebook los?." *Süddeutsche Zeitung*, April 5, 2018, available at: https://www.sueddeutsche.de/digital/datenmissbrauch-was-ist-eigentlich-gerade-bei-facebook-los-1.3932349.

Buchanan, Allen. *Better Than Human: The Promise and Perils of Biomedical Enhancement*. Oxford: Oxford University Press, 2017.

Burckhardt, Jacob. *Weltgeschichtliche Betrachtungen: Kritische Gesamtausgabe*. Vol. 10. München: Beck, 2000pp.

Butler, Judith. *Gender Trouble: Feminism and the Subversion of Identity*. London: Routledge, 2006.

Butler, Judith. *The Psychic Life of Power: Theories in Subjection*. Stanford: Stanford University Press, 1997.

Bykova, Alexandra and Roman Stöllinger. "Trade Balances and International Competitiveness in Cyber-physical, Digital task-intensive, ICT Capital intensive and Traditional Industries." *Research Report: The Vienna Institute for International Economic Studies* 468, May (2023): 5–38.

Canguilhem, Georges. *Le normal et le pathologique*. Paris: PUF, 2009 [1966].

Castelvecchi, Davide. "Is Facial Recognition too Biased to be let Loose? The Technology is Improving–but the Bigger Issue is How It's Used." *Nature* 587 (2020): 347–9.

Castoriadis, Cornelius. *The Imaginary Institution of Society*. Cambridge: MIT Press, 1997 [1975].

Cavell, Stanley. *Conditions Handsome and Unhandsome: The Constitution of Emersonian Perfectionism*. Chicago/London: Chicago University Press, 1990.

Chason, Rachel and Samantha Schmidt. "The Freedom To Assemble in Two Acts: Lafayette Square, Capitol Rallies Met Starkly Different Policing Response." *Washington Post*, January 14, 2021, available at: https://www.washingtonpost.com/dc-md-va/interactive/2021/ blm-protest-capitol-riot-police-comparison/.

Claessens, Dieter. *Rolle und Macht*. München: Juventa, 1974.

Collins, Patricia Hill. *Black Feminist Thought: Knowledge, Consciousness, and the Politics of Empowerment*. 2nd ed. New York/London: Routledge, 2000.

Colombo, Agustín and Edward McGushin. "Confessions of the Flesh–Guest Editors' Introduction." *Foucault Studies: Special Issue. Foucault's History of Sexuality Vol. 4, Confessions of the Flesh* 29 (2021): 1–5.

Conant, James. "Nietzsche's Perfectionism: A Reading of 'Schopenhauer as Educator.'" In *Nietzsche's Postmoralism: Essays on Nietzsche's Prelude to Philosophy's Future*. Ed. R. Schacht. 181–257. Cambridge: Cambridge University Press, 2001.

Conant, James. *Friedrich Nietzsche: Perfektionismus und Perspektivismus*. 2nd ed. Konstanz: Konstanz University Press, 2018.

Conant, James."The Dialectic of Perspectivism II." *Nordic Journal of Philosophy* 7, no. 1 (2006).

Conant, James. "The Dialectic of Perspectivism I." *Nordic Journal of Philosophy* 6, no. 2 (2005).

Connolly, William. "Spinoza and Us."*Political Theory* 29, no. 4 (2001): 583–94.

Connolly, William E. *The Ethos of Pluralization*. Minneapolis: University of Minnesota, 1995.

Cornell, Drucilla. *The Imaginary Domain: Abortion, Pornography and Sexual Harrassment*. London/New York: Routledge, 1995.

Couldry, Nick and Ulises A. Mejias. *The Costs of Connection: How Data is Colonizing Human Life and Appropriating it for Capitalism.* Stanford: Stanford University Press, 2019.

Crenshaw, Kimberlé. "Demarginalizing the Intersection of Race and Sex: A Black Feminist Critique of Antidiscrimination Doctrine, Feminist Theory and Antiracist Politics." *University of Chicago Legal Forum* 140, no. 1 (1989): 139–67.

Criado-Perez, Caroline. *Unsichtbare Frauen: Wie eine von Daten beherrschte Welt die Hälfte der Bevölkerung ignoriert*. Trans. S. Singh. München: btb, 2020.

Critchley, Simon. *Infinitely Demanding, Ethics of Commitment, Politics of Resistance*. London/New York: Verso, 2007.

Cruise, Tom. *Minority Report*. DVD. Dir. Steven Spielberg. Los Angeles: 20th Century Fox/DreamWorks Pictures, 2002.

Dabrock, Peter, and Tabea Ott. "Transparent human–Non-transparent Technology? The Janus-faced Call for Transparency in AI-Based Health Care Technology." *Frontiers in Genetics* 13 (2022).

Dahrendorf, Ralf. *Homo sociologicus: Ein Versuch zur Geschichte, Bedeutung und Kritik der Kategorie der sozialen Rolle*. 5th ed. Köln/Opladen: Westdeutscher Verlag, 1965.

Darc, Mireille. *The Weekend.* DVD. Dir. Jean-Luc Godard. Leipzig: Arthaus, 2013 [1967].

Davidson, Arnold I. "Ethics as Ascetics: Foucault, the History of Ethics, and Ancient Thought." In *The Cambridge Companion to Michel Foucault*. Ed. G. Gutting. 115–40. Cambridge: Cambridge University Press, 1994.

Davidson, Arnold I. *The Emergence of Sexuality: Historical Epistemology and the Formation of Concepts*. Cambridge: Harvard University Press, 2004.

Davidson, Helen. "Beijing Blocks Access to Clubhouse App after Surge in User Numbers." *The Guardian*, February 8, 2021, available at: https://www.theguardian

.com/world/2021/feb/08/bingeing-free-expression-popularity-of-clubhouse-app-soars-in-china.

"Deaths at Sea on Migrant Routes to Europe Almost Double, Year on Year." *UN News: Global Perspective Human Stories*, April 29, 2022, available at: https://news.un.org/en/story/2022/04/1117292.

Debord, Guy. *La société du spectacle*. Paris: Buchet/Chastel, 1967.

Deleuze, Gilles. "Postscript on the Societies of Control." *October* 59 (1996): 443–8.

Deleuze, Gilles. *Abécédaire*: *Gilles Deleuze von A bis Z.* Documentary. Dir. P.-A. Boutang. Fridolfing: absolut Medien, 1996.

Deleuze, Gilles. "Wunsch und Lust." In *Schizophrenie und Gesellschaft: Texte und Gespräche von 1975 bis 1995*. Ed. D. Lapoujade. Trans. E. Moldenhauer. 117–28, Frankfurt a.M.: Suhrkamp, 2005 [1994].

Deleuze, Gilles. "Ethology: Spinoza and Us." In *Incorporations*. Ed. J. Crary and S. Kwinter. 625–33. New York: Zone Books, 1992.

Deleuze, Gilles. "Das Leben als Kunstwerk." In *Gilles Deleuze. Unterhandlungen 1972–1990*. 136–46. 6th ed. Frankfurt a.M.: Suhrkamp, 1993 [1986].

Deleuze, Gilles. *Foucault*. Trans. H. Kocyba. 8th ed. Frankfurt a.M.: Suhrkamp, 2015 [1986].

Deleuze, Gilles and Felix Guattari. *Kafka: Toward a Minor Literature*. Trans. D. Polan. Minneapolis/London: University of Minnesota Press, 1986 [1975].

Deleuze, Gilles and Felix Guattari. *Anti-Oedipus: Capitalism and Schizophrenia*. 10th ed. London et al: Bloomsbury Academic, 2020 [1972].

Deleuze, Gilles. *Difference and Repetition*. New York: Columbia University Press, 1995 [1968].

Derrida, Jacques. *Schurken: Zwei Essays über die Vernunft*. Trans. H. Brühmann. Frankfurt a.M.: Suhrkamp, 2003.

Derrida, Jacques. *Politik der Freundschaft*. Trans. St. Lorenzer. 5th ed. Berlin: Suhrkamp, 2018 [1994].

Derrida, Jacques. "La mythologie blanche: La métaphore dans le texte philosophique." *Poétique* 5 (1971): 1–52.

Descartes, René. *Philosophische Schriften. In einem Band. Mit einer Einführung von Rainer Specht.* Hamburg: Meiner, 1996.

Detel, Wolfgang. *Foucault und die klassische Antike: Macht, Moral, Wissen*. Frankfurt a.M.: Suhrkamp, 1998.

Dhawan, Nikita. "Die Aufklärung vor den Europäer*innen retten." In *Normative Ordnungen*. Ed. R. Forst and K. Günther. 191–208. Berlin: Suhrkamp, 2021.

Dhawan, Nikita. *Impossible Speech: On the Politics of Silence and Violence.* Sankt Augustin: Academia, 2007.

Dickel, Sascha. "Der neue Mensch–ein (technik)utopisches Upgrade: Der Traum von Human Enhancement," available at: https://www.bpb.de/apuz/233464/der-neue-mensch-ein-technikutopisches-upgrade-der-traum-vom-human-enhancement.

Diefenbach, Katja et al., ed. *Encountering Althusser: Politics and Materialism in Contemporary Radical Thought.* London: Bloomsbury, 2013.

Drösser, Christoph. "Der Name HAL ist von IBM abgeleitet: Stimmt's?." *ZEIT*, January 27, 2000, available at: https://www.zeit.de/2000/05/200005.stimmts_hal_ibm_.xml.

Dübgen, Franziska. *Was ist gerecht? Kennzeichen einer transnationalen solidarischen Politik*. Frankfurt a.M./New York: Campus, 2014.

Dullea, Keir. *2001: A Space Odyssey*. DVD. Dir. Stanley Kubrick. Los Angeles: Warner Bros, 2004 [1968].

Dutton, William H. et al. *The Internet Trust Bubble: Global Values, Beliefs and Practices, World Economic Forum 2013*, available at: https://www3.weforum.org/docs/WEF_InternetTrustBubble_Report2_2014.pdf.

Eberlein, Undine. *Einzigartigkeit: Das romantische Individualitätskonzept der Moderne*. Frankfurt a.M./New York: Campus, 2000.

Ehrenberg, Alain. *Das erschöpfte Selbst: Depression und Gesellschaft in der Gegenwart*. Trans. M. Lenzen. Frankfurt a.M.: Campus, 2015.

Elden, Stuart. "The Problem of Confession: The Productive Failure of Foucault's 'History of Sexuality.'" *Journal for Cultural Research* 9, no. 1 (2005): 23–41.

Eligon, John. "Derek Chauvin Verdict Brings a Rare Rebuke of Police Misconduct." *New York Times*, April 20, 20021, available at: https://www.nytimes.com/2021/04/20/us/george-floyd-chauvin-verdict.html.

Emcke, Carolin. *Gegen den Hass*. 5th ed. Frankfurt a.M.: Fischer, 2020 [2016].

Emcke, Carolin. *Ja heißt ja und* .... Frankfurt a.M.: Fischer, 2019.

Eribon, Didier. *Gesellschaft als Urteil*: *Klassen, Identitäten, Wege*. Trans. T. Haberkorn. Frankfurt a.M.: Suhrkamp, 2017 [2013].

Eribon, Didier. *Returning to Reims*. Trans. M. Lucey. London: Penguin Books, 2019 [2009].

Eribon, Didier. *Michel Foucault*. 2nd ed. Paris: Éditions Flammarion, 2011 [1989].

Erling, Johnny. "So absurd ausgefeilt ist Chinas Überwachungssystem." *DIE WELT*, April 17, 2019, available at: https://www.welt.de/wirtschaft/article192029849/Social-Scoring-So-absurd-ausgefeilt-ist-Chinas-Ueberwachungssystem.html.

Ernaux, Annie. *Le jeune homme*. Paris: Gallimard, 2022.

Ernaux, Annie. *A Girl's Story*. Trans. A. L. Strayer. London: Fitzcarraldo Edition, 2020 [2016].

Ernaux, Annie. *Happening*. Trans. T. L. London: Fitzcarraldo Editions, 2019 [2000].

Ernaux, Annie. *The Years*. Trans. A. L. Strayer. London: Fitzcarraldo Editions, 2017 [2008].

Ernaux, Annie. *Shame*. Trans. T. Leslie. New York/Oakland/Liverpool: Seven Stories Press, 1999 [1997].

Ernaux, Annie. *A Man's Place*. Trans. T. Leslie. London: Fitzcarraldo Editions, 2012 [1983].

Ernaux, Annie. *A Woman's Story*. Trans. T. Leslie. New York: Seven Stories Press, 2003 [1987].

Erwin, Sean. "Political Technique, the Conflict of Umori, and Foucault's Reading of Machiavelli in 'Sécurité, Territoire, Population.'" *Foucault Studies* 9 (2015): 172–90.

Ewald, François. "Foucault: Ein vagabundierendes Denken." In *Michel Foucault: Dispositive der Macht*. 7–20. Berlin: Merve, 1978.

Ewald, François. "Omnes et singulatim: After Risk." *Carceral Notebooks* 7 (2011): 77–107.

Faden, Ruth and Tom L. Beauchamp. *The History and Theory of Informed Consent*. Oxford: Oxford University Press, 1986.

Fanon, Frantz. *Black Skin, White Masks*. Trans. R. Philcox. New York: Grove Press, 2008 [1954].

Faubion, James D. "Introduction." In *Michel Foucault. Aesthetics, Methods, And Epistemology*. Ed. J. D. Faubion. x-xxxvi. New York: Penguin Press, 2020.

Fenner, Dagmar. *Selbstoptimierung und Enhancement. Ein ethischer Grundriss*. Wien et al.: utb, 2019.

Feyerabend, Paul. *Against Method: Outline of an Anarchistic Theory of Knowledge*. London: Verso, 1975.

Fischel, Joseph J. *Screw Consent: A Better Politics of Sexual Justice*. Oakland, CA: University of California Press, 2019.

Fischel, Joseph J. *Sex and Harm in the Age of Consent*. Minneapolis: University of Minnesota Press, 2016.

Fisher, Marc et al. "The Four Hour Insurrection: How A Trump Mob Halted American Democracy." *Washington Post*, January 7, 2021, available at: https://www .washingtonpost. com/graphics/2021/politics/trump-insurrection-capitol/.

Fleig, Anne. "Automaten mit Köpfchen: Lebendige Maschinen und künstliche Menschen im 18. Jahrhundert." *Goethe-Zeit Portal*, January 19, 2004, available at: http://www.goethezeitportal.de/db/wiss/epoche/fleig_automaten.pdf.

Flügel, Oliver. "Jenseits von Prozedur und Substanz: Jacques Derrida und die normative Demokratietheorie." In *Politische Philosophie und Dekonstruktion. Beiträge zur politischen Theorie im Anschluss an Jacques Derrida*. Ed. A. Niederberger and M. Wolf. 119–42. Bielefeld: Transcript, 2007.

Forst, Rainer. "The Justification of Trust in Conflict: Conceptual and Normative Groundwork." *ConTrust Working Paper*, no. 2 (2022), available at: contrust.uni-frankfurt.de/wp-2.

Forst, Rainer. *Die noumenale Republik: Kritischer Konstruktivismus nach Kant*. Berlin: Suhrkamp, 2021.

Forst, Rainer. *Normativität und Macht: Zur Analyse sozialer Rechtfertigungsordnungen*. Berlin: Suhrkamp, 2015.

Forst, Rainer. "Noumenal Power." The Journal of Political Philosophy 23, no. 2 (2015): 111–27.

Forst, Rainer. *Das Recht auf Rechtfertigung: Elemente einer konstruktivistischen Theorie der Gerechtigkeit*. 3rd. Frankfurt a.M.: Suhrkamp, 2014.

Forst, Rainer. "Gerechtigkeit nach Marx." In *Nach Marx*. Ed. R. Jaeggi and D. Loick. 107–21. Berlin: Suhrkamp, 2013.

Forst, Rainer. "Die Ungerechtigkeit der Gerechtigkeit: Normative Dialektik nach Ibsen, Cavell und Adorno." In *Kritik der Rechtfertigungsverhältnisse: Perspektiven einer kritischen Theorie der Politik*. Ed. R. Forst. 181–95. Berlin: Suhrkamp, 2011.

Forst, Rainer. "Eine politische Theorie der Freiheit: Das Werk von James Tully." Introduction to *Politische Philosophie als kritische Praxis*, by James Tully. 7–11. Frankfurt a.M.: Campus, 2009.

Forst, Rainer. *Toleranz im Konflikt: Geschichte, Gehalt und Gegenwart eines umstrittenen Begriffs*. Frankfurt a.M.: Suhrkamp, 2003.

Forst, Rainer. "Endlichkeit, Freiheit, Individualität. Die Sorge um sich bei Heidegger und Foucault." In *Ethos der Moderne: Foucaults Kritik der Aufklärung*. Ed. E. Erdmann, R. Forst, and A. Honneth. 146–86. Frankfurt a.M./New York: Campus, 1990.

Foucault, Michel. *Schriften in vier Bänden: Dits et Écrits*. Ed. D. Defer and F. Ewald, trans. M. Bischoff, H.-D. Gondek and H. Kocyba. Frankfurt a.M.: Suhrkamp, 2001–2005.

Foucault, Michel. *Histoire de la sexualité*. Vol. 4: *Les aveux de la chair*. Ed. F. Gros. Paris: Gallimard 2018 [posthumous].

———English edition: Foucault, Michel. *The History of Sexuality*. Vol. 4: *Confessions of the Flesh*. Ed. F. Gros, trans. R. Hurley. New York: Pantheon Books, 2021 [posthumous].

Foucault, Michel. *The History of Sexuality*. Vol. 2: *The Use of Pleasure*. Trans. R. Hurley. New York: Vintage Books, 1990 [1984].

———German edition: Foucault, Michel. *Sexualität und Wahrheit*. Vol. 2: *Der Gebrauch der Lüste*. Trans. U. Raulff and W. Seitter. 13th ed. Berlin: Suhrkamp, 2017 [1984].

———French edition: Foucault, Michel. *Histoire de la sexualité*. Vol. 2: *L'usage des plaisirs*. Paris: Gallimard, 2014 [1984].

Foucault, Michel. *The History of Sexuality*. Vol. 3: *The Care of the Self*. Trans. R. Hurley. New York: Vintage Books, 1988 [1984].

Foucault, Michel. "What is Enlightenment?." In *Essential Works of Foucault 1954–1948*. Vol. 1: *Ethics: Subjectivity and Truth*. Ed. P. Rabinow, trans. R. Hurley, 303–20. New York/London: Penguin Books, 2000 [1984].

Foucault, Michel. "The Ethics of the Concern for Self as a Practice of Freedom."In *The Essential Work of Foucault (1954–1984)*. Vol. 1: *Ethics: Subjectivity and Truth*. Ed. P. Rabinow, trans. R. Hurley et al., 281–302. London et al.: Penguin Books, 2000 [1984].

———German edition: "Die Ethik der Sorge um sich als Praxis der Freiheit." In *Ästhetik der Existenz: Schriften zur Lebenskunst*. Ed. D. Defert and F. Ewald, trans. M. Bischoff et al., 253–79. 5th ed. Berlin: Suhrkamp, 2015 [1984].

Foucault, Michel. "On the Genealogy of Ethics: An Overview of the Work in Progress." In *Essential Works of Foucault (1954–1984)*. Vol. 1, *Ethics: Subjectivity and Truth*. Ed. P. Rabinow, trans. R. Hurley et al., 253–80. London et al.: Penguin Books, 2000 [1984].

———German edition: Foucault, Michel. "Zur Genealogie der Ethik: Ein Überblick über die laufende Arbeit." In *Ästhetik der Existenz: Schriften zur Lebenskunst*. Ed. D. Defert and F. Ewald, trans. M. Bischoff et al., 191–219. 5th ed. Berlin: Suhrkamp, 2015 [1984].

———French edition: Foucault, Michel. "À propos de la généalogie de l'éthique: Un aperçu du travail en cours." In *Un parcours philosophique: Au-delà de l'objectivité et de la subjectivité*. Ed. H. L. Dreyfus and P. Rabinow, trans. F. Durand-Bogaert, 323–46. Paris: Gallimard, 1984.

Foucault, Michel. "An Aesthetics of Existence." In *Foucault Live: Collected Interviews, 1961–1984*, ed. S. Lotringer, trans. L. Hochroth and J. Johnston, 450–54. 2nd ed. South Pasadena: Semiotext(e), 1996 [1984].

Foucault, Michel. "Friendship as a Way of Life." In *Essential Works of Foucault (1954–1984)*. Vol. 1: *Ethics: Subjectivity and Truth*. Ed. P. Rabinow, trans. R. Hurley, 303–20. London: Penguin Books, 2000 [1984].

Foucault, Michel. "The Return of Morality." In *Foucault Live: Collected Interviews 1961–1984*. Ed. S. Lotringer, trans. L. Hochroth and J. Johnston, 465–73. 2nd ed. South Pasadena: Semiotext(e), 1996 [1984].

———German edition: Foucault, Michel. "Die Rückkehr der Moral." In *Dits et Écrits*. Vol. 4: *1980–1988*. Ed. D. Defert and F. Ewald, trans. M. Bischoff et al., 859–72. Frankfurt a.M.: Suhrkamp, 2005 [1984].

Foucault, Michel. *The Government of Self and Others II: The Courage of the Truth. Lectures at the Collège de France 1983–1984*. Ed. F. Gros, trans. G. Burchell. New York: Palgrave Macmillan, 2011.

Foucault, Michel. *Discourse and Truth & Parrēsia*. Ed. H.-P. Fruchaud and D. Lorenzini, trans. N. Luxon. Chicago/London: The University of Chicago Press, 2019 [1982/83].

Foucault, Michel. *The Government of Self and Others: Lectures at the Collège de France 1982–1983*. Ed. F. Gros, trans. G. Burchell. New York: Palgrave Macmillan, 2011 [1982/1983].

———German edition: *Die Regierung des Selbst und der anderen: Vorlesung am Collège de France 1982/83*. Ed. F. Ewald and A. Fontana, trans. J. Schröder. Frankfurt a.M.: Suhrkamp, 2012.

Foucault, Michel. "The Subject and Power." In *Michel Foucault: Beyond Structuralism and Hermeneutics*. Ed. H. L. Dreyfus and P. Rabinow, 208–26. 2nd ed. Chicago: The University of Chicago Press, 1983 [1982].

Foucault, Michel. *The Hermeneutics of the Subject: Lectures at the Collège de France 1981–1982*. Ed. F. Gros, trans. Graham Burchell. 2nd ed. London: Picador, 2005 [1981/1982].

Foucault, Michel. "Subjectivity and Truth." In *Essential Works of Foucault 1954–1984*. Vol. 1: *Ethics: Subjectivity and Truth*. Ed. P. Rabinow, trans. R. Hurley, 87–92. New York/London: Penguin Books, 2000 [1981].

———German edition: Foucault, Michel. "Subjektivität und Wahrheit." In *Ästhetik der Existenz: Schriften zur Lebenskunst*. Ed. D. Defert and F. Ewald, trans. M. Bischoff et al., 74–80. 5th ed. Berlin: Suhrkamp, 2015 [1981].

Foucault, Michel. *Wrong-Doing, Truth-Telling: The Function of Avowal in Justice*. Ed. F. Brion and B. E. Harcourt, trans. S. W. Sawyer. Chicago/London: The University of Chicago Press, 2014 [1981].

Foucault, Michel. *On the Government of the Living: Lectures at the Collège de France 1979–1980*. Ed. M. Senellart, trans. G. Burchell. New York: Picador, 2012.

———German edition: Foucault, Michel. *Die Regierung der Lebenden*. Ed. F. Ewald and A. Fontana, trans. A. Hemminger. Berlin: Suhrkamp, 2014.

Foucault, Michel. "'Omnes et Singulatim': Zu einer Kritik der politischen Vernunft." In *Dits et Écrits*. Vol. 4: *1980–1988*. Ed. D. Defert and F. Ewald, trans. M. Bischoff, 165–98. Frankfurt a.M.: Suhrkamp, 2005 [1979].

Foucault, Michel. "Sicherheit, Territorium und Bevölkerung." In *Dits et Écrits*. Vol. 3: *1976–1979*. Ed. D. Defert and F. Ewald, trans. M. Bischoff, 900–5. 2nd ed. Berlin: Suhrkamp, 2016 [1978].

Foucault, Michel. "What is Critique?." In *The Politics of Truth*. Ed. S. Lotringer, trans. L. Hochroth and C. Porter, 41–82. South Pasadena: Semiotext(e), 2007 [1978].

———German edition: Foucault, Michel. *Was ist Kritik?*. Trans. W. Seitter. Berlin: Merve, 1992 [1978].

Foucault, Michel. *Security, Territory, Population. Lectures at the Collège de France 1977–1978*. Ed. M. Senellart, trans. G. Burchell. New York: Picador 2007 [2004].

Foucault, Michel. "The Confession of the Flesh." In *Power/Knowledge. Selected Interviews & Other Writings 1972–1977*. Ed. C. Gordon, trans. C. Gordon et al. 194–228. New York: Vintage Books, 1980 [1977].

———German edition: "Ein Spiel um die Psychoanalyse. Gespräch mit Angehörigen des Département de Psychoanalyse der Universität Paris VIII in Vincennes." In *Dispositive der Macht. Michel Foucault über Sexualität, Wissen und Wahrheit*. 118–75. Berlin: Merve, 1978 [1977].

Foucault, Michel. *The History of Sexuality*. Vol. 1: *An Introduction*. Trans. R. Hurley. New York: Vintage Books, 1990 [1976].

———German edition: Foucault, Michel. *Sexualität und Wahrheit*. Vol. 1: *Der Wille zum Wissen*. Trans. U. Raulff and W. Seitter. 21st ed. Frankfurt a.M.: Suhrkamp, 2017 [1976].

Foucault, Michel. *Society Must be Defended: Lectures at the Collège de France 1975–1976*. Ed. M. Bertani and A. Fontana, trans. D. Macey. London et al.: Penguin Books, 2004.

———German edition: Foucault, Michel. *In Verteidigung der Gesellschaft: Vorlesungen am Collège de France (1975–1976)*. Ed. F. Ewald and A. Fontana, trans. M. Ott. 5th ed. Frankfurt a.M.: Suhrkamp, 2016.

Foucault, Michel. *Discipline and Punish: The Birth of the Prison*. Trans. A. Sheridan. 2nd ed. New York: Vintage Books, 1995 [1975].

———German edition: Foucault, Michel. *Überwachen und Strafen: Die Geburt des Gefängnisses*. Trans. W. Seitter. 16th ed. Frankfurt a.M.: Suhrkamp, 2016 [1975].

———French original: Foucault, Michel. *Surveiller et punir: Naissance de la prison*. Paris: Gallimard, 2013 [1975].

Foucault, Michel. *Abnormal: Lectures at the Collège de France 1974–1975*. Ed. V. Marchetti and A. Salomoni, trans. G. Burchell. London/New York: Verso, 2016 [1974/1975].

Foucault, Michel. "Die Prosa der Welt." In *Dits et Écrits*. Vol. 1: *1954–1969*. Ed. D. Defert and F. Ewald, trans. M. Bischoff et al., 622–43. Frankfurt a.M.: Suhrkamp, 2001 [1974].

Foucault, Michel. "Truth and Juridical Forms." In *Essential Works 1954–1984*. Vol. 3: *Power*. Ed. J. D. Faubion, trans. R. Hurley et al., 39–88. 3rd ed. New York/London: Penguin Books, 2002 [1973].

———German edition: Foucault, Michel. *Die Wahrheit und die juristischen Formen: Mit einem Nachwort von Martin Saar*. Trans. M. Bischoff. 5th ed. Berlin: Suhrkamp, 2015 [1973].

Foucault, Michel. *The Punitive Society: Lectures at the Collège de France 1972–1973*. Ed. B. E. Harcourt, trans. G. Burchell. Basingstoke/New York: Palgrave Macmillan, 2015.

Foucault, Michel. "Der 'Anti-Ödipus': Eine Einführung in eine neue Lebenskunst." In *Dispositive der Macht. Michel Foucault über Sexualität, Wissen und Wahrheit*. 118–75. Berlin: Merve, 1978 [1972].

Foucault, Michel. *Penal Theories and Institutions: Lectures at the Collège de France 1971–1972*. Ed. B. E. Harcourt, trans. G. Burchell. New York: Palgrave Macmillan, 2019.

Foucault, Michel. "Nietzsche, Genealogy, History." In *The Foucault Reader: An Introduction to Foucault's Thought*. Ed. P. Rabinow, 76–100. London et al: Penguin Books, 1991 [1971].

Foucault, Michel. "Vorlesung über Nietzsche." In *Über den Willen zum Wissen: Vorlesungen am Collège de France 1970–1971. Gefolgt von Das Wissen des Ödipus*. Ed. F. Ewald and A. Fontana, trans. M. Bischoff. 259–81. Berlin: Suhrkamp, 2012.

Foucault, Michel. *Lectures on the Will to Know: Lectures at the Collège de France 1970–1971*. Ed. D. Defert, London: Palgrave Macmillan, 2013.

Foucault, Michel. *Die Ordnung des Diskurses*. Trans. W. Seitter. 10th ed. Frankfurt a.M.: Fischer, 2007 [1970].

Foucault, Michel. *The Archeology of Knowledge And the Discourse on Language*. Trans. A. M. S. Smith. New York: Vintage Books, 2010 [1969].

Foucault, Michel. *The Order of Things: An Archeology of the Human Sciences*. New York: Vintage Books, 1994 [1966].

Foucault, Michel. *Madness and Civilization: A History of Insanity in the Age of Reason*. Trans. R. Howard. New York: Vintage Books, 1988 [1961].

Frankenberg, Günter. "Staat als Begriff und Vorstellung." *Rechtsgeschichte* 15 (2009): 145–68.

Fraser, Nancy and Rahel Jaeggi. *Capitalism: A Conversation in Critical Theory*. Ed. B. Milstein. Cambridge: Polity, 2018.

Fraser, Nancy. *Unruly Practices: Power, Discourse, and Gender in Contemporary Social Theory*. Minneapolis: University of Minnesota Press, 1989.

Freud, Sigmund. *Triebe und Triebschicksale: Psychologie des Unbewussten*. Frankfurt a.M.: Fischer, 2000 [1915].

Frilot, Shari. "Sally Potter." *Bomb Magazine*, June 1, 1993, available at: https://bombmagazine.org/articles/sally-potter/.

Frisch, Max. *Homo faber: Ein Bericht*. Berlin: Suhrkamp, 2011 [1957].

Fuchs, Christian. *Soziale Medien und Kritische Theorie: Eine Einführung*. Trans. F. Kurz. München: UVK, 2019.

Fukuyama, Francis. *The End of History and the Last Man*. New York/London: Penguin Press, 1993.

Gatens, Moira and Genevieve Lloyd. *Collective Imaginings: Spinoza, Past and Present*. London: Routledge, 1999.

Gatens, Moira. *Imaginary Bodies: Ethics, Power and Corporeality*. London/New York: Routledge, 1996.

Gehring, Petra. "Eine Topo-Technologie der Gefährlichkeit: Digitale Einsperrtechniken und sozialer Raum." In *Techniken der Subjektivierung*. Ed. A. Gelhard, T. Alkemeyer and N. Ricken. 299–314. München: Fink, 2013.

Gehring, Petra. "The Inverted Eye: Panopticon and Panopticism. Revisited." *Foucault Studies*, no. 23 (2017): 42–62.

Geisenhanslüke, Achim. "Literatur und Diskursanalyse." In *Michel Foucault: Eine Einführung in sein Denken*. Ed. M. S. Kleiner. 60–71. Frankfurt.M./New York: Campus, 2001.

Gerhardt, Volker. *Vom Willen zur Macht: Anthropologie und Metaphysik der Macht am exemplarischen Fall Friedrich Nietzsche*. Berlin/New York: de Gruyter, 1996.

Geuss, Raymond. "Die Hoffnung." In *Auf Nietzsches Balkon III: Beiträge aus der Villa Silberblick*. Ed. S. Bianchi. 232–7. Weimar: asw-Verlag, 2018.

Geuss, Raymond. "Kritik, Aufklärung, Genealogie." In *Michel Foucault. Zwischenbilanz einer Rezeption: Frankfurter Foucault-Konferenz 2001*. Ed. A. Honneth and M. Saar, 145–56. Frankfurt a.M.: Suhrkamp, 2003.

Geuss, Raymond. "Nietzsche and Genealogy." In *Morality, Culture, and History: Essays on German Philosophy*. Ed. R. Geuss. 1–28. Cambridge: Cambridge University Press, 1999.

Geuss, Raymond. "Nietzsche's Philosophical Ethnology." *A Journal of Humanities and the Classics* 24, no. 3 (2017): 89–116.

Geuss, Raymond. *Changing the Subject: Philosophy from Socrates to Adorno*. Cambridge, MA: Harvard University Press, 2017.

Geuss, Raymond. *The Idea of Critical Theory: Habermas and the Frankfurt School*. Cambridge: Cambridge University Press, 1981.

Gilligan, Carol. *In a Different Voice: Psychological Theory and Women's Development*. Cambridge, MA: Harvard University Press, 1982.

Goethe, Johann W. *Faust: Der Tragödie erster und zweiter Teil*. Ed. E. Trunz. 16th ed. München: Beck, 1996 [1808].

Goethe, Johann W. *Sämtliche Werke. 2 Volumes, Hamburger Ausgabe*. München: Beck, 2000.

Goffman, Erving. *The Presentation of Self in Everyday Life*. New York: Doubleday, 1959.

———German edition: Goffman, Erving. *Wir alle spielen Theater: Die Selbstdarstellung im Alltag*. Trans. P. Weber-Schäfer. München: Piper, 2003.

Goldstein, Jürgen. *Nominalismus und Moderne: Zur Konstitution neuzeitlicher Subjektivität bei Hans Blumenberg und Wilhelm von Ockham*. Freiburg: Alber, 1998.

Gorris, Lothar. "'I can't breathe': Der Satz eines Landes, das zu zerreißen droht." *Der Spiegel*, June 5, 2020, available at: https://www.spiegel.de/ausland/george

-floyd-i-can-t-breathe-ein-satz-veraendert-die-welt-a-00000000-0002-0001-0000-000171426670.

Gramelsberger, Gabriele. "Digital Enhancement: Zum Konstitutionsverhältnis von Mensch, Computer und Welt," available at: https://idw-online.de/de/news666229.

Greely, Henry T. "Some First Steps Toward Responsible Use of Cognitive-Enhancing Drugs by the Healthy." *The American Journal of Bioethics* 13, no. 7 (2013): 39–41.

Greenwald, Glenn. *No Place to Hide: Edward Snowden, the NSA and the Surveillance State*. New York, NY: Picador, 2014.

Gregg, Melissa and Gregory J. Seigworth. *The Affect Theory Reader*. Durham: Duke University Press, 2010.

Gros, Frédéric. "Introduction." In *Michel Foucault. Discourse and Truth and Parrésia*, ed. H.-P. Fruchaud and D. Lorenzini. xvii–4. Chicago/London: Chicago University Press, 2019.

Grunwald, Armin. "Zwischen Fortschrittsoptimismus und Technikdämonisierung." In *Schöne neue Welt? Zwischen technischen Möglichkeiten und ethischen Herausforderungen*. Ed. M. Lindenau and M. Meier Kressig. 77–100. Bielefeld: transcript, 2020.

Guiblini, Alberto and Sagar Sanyal. "Challenging Human Enhancement." In *The Ethics of Human Enhancement: Understanding the Debate*. Ed. S. Clark et al. 1–26. Oxford: Oxford University Press, 2016.

Günther, Klaus. "Von normativen zu smarten Ordnungen." In *Normative Ordnungen*. Ed. R. Forst and K. Günther. 523–52. Berlin: Suhrkamp, 2021.

Habermas, Jürgen. *Der philosophische Diskurs der Moderne: Zwölf Vorlesungen*. 12th ed. Berlin: Suhrkamp, 2016 [1985].

Habermas, Jürgen. *Die Zukunft der menschlichen Natur: Auf dem Weg zu einer liberalen Eugenik?*. 5th ed. Berlin: Suhrkamp, 2018 [2001].

Hacking, Ian. *Historical Ontology*. Cambridge, MA: Harvard University Press, 2004.

Hadot, Pierre. *La philosophie comme manière de vivre: Entretiens avec Jeannie Carlier et Arnold I. Davidson.* Paris: Albin Michel, 2001.

Haggerty, Kevin D. and Richard V. Ericson. "The Surveillance Assemblage." *British Journal of Sociology* 51, no. 4 (2000): 605–22.

Hahn, Ulla. "Vernunft ist auch eine Herzenssache." *Frankfurter Allgemeine Zeitung*, March 10, 2019, available at: https://www.faz.net/aktuell/feuilleton/buecher/literatur-und-ki-vernunft-ist-auch-eine-herzenssache-16079038-p4.html.

Hahne, Nina. *Essayistik als Selbsttechnik: Wahrheitspraxis im Zeitalter der Aufklärung*. Berlin/Boston: de Gruyter, 2015.

Halliwell, Stephen. *Aristotle's Poetics*. Chicago: University of Chicago Press, 1998.

Hamacher, Werner, ed. *Nietzsche aus Frankreich: Essays von Maurice Blanchot, Jacques Derrida, Pierre Klossowski, Philippe Lacoue-Labarthe, Jean-Luc Nancy und Bernard Pautrat*. Frankfurt a.M./Berlin: Ullstein, 1986.

Hampe, Michael. *Die Dritte Aufklärung*. Berlin: Nicolai Verlag, 2018.

Hao, Karen. "This is How AI Bias Really Happens–And Why It's So Hard to Fix." *MIT Technology Review*, February 4, 2019, available at: https://www.technologyreview.com/ 2019/02/04/137602this-is-how-ai-bias-really-happensand-why-its-so-hard-to-fix/.

Harari, Yuval N. *Homo Deus: Eine Geschichte von Morgen*. Trans. A. Wirthensohn. München: Beck, 2017.

Harcourt, Bernard E. "Foucault's Keystone: Confessions of the Flesh. How the Fourth and Final Volume of the 'History of Sexuality' Completes Foucault's Critique of Modern Western Societies." *Columbia Public Law Research Paper*, No. 14–647 (2019): 1–24.

Harcourt, Bernard E. "The Fight Ahead." *Boston Review*, January 7, 2021, available at: http://bostonreview.net/race-politics/bernard-e-harcourt-fight-ahead.

Harcourt, Bernard E. "The Illusion of Influence: On Foucault, Nietzsche, and a Fundamental Misunderstanding." *Columbia Public Law Research Paper* 14–627 (2019).

Harcourt, Bernard E. *Against Prediction: Profiling, Policing, and Punishing in an Actual Age*. Chicago: The University of Chicago Press, 2007.

Harcourt, Bernard E. *Exposed: Desire and Disobedience in the Digital Age*. London/Cambridge, MA: Harvard University Press, 2015.

Harcourt, Bernard E. *Language of the Gun: Youth, Crime and Public Policy*. Chicago/London: The University of Chicago Press, 2006.

Harcourt, Bernard E., Martin Saar, and Sarah Bianchi. "The Critique and Politics of Identity: On the Affinities between Critical Theory and Poststructuralism. A Conversation with Bernard E. Harcourt and Martin Saar Conducted by Sarah Bianchi." *Journal for the Study of Contemporary Power: The Coils of the Serpent* 10 (2022): 118–30.

Hark, Sabine. *deviante Subjekte: Die paradoxe Politik der Identität*. Opladen: Leske+Budrich, 1996.

Hark, Sabine. *Dissidente Partizipation: Eine Diskursgeschichte des Feminismus*. Frankfurt a.M.: Suhrkamp, 2005.

Hark, Sabine. *Koalitionen des Überlebens: Queere Bündnispolitiken im 21. Jahrhundert*. Göttingen: Wallstein Verlag, 2017.

Harris, John. *Enhancing Evolution: The Ethical Case for Making Better People*. Princeton: Princeton University, 2007.

Harwell, Drew and Craig Timberg. "How America's Surveillance Networks Helped the FBI Catch the Capitol Mob." *Washington Post*, April 2, 2021, available at: https://www.washingtonpost.com/technology/2021/04/02/capitol-siege-arrests-technology-fbi-privacy/.

Haug, Frigga. "Sozialistischer Feminismus: Eine Verbindung im Streit." In *Handbuch Frauen- und Geschlechterforschung. Theorie, Methoden, Empirie*. Ed. R. Becker and B. Kortendiek. 49–55. Wiesbaden: VS Verlag für Sozialwissenschaften, 2004.

Haugaard, Mark and Matthias Kettner, ed. *Theorising Noumenal Power. Rainer Forst and his Critiques.* London: Routledge, 2020.

Hediger, Vinzenz and Rembert Hüser, ed. *Jean-Luc Godard: Film denken nach der Geschichte des Kinos*. Paderbon: Fink, 2023.

Hegel, Georg W. F. *Lectures on the History of Philosophy*. Vol. 2. Trans. E. S. Haldane and F. H. Simson. London 1896 [1832]).

Heilinger, Jan-Christoph. *Anthropologie und Ethik des Enhancement*. Berlin/New York: de Gruyter, 2010.

Heine, Heinrich. *Zur Geschichte der Religion und Philosophie in Deutschland*. Stuttgart: Reclam, 1997 [1834].

Hill, Evan et al. "How George Floyd Was Killed in Police Custody." *New York Times*, May 31, 2020, available at: https://www.nytimes.com/2020/05/31/us/george-floyd-investigation.html.

Hintze, Arend. "Understanding the Four Types of Artificial Intelligence, from Reactive Robots to Self-Aware Beings." *The Conversation*, November 14, 2016, available at: https://theconversation.com/understanding-the-four-types-of-ai-from-reactive-robots-to-self-aware-beings-67616.

Hobbes, Thomas. *Leviathan oder Stoff, Form und Gewalt eines kirchlichen und bürgerlichen Staates*. Trans. W. Euchner. Frankfurt a.M.: Suhrkamp, 1984 [1651].

Hoffmann, Ernst T. A. *Der Sandmann*. Berlin: Henricus Edition Deutsche Klassik, 2019 [1817].

Hofmann, Jeanette. "Digitale Infrastrukturen im Wandel." *Bürger & Staat* 72, no. 1–2 (2022): 56–62.

Honig, Bonnie. *Political Theory and the Displacement of Politics*. New York: Cornell University Press, 1993.

Honneth, Axel, "Rekonstruktive Gesellschaftskritik unter genealogischem Vorbehalt: Zur Idee der 'Kritik' in der Frankfurter Schule." In *Pathologien der Vernunft: Geschichte und Gegenwart der Kritischen Theorie*. Ed. A. Honneth. 57–69. 5th ed. Frankfurt a.M.: Suhrkamp, 2020.

Honneth, Axel. *Verdinglichung: Eine anerkennungstheoretische Studie*. Berlin: Suhrkamp, 2015.

Honneth, Axel: *Das Recht der Freiheit: Grundriss einer demokratischen Sittlichkeit*. Berlin: Suhrkamp, 2011.

Honneth, Axel. "Einleitung: Genealogie als Kritik." In *Michel Foucault: Zwischenbilanz einer Rezeption: Frankfurter Foucault-Konferenz 2001*. Ed. A. Honneth and M. Saar. 117–22. Frankfurt a.M.: Suhrkamp, 2003.

Honneth, Axel. "'Angst und Politik.' Stärken und Schwächen der Pathologiediagnose von Franz Neumann." In *Kritische Theorie der Politik. Franz L. Neumann–eine Bilanz*. Ed. M. Iser and D. Strecker. 200–7. Baden-Baden: Nomos, 2002.

Honneth, Axel. *Kampf um Anerkennung: Zur moralischen Grammatik sozialer Konflikte*. Frankfurt a.M.: Suhrkamp, 1994.

Honneth, Axel. "Soziologie: Eine Kolumne. Ästhetisierung der Lebenswelt." *Merkur* 46, no. 519 (1992): 522–7.

Honneth, Axel. *Kritik der Macht. Reflexionsstufen einer kritischen Gesellschaftstheorie*. Frankfurt a.M.: Suhrkamp, 1988.

———English edition: Honneth, Axel. *The Critique of Power: Reflective Stages in a Critical Social Theory*. Trans. K. Baynes. Cambridge, MA: MIT Press, 1993 [1988].

Horkheimer, Max and Theodor W. Adorno. *Dialektik der Aufklärung: Philosophische Fragmente*. 23rd ed. Frankfurt a.M.: Fischer, 2017 [1947].

Hudson, Tymber. "Garden of Vitality." *Abolition 13/13*. February 25, 2021, available at: http://blogs.law.columbia.edu/abolition1313/10-13-abolish-family-policing/.

Huizinga, Johan. *Homo ludens: Vom Ursprung der Kultur im Spiel*. Hamburg: Rowohlt, 2004 [1938].

Huxley, Aldous. *Brave New World*. London: Vintage Penguin Books, 2007 [1932].

IKEA. https://www.ikea.com/gb/en/product-guides/ikea-home-smart-system/.

"Intersektionalität: Eine kurze Einführung." In *Heinrich Böll Stiftung*, April 12, 2019, available at: https://www.gwi-boell.de/de/2019/04/12/intersektionalitaet-eine-kurze-einfuehrung.

Israel, Jonathan I. *Radical Enlightenment: Philosophy and the Making of Modernity 1650–1750*. Oxford: Oxford University Press, 2001.

Jaeggi, Rahel and Robin Celikates. *Sozialphilosophie: Eine Einführung*. München: Beck, 2017.

Jaeggi, Rahel. *Kritik von Lebensformen*. Berlin: Suhrkamp, 2014.

Jaeggi, Rahel. *Entfremdung: Zur Aktualität eines sozialphilosophischen Problems*. Frankfurt a.M./New York: Campus, 2005.

Jay, Martin. *Downcast Eyes: The Denigration of Vision in Twentieth-Century French Thought*. Berkeley/Los Angeles/London: University of California Press, 1994.

Jay, Martin. *The Dialectical Imagination: The History of The Frankfurt School and the Institute of Social Research 1923–1950*. Boston and Toronto: Little, Brown and Company, 1973.

Jefferson, Thomas et al. "The Constitution of the United States." In *America's Founding Documents*, available at: https://www.archives.gov/founding-docs/constitution.

"Joe Biden: There is Institutional Racism in America." *CNBC*, October 22, 2020, available at: https://www.cnbc.com/video/2020/10/22/joe-biden-there-is-institutional-racism-in-america.html.

John Earl, Joseph, ed. *Ferdinand de Saussure: Critical Assessment of Leading Linguists. Volume 3: Saussure, Structuralism and Post-Structuralism*. London: Routledge, 2013.

Kafka, Franz. *Der Prozess und ausgewählte Parabeln*. Paderborn: Schöningh, 2001 [1925].

Kaiser, Britanny. *Targeted: My Inside Story of Cambridge Analytica and How Trump, Brexit and Facebook Broke Democracy*. London: HarperCollinsPublishers, 2019.

Kaiser, Britanny. *The Great Hack*. Documentary. Dir. K. Amer and J. Noujaim. Netflix: Los Gatos, 2019.

Kammler, Clemens et al., eds. *Foucault Handbuch: Leben–Werk–Wirkung*. Stuttgart: Metzler, 2014.

Kant, Immanuel. "Was ist Aufklärung?." In *The Politics of Truth*. Ed. S. Lotringer. Trans. L. Hochroth and C. Porter, 29–38. South Pasadena: Semiotext(e), 2007.

Kant, Immanuel. *Kritik der Urteilskraft*. AA V. Berlin/New York: de Grutyer, 1971 [1790]. [Abbreviation: KU]

Karg, Stefanie. "The Next Rembrandt: Kunstwerke aus dem 3 D-Drucker?!." *Kultur Hoch N*, available at: https://www.kulturhochn.de/the-next-rembrandt-kunstwerke-aus-dem-3d-drucker/.

Kass, Leon et al. *Beyond Therapy: Biotechnology and the Pursuit of Happiness: A Report by the President's Council on Bioethics*. 2nd ed. New York: HarperCollins Publishers, 2003.

Keller, Maren. "Selbstoptimierung: Die Arbeit am Ich. Fitter, gebildeter, schöner–die Soziologien Anja Röcke erklärt, warum wir uns selbst als eine Dauerbaustelle betrachten." *Der Spiegel*, April 25, 2022, available at: https://www.spiegel.de/wissenschaft/mensch/selbstoptimierung-warum-wir-uns-selbst-als-eine-dauerbaustelle-betrachten-a-0c64a4f6-0002-0001-0000-000200446780.

Kelly, Mark G. E. *For Foucault: Against Normative Political Theory.* Albany: SUNY Press, 2018.

Kerner, Ina. "Gayatri Chakravorty Spivak, Can the Subaltern Speak?." In *Schlüsselwerke der Kulturwissenschaften*. Ed. C. Leggewie et al. 134–8. Bielefeld: transcript, 2012.

Kerner, Ina. "Geschlecht." In *Politische Theorie: 25 umkämpfte Begriffe zur Einführung*. Ed. G. Göhler et al. 126–41. 2nd ed. Wiesbaden: VS Verlag für Sozialwissenschaften, 2011.

Kerner, Ina. "Konstruktion und Dekonstruktion von Geschlecht: Perspektiven für einen neuen Feminismus." *gender . . . politik . . . online* (2007), available at: https://www.fu-berlin.de/sites/gpo/pol_theorie/ Zeitgenoessische_ ansaetze/KernerKonstruktion_ und_ Dekonstruktion/kerner.pdf.

Kersting, Wolfgang. "Einleitung: Die Gegenwart der Lebenskunst." In *Kritik der Lebenskunst*. Ed. W. Kersting and C. Langbehn. 10–90. 2nd ed. Frankfurt a.M.: Suhrkamp, 2016.

King Jr., Martin Luther. *Stride Toward Freedom: The Montgomery Story*. New York, Evanstone and London: Harper & Row Publishers, 1958.

Kirloskar-Steinbach, Monika, ed. *Dialogue and Decolonization: Historical, Philosophical, and Political Perspectives. Garrick Cooper, Sudipta Kavirai, Charles W. Mills, Sor-Hoon Tan and James Tully*. London et al: Bloomsbury Press, forthcoming 2023.

Kitcher, Philip. *Genetik und Ethik: Die Revolution der Humangenetik und ihre Folgen*. München: Luchterhand, 1998.

Koopman, Colin. *Genealogy as Critique: Foucault and the Problems of Modernity.* Bloomington: Indiana University Press, 2012.

Koopman, Colin. *How We Became Our Data: A Genealogy of the Informational Person.* Chicago/London: The University of Chicago Press, 2019.

Kracauer, Siegfried. *The Mass Ornament: Weimar Essays*. Trans. T. Y. Levin. Cambridge, Mass./London: Harvard University Press, 1995 [1963/1927].

Krasmann, Susanne et al., eds. *Sichtbarkeitsregime: Überwachung, Sicherheit und Privatheit im 21. Jahrhundert*. Wiesbaden: Springer, 2011.

Kubrick, Stanley and Arthur C. Clark. *Screenplay: 2001. A Space Odyssey*, available at: http://www.archiviokubrick.it/opere/film/2001/script/2001-originalscript.pdf.

Kumkar, Nils. *Alternative Fakten: Zur Praxis der kommunikativen Erkenntnisverweigerung*. Berlin: Suhrkamp, 2022.

Kuo, Lily. "China Bans 23m from Buying Travel Tickets as Part of 'Social Credit' System." *The Guardian*, March 1, 2019, available at: https://www

.theguardian. com/ world/ 2019/ mar/01/china-bans-23m-discredited-citizens-from-buying-travel-tickets-social-credit-system.

Lacan, Jacques. *Das Seminar.* Buch II: *Das Ich in der Theorie Freuds und in der Technik der Psychoanalyse*. Trans. H.-J. Metzger. Weinheim/Berlin: Quadriga, 1991 [1954/1955].

Lacan, Jacques. *Das Spiegelstadium als Bildner der Ichfunktion, wie sie uns in der psychoanalytischen Erfahrung erscheint*. Schriften I. Trans. P. Stehlin. Weinheim/Berlin: Quadriga, 1986 [1936/1949].

Lagasnerie, Geoffroy de. *L'art de la révolte. Snowden, Assange, Manning*. Paris: Fayard, 2015.

Lagasnerie, Geoffroy de. *La dernière leçon de Michel Foucault? Sur le néoliberalisme, la théorie et la politique*. Paris: Fayard, 2012.

Lagasnerie, Geoffroy de. *Penser dans un monde mauvais*. Paris: PUF, 2017.

Large, Duncan. Translator's Introduction to *Nietzsche and Metaphor*, by Sarah Kofman. vii–xxii. Stanford: Stanford University Press.

Latour, Bruno. "On Actor-Network Theory: A Few Clarifications." *Soziale Welt* 47, no. 4 (1996): 369–81.

Le Blanc, Guillaume. "Why Read Foucault's 'Confessions of the Flesh' Today?." *Critique 13/13*, December 2, 2019, available at: http://blogs.law.columbia.edu/critique1313/guillaume-le-blanc-why-read-foucaults-confessions-of-the-flesh-today-english-version/.

Le Blanc, Guillaume. *La pensée Foucault*. Paris: Ellipses, 2014.

Leatherby, Lauren et al. "How a Presidential Rally Turned into a Capitol Rampage." *New York Times*, January 12, 2021, available at: https://www.nytimes.com/interactive/ 2021/ 01/12/us/capitol-mob-timeline.html.

Lee, Gavin. "Corona-Virus: What Went Wrong at Germany's Gütersloh Meat Factory?." *BBC News*, June 25, 2020, online: https://www.bbc.com/news/world-europe-53177628.

Lemke, Thomas, Susanne Krasmann and Ulrich Bröckling. "Gouvernementalität, Neoliberalismus und Selbsttechnolgien." In *Gouvernementalität der Gegenwart: Studien zur Ökonomisierung des Sozialen*. Ed. T. Lemke, S. Krasmann and U. Bröckling. 7–40. 8th ed. Frankfurt a.M.: Suhrkamp, 2019.

Lemm, Vanessa. *Homo Natura. Nietzsche, Philosophical Anthropology and Biopolitics*. Edinburgh: Edinburgh University Press, 2020.

Locke, John. *Zweiter Kommentar über die Regierung*. Trans. H. J. Hoffmann. Frankfurt a.M.: Suhrkamp, 2008.

Locker, Anatol. "Kunst oder Kokolores: Wie Roboter Ai-Da funktioniert." *ZDF*, June 12, 2019, available at: https://www.zdf.de/nachrichten/heute/kunst-oder-kokolores-wenn-roboter-zeichnen-100.html.

Loh, Janina. *Trans- und Posthumanismus zur Einführung*. 3rd ed. Hamburg: Junius, 2020.

Loh, Janina. *Roboterethik: Eine Einführung*. Berlin: Suhrkamp, 2019.

Lorenzini, Daniele. "On Possibilising Genealogy." *Inquiry: An Interdisciplinary Journal of Philosophy* (2020).

Lorenzini, Daniele. "The Emergence of Desire: Notes Toward A Political History of the Will." *Critical Inquiry* 45 (2019): 448–70.

Lorenzini, Daniele. *Éthique et politique de soi: Foucault, Hadot, Cavell et les techniques de l'ordinaire*. Paris: Vrin, 2015.

Louis, Édouard. *Changer: méthode*. Paris: Seuil, 2021.

Louis, Édouard. *Who killed my father?*. Trans. L. Stein. London: Harvill Secker, 2019 [2018].

Louis, Édouard. *The End of Eddy*. Trans. M. Lucey. London: Harvill Secker, 2017 [2014].

Loukidelis, Nikolaos, ed. *Nietzsche und die Lebenskunst: Ein philosophisch-psychologisches Kompendium*. Stuttgart: Metzler, 2016.

Lukács, Georg. *Die Zerstörung der Vernunft*. Bielefeld: Aisthesis Verlag, 2022 [1954].

Lukács, Georg. *History and Class Consciousness*. Cambridge: MIT Press, 1972 [1923].

Lukács, Georg. "Über Form und Wesen des Essay." In *Georg Lukács: Ästhetik, Marxismus, Ontologie. Ausgewählte Texte*. Ed. R. Dannemann and A. Honneth. 41–62. Berlin: Suhrkamp, 2021 [1910].

Lukes, Steven. *Power: A Radical View*. Basingstoke and London: Palgrave Macmillan, 1974.

Lukrez, *Über die Natur der Dinge oder vom Wesen des Weltalls*. Trans. E. M. Noller. Stuttgart: Reclam, 2021 [around 55 BCE].

Lutz-Bachmann, Matthias. "Die Zukunft mit Algorithmen gestalten." *Goethe-Uni online*, September 12, 2019, available at: https://aktuelles.uni-frankfurt.de/veranstaltungen/die-zukunft-mit-algorithmen-gestalten-interview-mit-prof-lutz-bachmann/.

Lyon, David. *Theorizing Surveillance: The Panopticon and Beyond*. Cullompton/Portland: Willan Publishing, 2006.

Maak, Niklas. "Google-Stadt ist abgebrannt. Keine Smart-City in Toronto." *Frankfurter Allgemeine Zeitung*, May 11, 2020, available at: https://www.faz.net/aktuell/feuilleton/debatten/keine-smart-city-in-toronto-google-stadt-ist-abgesagt-16763217.html.

Machiavelli, Niccolò. *Der Fürst*. Trans. E. Rudolph. Hamburg: Felix Meiner, 2019 [1513].

———English edition: Machiavelli, Niccolo. *The Prince*. Trans. W. K. Marriott. Berlin: epubli, 2018.

Magnus, Bernd. *Nietzsche's Existential Imperative*. Bloomington: Indiana University Press, 1978.

Marchart, Oliver. *Das unmögliche Objekt: Eine postfundamentalistische Theorie der Gesellschaft*. Berlin: Suhrkamp, 2013.

Marcuse, Herbert. *Der eindimensionale Mensch: Studien zur Ideologie der fortgeschrittenen Industriegesellschaft*. 21th ed. Darmstadt et al.: Luchterhand, 1967 [1964].

Marcuse, Herbert. *Triebstruktur und Gesellschaft. Ein philosophischer Beitrag zu Sigmund Freud*. Frankfurt a.M.: Suhrkamp, 1965 [1955].

Marx, Karl. *The Eighteenth Brumaire of Louis Bonaparte*. Without location: E-art now, 2020 [1852].

Massumi, Brian, ed. *Politics of Everyday Fear*. Minneapolis: The University of Minnesota Press, 1993.

Massumi, Brian. *Politics of Affect*. Cambridge/Malden: Polity Press, 2015.

Maupassant, Guy de. *Bel-Ami*. Trans. D. Parmee. London/New York: Penguin Classics, 1975 [1885].

McCormick, John. *Machiavellian Democracy*. Cambridge: Cambridge University Press, 2011.

Meißner, Hanna. *Jenseits des autonomen Subjekts: Zur gesellschaftlichen Konstitution von Handlungsfähigkeit im Anschluss an Butler, Foucault und Marx*. Bielefeld: transcript, 2010.

Menke, Christoph. "Zweierlei Übung. Zum Verhältnis sozialer Disziplinierung und ästhetischer Existenz." In *Michel Foucault. Zwischenbilanz einer Rezeption. Frankfurter Foucault-Konferenz 2001*. Ed. A. Honneth and M. Saar, 199–210. Frankfurt a.M.: Suhrkamp, 2003.

Menke, Christoph. *Kraft: Ein Grundbegriff ästhetischer Anthropologie*. Berlin: Suhrkamp, 2017.

"Merkel sieht Twitter-Sperre kritisch." *Tagesschau*, January 11, 2021, available at: https://www.tagesschau.de/inland/merkel-trump-twitter-103.html.

Milde, Nadine. "Es spielt (k)eine Rolle: Orlando 'revisited' vom Buch zum Film." *Freiburger Zeitschrift für Geschlechterstudien* 24 (2010): 251–70.

Moellendorf, Darrel. *The Moral Challenge of Dangerous Climate Change: Values, Poverty, and Policy*. Cambridge: Cambridge University Press, 2014.

Moeller, Hans-Georg and Paul J. D'Ambrosio. *You and Your Profile: Identity after Authenticity*. New York: Columbia University Press, 2021.

Montag, Warren, ed. *Althusser and his Contemporary: Philosophy's Perpetual War.* Durham: Duke University Press, 2020.

Montag, Warren. "The Soul is the Prison of the Body: Althusser and Foucault 1970–1975." *Yale French Studies*, no. 88 (1995): 53–77.

Montaigne, Michel de. *Essais*. Ed. R.-R. Wuthenow. Frankfurt a.M.: Suhrkamp 2001 [1580].

Mouffe, Chantal. *Agonistics: Thinking the World in Plurality*. London: Verso, 2013.

Mouffe, Chantal. *Über das Politische: Wider die kosmopolitische Illusion*. Frankfurt a.M.: Suhrkamp, 2007 [2005].

Muhle, Maria. *Eine Genealogie der Biopolitik: Zum Begriff des Lebens bei Foucault und Canguilhem*. München: Fink, 2013.

Mühlhoff, Rainer et al., eds. *Affekt Macht Netz: Auf dem Weg zu einer Sozialtheorie der digitalen Gesellschaft*. Bielefeld: Transcript, 2019.

Mullen, Jethro. "Google Rushes to Fix Software that Tagged Photo with Racial Slur." *CNN*, July 2, 2015, available at: https://edition.cnn.com/2015/07/02/tech/google-image-recognition-gorillas-tag/index.html.

Mulvey, Laura. "Visual Pleasure and Narrative Cinema." *Screen* 16, no. 3 (1975): 6–18.

Münchhausen, Marco von. *Das Münchhausen-Prinzip: Wie Sie sich am eigenen Schopf aus dem Sumpf ziehen*. München: Kösel, 2011.

Munro, Kelsey. "China's Social Credit System Could Interfere in Other Nations' Sovereignity." *The Guardian*, June 27, 2018, available at: https://www.theguardian.com/ world/2018/jun/28/chinas-social-credit-system-could-interfere-in-other-nations-sovereignty.

Nagel, Thomas. *The View from Nowhere*. Oxford et al.: Oxford University Press, 1986.

Nassehi, Armin. *Muster: Theorie der digitalen Gesellschaft*. München: Beck, 2019.

Nehamas, Alexander. "How one Becomes What One is." In *The Existentialists: Critical Essays on Kierkegaard, Nietzsche, Heidegger, and Sartre*. Ed. C. B. Guignon. 73–100. Lanham et al: Rowman & Littlefield Publishers, 2004.

Nehamas, Alexander. "The Genealogy of Genealogy: Interpretation in Nietzsche's Second Untimely Meditation and On the Genealogy of Morals." In *Nietzsche's On the Genealogy of Morals: Critical Essays*. Ed. C. D. Acampora. 57–66. Lanham: Rowman & Littlefield Publishers, 2006.

Nehamas, Alexander. *The Art of Living: Socratic Reflections from Plato to Foucault*. Berkeley et al.: University of California Press, 1998.

Nguyen, C. Thi. "Echo Chambers and Epistemic Bubbles." *Episteme* 17, no. 2 (2020): 141–61.

Nichols, Robert. *The World of Freedom: Heidegger, Foucault and the Politics of Historical Ontology*. Stanford: Stanford University Press, 2014.

Nida-Rümelin, Julian and Nathalie Weidenfeld. *Digitaler Humanismus: Eine Ethik für das Zeitalter der Künstlichen Intelligenz*. München: Piper, 2018.

Nietzsche, Friedrich. *Kritische Studienausgabe Briefe*. Ed. G. Colli and M. Montinari. 2nd ed. München/Berlin: dtv/de Gruyter, 1986.

Nietzsche, Friedrich. *Kritische Studienausgabe*. Ed. G. Colli and M. Montinari. 2nd ed. München/Berlin: dtv/de Gruyter, 1986.

Nietzsche, Friedrich. *Nachgelassene Fragmente* (1869–1889). In *KSA* 7–13.

Nietzsche, Friedrich. *Twilight of the Idols*. Ed. A. Ridley and J. Norman, trans. J. Norman. Cambridge: Cambridge University Press, 2005.

———German edition: "Götzendämmerung (1889)." In *KSA* 5.

Nietzsche, Friedrich. *Ecce Homo*. Ed. N. Ridley and J. Norman, trans. J. Norman. Cambridge: Cambridge University Press, 2005.

———German edition: "Ecce Homo (1888/89)." In *KSA* 6.

Nietzsche, Friedrich. *On the Genealogy of Morals*. Ed. K. Ansell-Pearson, trans. C. Diethe. 3rd ed. Cambridge: Cambridge University Press, 2017.

———German edition: "Zur Genealogie der Moral (1887)." In *KSA* 5.

Nietzsche, Friedrich. *Beyond Good and Evil.* Ed. R.-P. Horstmann and J. Norman, trans. J. Norman. 7th ed. Cambridge: Cambridge University Press, 2009.

———German edition: "Jenseits von Gut und Böse (1886)." In *KSA* 5.

Nietzsche, Friedrich. *Thus Spoke Zarathustra*. Ed. A. del Caro and R. Pippin, trans. A. del Caro. Cambridge: Cambridge University Press, 2006.

———German edition: "Also sprach Zarathustra (1883–1885)." In *KSA* 4.

Nietzsche, Friedrich. *The Gay Science*. Ed. B. Williams, trans. J. Nauckhoff and A. del Caro. Cambridge: Cambridge University Press, 2001.

———German edition: "Die Fröhliche Wissenschaft (1882)." In *KSA* 3.

Nietzsche, Friedrich. *Daybreak. Thoughts on Prejudices of Morality*. Ed. M. Clark and B. Leiter, trans. R. J. Hollingdale. Cambridge: Cambridge University Press, 1997.
———German edition: "Morgenröthe. Gedanken über moralische Vorurtheile (1881)." In *KSA* 3.
Nietzsche, Friedrich. *Human, All too Human. A Book for Free Spirits*, ed. R. Schacht, trans. R. J. Hollingdale. 14th ed. Cambridge: Cambridge University Press, 2010.
———German edition: "Menschliches, Allzumenschliches. Ein Buch für freie Geister I (1878) and II (1879)." In *KSA* 2.
Nietzsche, Friedrich. "The Wanderer and his Shadow (part two, *Human All too Human*)." In *Human, All too Human: A Book for Free Spirits*. Ed. R. Schacht, trans. R. J. Hollingdale. 14th ed. Cambridge: Cambridge University Press, 2010.
———German edition: "Der Wanderer und sein Schatten (part two, *Menschliches, Allzumenschliches*, 1879)." In *KSA* 2.
Nietzsche, Friedrich. *Untimely Meditations*. Ed. D. Breazeale, trans. R. J. Hollingdale. Cambridge: Cambridge University Press, 1997.
———German edition: "Unzeitgemäße Betrachtungen (1873pp)." In *KSA* 1.
Nietzsche, Friedrich. "On the Use and Disadvantages of History for Life." In: *Untimely Meditations*. Ed. D. Breazeale, trans. R. J. Hollingdale. Cambridge: Cambridge University Press, 1997.
———German edition: "Vom Nutzen und Nachteil der Historie für das Leben (1874)." In *KSA* 1.
Nietzsche, Friedrich. "Über Wahrheit und Lüge im außermoralischen Sinn (1873)." In *KSA* 1.
Nietzsche, Friedrich. "Über die Zukunft unserer Bildungsanstalten (1872)." In *KSA* 1.
Nietzsche, Friedrich. *Birth of Tragedy out of the Spirit of Music*. Ed. M. Tanner, trans. S. Whiteside. New York/London: Penguin Classics, 1993.
———German edition: "Die Geburt der Tragödie aus dem Geist der Musik (1872)." In *KSA* 1.
Oksala, Johanna. "Foucault's Politicization of Ontology." *Continental Philosophy Review* 43, no. 4 (2010): 445–66.
Oksala, Johanna. *Foucault on Freedom*. Cambridge: Cambridge University Press, 2005.
Orwell, George. *1984*. London: Penguin Classics, 2013 [1949].
Ovid. *Metamorphosen*. Trans. M. v. Albrecht. Stuttgart: Reclam, 2019 [around 8 AD].
Owen, David. "Perfectionism, Parrhesia, and the Care of the Self: Foucault and Cavell." In *The Claim to Community. Essays on Stanley Cavell and Political Philosophy*. Ed. A. Norris. 128–55. Stanford: Stanford University Press, 2006.
Owen, David. "Kritik und Gefangenschaft: Genealogie und Kritische Theorie." In *Michel Foucault: Zwischenbilanz einer Rezeption. Frankfurter Foucault-Konferenz 2001*. Ed. A. Honneth and M. Saar. 122–45. Frankfurt a.M.: Suhrkamp, 2003.
Owen, David. *Nietzsche, Politics & Modernity*. London et al.: Sage Publications, 1995.
Owen, David. *Maturity and Modernity: Nietzsche, Weber, Foucault and the Ambivalence of Reason*, London/New York: Routledge, 1994.
Packard, Vans. *The Hidden Persuaders: The Naked Society and the Status Seekers*. New York: Pocket Books, 1957.

Panagia, Davide. "#datapolitics. An Interview with Davide Panagia." *Contrivers' Review* (2017) 11.

Patton, Paul. "Nietzsche and Metaphor." In *Enigmas: Essays on Sarah Kofman*. Ed. P. Deutscher and K. Oliver. 97–108. Ithaca: Cornell University Press, 1999.

Paxton, Marie. *Agonistic Democracy: Rethinking Political Institutions in Pluralist Times*. New York/Abingdon: Routledge, 2020.

Penz, Otto and Birgit Sauer. *Affektives Kapital: Die Ökonomisierung der Gefühle im Arbeitsleben*. Frankfurt a.M./New York: Campus, 2016.

Persson, Ingmar and Julian Savulescu. "The Duty to be Morally Enhanced." *Topoi* 38 (2019): 7–14.

Phoenix, Joaquin. *Her*. DVD. Dir. Sp. Jonze. Los Angeles: Warner Bros, 2013.

Pitkin, Hanna Fenichel. *Wittgenstein and Justice: On the Significance of Ludwig Wittgenstein for Social and Political Thought*. Berkeley: University of California Press, 1972.

Plato. *Phaidros*. Trans. T. Paulsen and R. Rehn. Hamburg: Meiner, 2019 [around 370 BCE].

Plato, *Protagoras*. Trans. H.-W. Krautz. Stuttgart: Reclam, 1986 [380 BCE].

Politi, Daniel and Ernesto Londoño. "How Support for Legal Abortion Went Mainstream in Argentina." *New York Times*, January 1, 2021, available at: https://www.nytimes.com/ 2021/01/01/world/americas/argentina-abortion.html.

Propris, Lisa de and David Bailey, eds. *Industry 4.0 and Regional Transformation*. London: Routledge, 2020.

Protevi, John. "Affect and Life in Spinoza, Nietzsche, and Bergson." In *Affect and Literature*. Ed. A. Houen. 66–82. Cambridge: Cambridge University Press, 2020.

Quinn, Tristan. "Horizon: The Immortalist. Uploading the Mind to a Computer." *BBC News*, March 14, 2016, available at: https://www.bbc.com/news/magazine-35786771, last accessed: 29.03.2020.

Rajchman, John. "Foucault's Art of Seeing." *October* 44 (1988): 88–117.

Rebentisch, Juliane. *Die Kunst der Freiheit: Zur Dialektik demokratischer Existenz*. Berlin: Suhrkamp, 2012.

Reckwitz, Andreas. "Die Transformation der Sichtbarkeitsordnung: Vom disziplinären Blick zu den kompetitiven Singularitäten." In *Vierzig Jahre 'Überwachen und Strafen': Zur Aktualität der Foucault'schen Machtanalyse*. Ed. M. Rölli and R. Nigro. 197–212. Bielefeld: Transcript, 2017.

Reckwitz, Andreas. *Die Gesellschaft der Singularitäten: Zum Strukturwandel der Moderne*. 3rd ed. Berlin: Suhrkamp, 2017.

Redecker, Eva von. *Zur Aktualität von Judith Butler: Einleitung in ihr Werk*. Wiesbaden: VS Verlag für Sozialwissenschaften, 2011.

Reich, Wilhelm. *Die sexuelle Revolution*. Frankfurt a.M.: Fischer, 1993 [1936].

Reschke, Renate. *Denkumbrüche mit Nietzsche: Zur anspornenden Verachtung der Zeit*. Berlin: Akademie-Verlag, 2000.

Richter, Gerhard. *Text. 1961 bis 2007 Schriften, Interviews, Briefe*. Köln: Walther König, 2008.

Richter, Norbert Alexander. *Grenzen der Ordnung: Bausteine einer Philosophie des politischen Handelns nach Plessner und Foucault*. Frankfurt a.M./New York: Campus, 2005.

Rieger, Jenni. "Mein Algorithmus und ich: Daniel Kehlmann testet KI." *Tagesschau*, February 10, 2021, available at: https://www.tagesschau.de/wirtschaft/daniel-kehlmann-kuenstliche-intelligenz-versuch-101.html.

Rilke, Rainer Maria. "Archaischer Torso Apollos." In *Der neue Gedichte anderer Teil*. Ed. E. Zinn, 503. 7th ed. Frankfurt a.M.: Insel, 1995 [1957, 1908].

Rousseau, Jean-Jacques. *A Discourse on Inequality*. London/New York: Penguin Classics, 1984 [1755].

Russell, Bertrand. *The History of Western Philosophy: And its Connection With Political and Social Circumstances From the Earliest Times to the Present Day*. London/New York: Routledge, 1967.

Saar, Martin. "Ordnung–Praxis–Subjekt: Oder: Was ist Sozialphilosophie?." *WestEnd: Neue Zeitschrift für Sozialforschung*, no. 2 (2019): 161–74.

Saar, Martin. "Philosophie in ihrer (und gegen ihre) Zeit." *Deutsche Zeitschrift für Philosophie* 67, no. 1 (2019): 1–22.

Saar, Martin. "Critical Theory and Poststructuralism." In *The Routledge Companion to the Frankfurt School*. Ed. P. E. Gordon, E. Hammer and A. Honneth. 323–35. New York/Abingdon: Routledge, 2018.

Saar, Martin. "Nachwort: Die Form des Lebens. Künste und Techniken des Selbst beim späten Foucault." In *Ästhetik der Existenz: Schriften zur Lebenskunst*. Ed. D. Defert and F. Ewald. 319–43. 5th ed. Berlin: Suhrkamp, 2015.

Saar, Martin. *Immanenz der Macht: Politische Theorie nach Spinoza*. Berlin: Suhrkamp, 2013.

Saar, Martin. "Analytik der Subjektivierung: Umrisse eines Theorieprogramms." In *Techniken der Subjektivierung*. Ed. A. Gelhard et al. 17–28. Paderborn: Wilhelm Fink, 2013.

Saar, Martin. "Macht und Kritik." In *Sozialphilosophie und Kritik*. Ed. R. Forst et al. 567–87. Frankfurt a.M.: Suhrkamp, 2009.

Saar, Martin. *Genealogie als Kritik: Geschichte und Theorie des Subjekts nach Nietzsche und Foucault*. Frankfurt a.M./New York, NY: Campus, 2007.

Safranski, Rüdiger. *Goethe: Das Kunstwerk des Lebens*. München: Carl Hanser, 2013.

Sandel, Michael J. *The Case Against Perfection: Ethics in the Age of Genetic Engineering*. Cambridge, Mass.: Harvard University Press, 2007.

Sander, Matthias. "Wem gehört die Smart City? Toronto ringt mit dem Google-Konzern." *Neue Zürcher Zeitung*, October 10, 2019, available at: https://www.nzz.ch/international/wem-gehoert-die-smart-city-toronto-ringt-mit-dem-google-konzern-ld.1513002.

Sartre, Jean-Paul. *Les jeux sont faits*. Paris: Gallimard, 1998 [1947].

Savulescu, Julian and Nick Bostrom, eds. *Human Enhancement*. Oxford: Oxford University Press, 2009.

Schiller, Friedrich. *On the Aesthetic Education of Man*. Mineola/New York: Dover Publications, 2004 [1794]).

Schmid, Wilhelm. *Mit sich selbst befreundet sein: Von der Lebenskunst im Umgang mit sich selbst*. 9th ed. Frankfurt a.M.: Suhrkamp, 2016.

Schmid, Wilhelm. *Philosophie der Lebenskunst: Eine Grundlegung*. Frankfurt a.M.: Suhrkamp, 1998.

Schnädelbach, Herbert. "Das Gesicht im Sand." In *Zwischenbetrachtungen im Prozess der Aufklärung: Jürgen Habermas zum 60. Geburtstag*. Ed. A. Honneth et al. 231–61. Frankfurt a.M.: Suhrkamp, 1989.

Schneider, Ulrich Johannes and Aaron Sabellek. "Szenographie als Philosophiegeschichte: Die Regierung des Selbst und der Anderen 1982/83." In *Fragmente eines Willens zum Wissen: Michel Foucaults Vorlesungen 1970–1984*. Ed. F. Vogelmann. 209–23. Stuttgart: Metzler, 2020.

Schneider, Ulrich Johannes. *Michel Foucault*. Darmstadt: Primus, 2004.

Schubert, Karsten. *Freiheit als Kritik: Sozialphilosophie nach Foucault*. Bielefeld: Transcript, 2018.

Schweiker, Willliam. *Dust that Breathes: Christian Faith and the New Humanisms*. Oxford: Wiley-Blackwell, 2010.

Searle, John R. "'Las Meninas' and the Paradoxes of Pictoral Representation." *Critical Inquiry* 6, no. 3 (1980): 477–88.

Sedgwick, Eve K. and Adam Frank. "Shame in the Cybernetic Fold: Reading Silvan Tomkins." In *Shame and Its Sisters: A Silvan Tomkins Reader*. Ed. E. K. Sedgwick and A. Frank. 1–28. Durham: Duke University Press, 1995.

Seel, Martin. *Ethisch-ästhetische Studien*. 3rd ed. Frankfurt a.M.: Suhrkamp, 2019.

Shakespeare, William. *As You Like It*. Leipzig: Amazon Distributions, without year [1603].

Shaw, Bernard. *Pygmalion: Romanze in 5 Akten*. Trans. H. Mueller. Berlin: Suhrkamp, 2012.

Shieber, Stuart M., ed. *The Turing Test: Verbal Behavior as the Hallmark of Intelligence*. Cambridge: MIT Press, 2004.

Simanowski, Roberto. *Data Love: The Seduction and Betrayal of Digital Technologies*. Trans. B. Pichon et al. New York: Columbia University Press, 2018.

Simanowski, Roberto. *Facebook Society: Losing Ourselves in Sharing Ourselves*. Trans. S. H Gillespie. New York: Columbia University Press, 2018.

Simon, Felix. "Google baut in Toronto einen vernetzten, perfekten Stadtteil–der unsere Freiheiteinschränken könnte." *Neue Zürcher Zeitung*, December 8, 2017, available at: https://www.nzz.ch/feuilleton/wie-wollen-wir-leben-ld.1336803?reduced=true.

Slaby, Jan. *Gefühl und Weltbezug: Die menschliche Affektivität im Kontext einer neo-existentialistischen Konzeption von Personalität*. Paderborn: mentis, 2008.

Sloterdijk, Peter. *Du musst dein Leben ändern: Über Anthropotechnik*. Frankfurt a.M.: Suhrkamp, 2009.

———English edition: Sloterdijk, Peter. *You Must Change Your Life: On Anthropotechnics*. Trans. W. Hoban. Cambridge: Polity Press, 2013.

Sloterdijk, Peter. *Regeln für den Menschenpark: Ein Antwortschreiben zu Martin Heideggers Brief über den Humanismus*. Frankfurt a.M.: Suhrkamp, 1999.

Snowden, Edward. *Permanent Record*. London: Pan Macmillan, 2019.

"So können Daten den Wahlkampf-Planern helfen." *Süddeutsche Zeitung*, March 23, 2018, available at: https://www.sueddeutsche.de/politik/wahlkampf-in-den-usa-daten-sammeln-fuer-den-wahlsieg-das-machen-alle-1.3918504-2.

Sommer, Andreas Urs. *Kommentar zu Nietzsches "Zur Genealogie der Moral."* Berlin/Boston: de Gruyter, 2019.

Sommer, Andreas Urs. *Kommentar zu "Der Anti-Christ," "Ecce Homo," "Dionysos-Dithyramben" und "Nietzsche contra Wagner."* Berlin/Boston: de Gruyter, 2013.

Sonderegger, Ruth. *Für eine Ästhetik des Spiels: Hermeneutik, Dekonstruktion und der Eigensinn der Kunst*. Frankfurt a.M.: Suhrkamp, 2000.

Sorgner, Stefan Lorenz. *Übermensch: Plädoyer für einen Nietzscheanischen Transhumanismus*. Basel: Schwabe, 2019.

Spinoza, Baruch de. *Politischer Traktat*. Ed. W. Bartuschat. Hamburg: Meiner, 1994 [1677]. [Abbreviation: TTP]

Spivak, Gayatri Chakravorty. "Can the Subaltern Speak?." In *Marxism and the Interpretation of Culture*. Ed. C. Nelson and L. Grossberg. 271–313. Urbana: University of Illinois Press, 1988.

Stendhal. *The Red and the Black*. Trans. R. Gard. London/New York: Penguin Classics, [1830].

Strauss, Leo. *Persecution and the Art of Writing*. Chicago/London: The University of Chicago Press, 1988 [1952].

Sturma, Dieter and Bert Heinrichs, eds. *Handbuch Bioethik*. Stuttgart/Weimar: Metzler, 2015.

Swift, Laura. *Euripides: Ion*. Bristol: Classical Press, 2008.

Tanriverdi, Hakan. "Aufregung um Spracherkennung. Samsung hört mit—aber nur manchmal." *Süddeutsche Zeitung*, February 9, 2015. https://www.sueddeutsche.de/digital/aufregung-um-spracherkennung-samsung-hoert-mit-aber-nur-manchmal-1.2341288.

Taylor, Charles. *A Secular Age*. Cambridge, MA/London: Harvard University Press, 2007.

Taylor, Charles. *Philosophy in an Age of Pluralism: The Philosophy of Charles Taylor in Question*. Ed. James Tully. Cambridge: Cambridge University Press, 1994.

Taylor, Charles. *Human Agency and Language: Philosophical Papers I*. Cambridge: Cambridge University Press, 1985.

Temple-Raston, Dina. "Lawyers for 18-years-old Capitol Rioter Want him Released to his Parents." *NPR*, February 26, 2021, available at: https://www.npr.org/sections / insurrection-at-the-capitol.

Terranova, Tiziana. *Network Culture: Politics for the Information Age*. London: Pluto Press, 2004.

Thomas, Günter. "Enhancement: Evangelisch theologische Optionen in der gegenwärtigen Debatte." In *Sind Sie gut genug? Zur (Selbst-) Optimierung und Vervollkommnung des Menschen*. Ed. Monika C. M. Müller et al. 25–38. Loccum: Loccumer Protokolle, 2011.

Toonders, Joris. "Data is the New Oill of the Digital Economy." *Wired* 7 (2014), available at: https://www.wired.com/insights/2014/07/data-new-oil-digital-economy/.

Tugendhat, Ernst. *Vorlesungen über Ethik*. Frankfurt a.M.: Suhrkamp, 1993.

Tully, James. *Politische Philosophie als kritische Praxis*. Trans. E. Engels. Frankfurt a.M.: Campus, 2009.

Vatter, Miguel. *The Republic of the Living: Biopolitics and the Critique of Civil Society*. New York: Fordham University Press, 2014.

Veyne, Paul. "Foucault Revolutionizes History." In *Foucault and his Interlocutors*. Ed. A. I. Davidson. Trans. C. Porter. 146–82. Chicago, IL/London: The University of Chicago Press, 1997.

Vikander, Alicia. *Ex Machina*. DVD. Dir. A. Garland. London: DNA Films, 2015.

Villa, Paula Irene, ed. *Schön normal: Manipulationen am menschlichen Körper als Technologien des Selbst*. Bielefeld: Transcript, 2008.

Villa, Paula Irene. *Judith Butler*. Frankfurt a.M./New York, Campus, 2003.

Vogelmann, Frieder, ed. *"Fragmente eines Willens zum Wissen"*: *Michel Foucaults Vorlesungen 1970–1984*. Stuttgart: Metzler, 2020.

Wagner, Greta. *Selbstoptimierung: Praxis und Kritik von Neuroenhancement*. Frankfurt a.M./New York: Campus, 2014.

Warren, Mark E. *Nietzsche and Political Thought*. Cambridge, MA/London: MIT Press, 1991.

Warren, Mark E., ed. *Democracy and Trust*. Cambridge: Cambridge University Press, 1999.

Weber, Max. *Wirtschaft und Gesellschaft: Grundriss einer verstehenden Soziologie*. Ed. J. Winckelmann. Tübingen: Mohr, 1976 [1921/22].

———English edition: Weber, Max. *Economy and Society*. Trans. K. Tribe. Cambridge, Mass./London: Harvard University Press, 2019 [1921/22].

Weingarten, Susanne and Marianne Wellershoff, eds. *Die widerspenstigen Töchter: Für eine neue Frauenbewegung*. Köln: Kiepenheuer und Witsch, 1999.

Welsch, Wolfgang. *Ästhetisches Denken*. 8th ed. Stuttgart: Reclam, 1990.

Williams, Bernard. *Truth and Truthfulness: An Essay on Genealogy*. Princeton: Princeton University Press, 2002.

Williams, Melissa ed. *Deparochializing Political Theory.* Cambridge: Cambridge University Press, 2020.

Woolf, Virginia. *A Room of One's Own*. Chicago: Albatross Publishers, 2015 [1929].

Woolf, Virginia. *Orlando*. London: HarperCollins Publishers, 2014 [1928].

Woolf, Virginia. *The Diary of Virginia Woolf.* Vol. 3: *1925–1930*. New York: Mariner Books, 1981.

Wylie, Christopher. *Mindf*ck: Cambridge Analytica and the Plot to Break America*. New York: Random House, 2019.

Zamora, Daniel and Michael C. Behrent, eds. *Foucault and Neoliberalism*. Cambridge/Malden: Polity Press, 2016.

Zeh, Juli. *Corpus Delicti: Ein Prozess*. München: btb, 2010.

Zerilli, Linda M. G. *Feminism and the Abyss of Freedom.* Chicago: Chicago University Press, 2005.

Zimmerli, Walther Ch. *Künstliche Intelligenz: Philosophische Probleme.* Stuttgart: Reclam, 1993.

Zuboff, Shoshana. *The Age of Surveillance Capitalism: The Fight for A Human Future at the Frontier of Power*. New York: Public Affairs, 2019.

Index

About the Author

Dr. Sarah Bianchi is postdoc at Goethe-University. Previously, she was visiting professor at Technical University Dresden.

www.ingramcontent.com/pod-product-compliance
Ingram Content Group UK Ltd.
Pitfield, Milton Keynes, MK11 3LW, UK
UKHW012118150926
475560UK00002B/164

9 781666 928310